W9-AXO-411

Geography

Second Edition

by Joseph Gonzalez and Thomas E. Sherer Jr.

RETURN TO:
The Media Center
Ben Davis High School
1200 N. Girls School Road
Indianapolis, IN 46214

ALPHA

A member of Penguin Group (USA) Inc.

910
GONZ

BEN DAVIS HIGH SCHOOL
MEDIA CENTER
The complete idiot's guide to

300096120

9/06

ALPHA BOOKS

Published by the Penguin Group

Penguin Group (USA) Inc., 375 Hudson Street, New York, New York 10014, U.S.A.

Penguin Group (Canada), 10 Alcorn Avenue, Toronto, Ontario, Canada M4V 3B2 (a division of Pearson Penguin Canada Inc.)

Penguin Books Ltd, 80 Strand, London WC2R 0RL, England

Penguin Ireland, 25 St Stephen's Green, Dublin 2, Ireland (a division of Penguin Books Ltd)

Penguin Group (Australia), 250 Camberwell Road, Camberwell, Victoria 3124, Australia (a division of Pearson Australia Group Pty Ltd)

Penguin Books India Pvt Ltd, 11 Community Centre, Panchsheel Park, New Delhi—110 017, India

Penguin Group (NZ), cnr Airborne and Rosedale Roads, Albany, Auckland 1310, New Zealand (a division of Pearson New Zealand Ltd)

Penguin Books (South Africa) (Pty) Ltd, 24 Sturdee Avenue, Rosebank, Johannesburg 2196, South Africa

Penguin Books Ltd, Registered Offices: 80 Strand, London WC2R 0RL, England

Copyright © 2004 by Penguin Group (USA) Inc. and Thomas E. Sherer Jr.

All rights reserved. No part of this book shall be reproduced, stored in a retrieval system, or transmitted by any means, electronic, mechanical, photocopying, recording, or otherwise, without written permission from the publisher. No patent liability is assumed with respect to the use of the information contained herein. Although every precaution has been taken in the preparation of this book, the publisher and author assume no responsibility for errors or omissions. Neither is any liability assumed for damages resulting from the use of information contained herein. For information, address Alpha Books, 800 East 96th Street, Indianapolis, IN 46240.

THE COMPLETE IDIOT'S GUIDE TO and Design are registered trademarks of Penguin Group (USA) Inc.

International Standard Book Number: 1-59257-188-3
Library of Congress Catalog Card Number: 2003116930

07 06 05   8 7 6 5 4 3

Interpretation of the printing code: The rightmost number of the first series of numbers is the year of the book's printing; the rightmost number of the second series of numbers is the number of the book's printing. For example, a printing code of 04-1 shows that the first printing occurred in 2004.

Printed in the United States of America

Note: This publication contains the opinions and ideas of its authors. It is intended to provide helpful and informative material on the subject matter covered. It is sold with the understanding that the authors and publisher are not engaged in rendering professional services in the book. If the reader requires personal assistance or advice, a competent professional should be consulted.

The authors and publisher specifically disclaim any responsibility for any liability, loss, or risk, personal or otherwise, which is incurred as a consequence, directly or indirectly, of the use and application of any of the contents of this book.

Most Alpha books are available at special quantity discounts for bulk purchases for sales promotions, premiums, fund-raising, or educational use. Special books, or book excerpts, can also be created to fit specific needs.

For details, write: Special Markets, Alpha Books, 375 Hudson Street, New York, NY 10014.

Publisher: *Marie Butler-Knight*
Product Manager: *Phil Kitchel*
Senior Managing Editor: *Jennifer Chisholm*
Acquisitions Editor: *Mikal E. Belicove*
Development Editor: *Jennifer Moore*
Production Editor: *Megan Douglass*
Copy Editor: *Molly Schaller*
Illustrator: *Jody Schaeffer*
Cover/Book Designer: *Trina Wurst*
Indexer: *Julie Bess*
Layout/Proofreading: *Angela Calvert, Mary Hunt*

Contents at a Glance

Contents

Foreword

There once was a time when men would climb aboard a ship and set sail with only a compass direction to guide them. They were true explorers, hunting for knowledge of far-off places that they only hoped existed. Brave and courageous, these men left their families and homes behind, knowing they might not survive the journey. Those who stumbled upon new land became heroes, for their stories of new places, people with different customs, and exciting experiences enthralled those back home.

Many of the explorers' exotic stories of adventure and new lands were hastily printed and sold—instantly becoming bestsellers. However, those stuck at home were so hungry for knowledge of the outside world, that explorers often could not find and describe new lands fast enough to satisfy them. So a few creative, but deceptive, authors wrote fake travelogues that included tales of giants, magic, and fantastical adventures. The public, whose thirst for knowledge of the world was unbelievably strong, bought both the genuine and fictitious travelogues.

It is incredible that during a time when there was so little known about the world, people desperately craved knowledge. Yet now that information about the world is readily available, individuals seem to have stopped their quest for geographic wisdom. Not only have we lost our enthusiasm for knowledge of the world around us, most residents of the United States lack even basic knowledge of world geography. In a recent survey by the National Geographic Society, 30 percent of Americans age 18 to 24 who were polled could not even locate the Pacific Ocean on a map. This geographic illiteracy is distressing, especially with the ever-growing interdependency of our world.

The inventions of the phone, television, and the Internet plus the ability to travel the world via plane should have taken away our excuses for ignorance. We no longer have to risk life and limb to learn about the unknown, the information can be at our fingertips in minutes. But where do you start to learn about the world? There is so much to know, it can be overwhelming to begin. *The Complete Idiot's Guide to Geography, Second Edition*, will give you a foundation from which to launch your own exploration.

Though this book may be missing the giants and magic of the fictitious travelogues of the past, it is still filled with exciting information about far-off places and people with different cultures. Inside you will learn about the Economic Tigers of Asia, the world's largest river, why Greenland is not considered a continent, the difference between Britain and the United Kingdom, and much more. You may also be surprised by what you learn about your own country—a place of incredible geographic diversity.

The Complete Idiot's Guide to Geography, Second Edition, will introduce you to our world and help you become a knowledgeable member of global society. So grab a comfortable chair and get ready for a geographical adventure.

—Matthew Rosenberg

Matthew Rosenberg is the editor and guide for About.com's Geography Site at http://geography.about.com. He is also the author of *The Handy Geography Answer Book* and *The Geography Bee Complete Preparation Handbook*.

Introduction

Thank you for purchasing the new, revised edition of *The Complete Idiot's Guide to Geography, Second Edition*. Please don't be put off by its title: We're all idiots when it comes to geography! Especially in a world that is changing so quickly, even professional geographers and mapmakers get stumped every so often and have to look things up in an atlas, just like everyone else! There's simply too much to geography and too much going on in the world today to be able to get a handle on it all—but that doesn't mean, of course, that you can't try.

This book is designed to give you the basics. It tells you what geography is all about and how it applies to your life. You will learn a good deal about the natural world and get a feel for the people who inhabit the many environments of our planet. You'll also learn a little about the history, culture, and politics of the world's regions. Best of all, you won't feel like an idiot the next time someone drops an exotic place name at a party or refers to a current global hot spot.

There is nothing more fascinating than the world we live in and the many different peoples we share it with. I hope that this book will set you on a lifelong quest for geographic knowledge. The Internet has become a terrific source of quick information on all aspects of our planet—its geography, resources, environments, peoples, history, cultures, current events, and more. To continue your geographic quest, check out such websites as geography.about.com, www.geographic.org, www.geohive.com, and the invaluable CIA factbook site, www.cia.gov. Subscribe to a geographic magazine, buy yourself a good atlas (no home should be without one!), but most of all—travel! Nothing brings geography to life as much as actually seeing places you've read about up close and in person.

The goal of this book is to give you a basic understanding of geography and an overview of the world's lands and peoples. But it is just an overview.

What You'll Learn in This Book

The Complete Idiot's Guide to Geography, Second Edition, divides your geographic journey into five parts.

Part 1, "An Overview of Geography," defines what geography is and introduces the nuts and bolts of the discipline. It lays out the basic terminology and principles you'll come across throughout the rest of the book. Part 1 is the foundation on which your geographic knowledge will be built.

In Part 2, "A Regional Look: The Developed World," you begin your journey around the globe, starting with the more economically developed nations of the world. Eight regions and more than two dozen countries are included in this section. Each regional chapter surveys the terrain, environments, climates, history, peoples, and cultures of the countries that make up the region. In each chapter, you'll find the basic facts about a country or group of countries—capitals, largest cities, tallest mountains, longest rivers, ethnic breakdowns, economic activities, and current events—plus a wealth of surprising statistics and interesting background information. Don't get overwhelmed or feel obliged to memorize it all—that's what atlases and almanacs are for.

Part 3, "A Regional Look: Nations in Transition," is new to this revised edition. Its content and format are the same as in Part 2, but its focus is on three regions that don't fit neatly into the group of developed countries covered in Part 2 or the developing countries surveyed in Part 4. The regions you'll explore in this section of the book—Eastern Europe, Russia, and China—have one basic feature in common. They are all in various stages of a transition from a state-controlled communist economy to an open, free-market capitalist economy.

Part 4, "A Regional Look: The Developing World," takes you on a whirlwind tour through nine regions and dozens of countries whose economies are on the lower rungs of the economic development scale. The countries you'll explore in this section range from the very poor and troubled to some of the most exciting and vibrant places on the planet. Brace yourself for lots of color, contrast, and controversy!

The shortest part of the book, **Part 5, "A Global Overview,"** looks at problems that concern every region on earth—environmental and economic issues that have an impact on all of us and call for global cooperation to resolve.

Extras

Throughout the five parts of this book, you'll find a variety of features designed to make your journey around the world easier and more fun. Each chapter includes several simplified maps of the region being surveyed. Although you won't want to navigate by them, the maps will give you a very good idea of what is located where. In each chapter you'll also find sidebars and feature boxes with extra information that clarifies or expands on subjects covered in the main text. Here are the road signs you'll encounter along the way:

GeoJargon

In these boxes you'll find handy definitions of obscure or unfamiliar geographic terms.

Geographically Speaking

These boxes examine geographic, historical, or cultural topics in greater depth than is possible in the main text. Everything is fair game in these sidebars from an explanation of plate tectonics to a closer look at India's caste system.

Terra-Trivia

Impress family and friends with these fascinating facts and bits of interesting information about people, places, and events around the world.

GeoRecord

Highest, widest, deepest, longest—here are some of the superlatives of world geography.

Eye on the Environment

The health of our planet should be a top concern of everyone who lives on it. Environmental issues are the focus of these sidebars, from the unintended consequences of Egypt's Aswan Dam to New Zealand's pioneering role in the use of earth-friendly geothermal energy.

Acknowledgments

Without a first edition there would be no revised edition. As the author of the new revised edition of *The Complete Idiot's Guide to Geography*, I would like to thank the author of the original edition, Thomas E. Sherer, Jr., for laying down such a solid foundation for this book. I would also like to thank Hildegard B. Anderson for her demographic research and my partner at FairWinds Antique Maps (www.fairmaps.com), Ralph E. Magnus, for help with the chapter on maps and mapmaking.

The maps in this book were created quickly and beautifully by Nicole and Mario Antonetti, under the direction of Ortelius Design. My thanks to them for a great job on short notice. For information about Ortelius Design, please contact them at 608-835-8301 or gary@orteliusdesign.com.

Part 1

An Overview of Geography

One of the most difficult challenges geographers face is defining exactly what geography means. Part 1 of this book tries to do just that. It fleshes out the meaning and scope of the term *geography* and gives you some basic background in physical, cultural, and regional geography. Along the way, you even get to refresh your map skills.

Although all this might sound like a great deal of heavy-duty ground to cover in just a few chapters, don't let that daunt you. The knowledge and skills you'll gain will be very useful as you set forth on your global explorations later in this book. With maps in hand and a newfound understanding of what geography really is, you'll be ready to take on the world.

Chapter 1

What Is Geography?

In This Chapter

- Geography defined
- How to study geography
- Why you need to know this stuff

You've probably picked up this book because you have some interest in geography. Perhaps you you're a map buff and spend countless hours poring over the atlas, or maybe you simply want to brush up your geography skills to enhance your reputation as a Trivial Pursuit ringer.

More likely, you believe that you know nothing about geography and have decided to rectify your hideous state of geographic illiteracy! If that's your situation, you might not even know what an understanding of geography entails—to be truthful, most people don't. Geography is a difficult discipline to define because it doesn't fit neatly into some academic box. This chapter describes what geography is all about, what it is not, and why the subject should matter to you.

The Mother of All Sciences

Geography has been called "the mother of all sciences" and "the science of place." However you describe it, geography involves the study of the physical and cultural factors that interact to make up the diversity of the earth's places and environments.

Is this scope somewhat broad? Yes, it is. Almost anything can fall into the realm of geographic study. If you want to learn about the earth, its peoples, and why the two interact as they do, you have the makings of a geographer. Although this book doesn't really put limitations on *what* geographers study, it's more specific about the *way* geographers look at things.

Terra-Trivia

The Greek scientist Eratosthenes was the first to use the term *geography*, which means literally "to describe the earth."

Geography is a *spatial* discipline, which means that geographers are concerned not only with what something is but also with the way it is distributed in space. Although geographers don't have the medical expertise to develop a cure for cancer, for example, they can still study the distribution of cancer cases and suggest possible causes based on that spatial information.

Our favorite definition of geography is "anything that can be mapped." We use this definition again later in this book because it gets at the heart of geography. If something can be mapped, it has a spatial component. Maps are the tools of geographers: If something can be mapped, it's geography.

What Geography Is Not: The Dread Place-Name Quiz

Many people, because of their experiences in school, incorrectly believe that the study of geography involves nothing more than studying for and taking place-name exams—you know, those outline maps where you have to fill in the names of countries, cities, rivers, and so on? If you mention geography to most people, you get a sour look. To them, geography means no more than an outline map and the memorization of country names and capitals.

Why is the place-name map so entrenched? Because, for decades, geography has been taught that way. Spatial thinking and investigation have been left in the shadows as teachers have focused on the place-name map. Although outline maps are easy to display in a classroom, they present only the most superficial geographic knowledge and leave the more important issues out of the discussion.

The importance of knowing where things are in the world can't be denied. As the world increasingly becomes a smaller, more interdependent place, knowing the names of the places around you is more necessary than ever, but geography is about much more than this single component. Would you grant a medical doctor a license based on his or her ability to name the parts of the body? Of course not. In addition to knowing medical terminology, doctors must be able to diagnose illnesses, prescribe medication, and treat patients.

Despite what we've said about place-name maps, after reading this book, you probably will gauge your newfound geographic literacy by your ability to identify places on a map.

Why We're Geographically Illiterate

If you live in the United States and took a geography class in school, you're in the minority. Most people in this country got their limited dose of geography in a social studies class because most states ditched geography as a distinct course decades ago. Geography was absorbed into the more general social studies curriculum along with history, economics, sociology, and anthropology.

One result of this amalgam called social studies is that Americans don't know their geography. Article after article in the press cites the woeful state of our country's geographic education. Although a "geographically illiterate" nation is good news when you're writing a book like this one, it's bad news for U.S. business.

Virtually every other industrialized nation on earth features geographic education as a primary component of its academic instruction. In most countries, geography is taught—from elementary school through high school—in a separate class. When graduates in these countries enter the global marketplace, they're familiar with resources, transportation systems, landforms, foreign languages, religions and customs, climate, demographics, and political systems. How many U.S. graduates can say the same? Fortunately, the study of geography is making something of a comeback in U.S. classrooms. Prompted in part by a famous National Geographic Society survey in 1988 that found that one in seven young adults in the United States could not locate their own country on a map, educators are placing greater emphasis on geography and are teaching it once again as a stand-alone course. Still, geography instruction faces considerable obstacles in the United States, where most schools stress the preparation of students to pass standardized tests in reading, writing, and math, often at the expense of other subjects.

Why Study Geography?

One of the most frequent complaints from kids in school is, "Why do I have to learn this stuff? I'll never need to use it." We never know what life will bring, of course, nor what we might someday need. But try to tell that to a 15-year-old! Nonetheless, people seem to focus better when they feel that what they're doing has some purpose.

So what purpose does geography have in your life—why learn about this stuff? In a nutshell, the study of geography places the whole world at your fingertips. It takes you on fantastic voyages to faraway environments and exotic landscapes and introduces you to our planet's kaleidoscope of peoples, cultures, and ways of life. To a great extent geography shapes history and current events. In our increasingly globalized and interconnected society, knowledge of geography is more important than ever to understanding the world and your place in it.

The Least You Need to Know

- ◆ Geography is a spatial discipline that encompasses "anything that can be mapped."

- ◆ There's more to geography than place names.

- ◆ Long buried in the mix of disciplines known as social studies, geography is making a comeback as a valued stand-alone subject.

- ◆ Geography is important in your life in ways you might not guess. Knowledge of geography opens your life to new frontiers.

Physical Geography: A Look at Mother Earth

In This Chapter

- ◆ How the sun is responsible for day and night and the seasons

- ◆ How geography affects the climate

- ◆ Whether the earth is really on solid ground

- ◆ The earth's moving surface

- ◆ How plate movement and volcanoes build up the earth's surface

- ◆ How water and ice wear down the earth's surface

Physical geography is the study of the physical features and phenomena of the earth. The study of physical geography encompasses so many subjects that geographers often wind up stepping on other scientists' toes. Geologists, hydrologists, meteorologists, biologists, and virtually every "ologist" you can name has some interest in the study of the earth. Because geographers are concerned with the earth primarily as it affects the human experience, they spend less time on isolated details; they're the "big picture" people when it comes to earth study.

This chapter looks at the geographical big picture (think broad strokes on an extremely large canvas). Even a college course in physical geography gives only the condensed version of the subject: Many people spend their entire lives trying to understand the subtleties of the earth and its systems. After a lifetime of effort, these people would report that they had only begun to scratch the surface. Complex? Yes. Daunting? Perhaps. Intimidating? Not on your life.

It All Revolves Around the Sun

The source of all energy on earth is the sun. Without it, there would be no day and night and no seasons. *No* life at all, in fact. Throughout human history, people have observed the sun's path through the sky with fascination and awe. They have worshipped the sun as a god and feared its demise during solar eclipses. Just how does the sun influence people's lives today?

Spin Control: What Causes Day and Night

Most people are aware that the sun is responsible for day and night and the changing seasons. When it comes to explaining the nuts and bolts of how it all happens, though, our knowledge begins to falter.

As shown in the following figure, a flashlight aimed at a tennis ball sums up the relationship between the sun and the earth. When the light (the sun) shines on one side of the tennis ball (the earth), the other side of the ball is dark. Therefore, half the earth is always illuminated, and half the earth is always in the dark.

How the sun causes day and night.

We know what you're thinking: "Thank goodness I live on the illuminated side!" But of course everyone has a little of the light and a little of the dark. The reason that one side isn't always in the light or always in the dark is that the earth rotates; it always spins one full rotation every 24 hours. The result is that about half the time you're in the light (daytime) and half the time you're in the dark (nighttime).

Naturally, this process is a little more complicated than my simple description. You know that during the summer the days are longer and the nights are shorter and that during the winter the days are shorter and the nights are longer. You might even know that at certain times of the year, near the North Pole or the South Pole, the sun never rises nor sets. We talk more about this subject later in this book, when we describe Northern Europe and the North and South poles. For now, however, remember that day and night are caused by both the sun's illumination of half the globe and the spinning of the earth.

Why the Seasons Change: Talking About a Revolution

The seasons are also caused by the earth's relationship to the sun. In this case, the spinning of the earth is not the important factor; rather, it's that the earth *revolves* around the sun. Although Copernicus had a tough time selling the idea to the Pope, the earth does orbit the sun. Because the earth's axis is inclined, or tilted (23^1/$_2$ degrees, to be exact), sometimes the Northern Hemisphere tilts toward the sun (in June), and sometimes the Southern Hemisphere does (in December).

As the next figure shows, the hemisphere tilted toward the sun experiences summer, and the hemisphere tilted away from the sun experiences winter. Although residents of the Northern Hemisphere assume that Christmas and New Year's are synonymous with winter, half the world would disagree. While New Englanders are bundling up their kids for a sleigh ride to grandmother's house for a Christmas ham, Australians are throwing another shrimp on the "barbie" for a Christmas dinner on the beach.

GeoJargon

Rotation is the spinning of the earth on its axis; a complete rotation occurs once every 24 hours. **Revolution** describes the earth's orbit around the sun; the earth completes a full revolution once every year (365 days, 5 hours, 49 minutes, and 12 seconds)—except in leap years, but that's another matter.

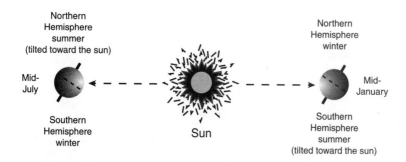

How the earth's tilt causes the seasons.

Considering Climate

Although people usually use the terms *weather* and *climate* interchangeably, they mean different things. Weather refers to the temporal, or the here and now. When people speak about the weather (which many people do frequently), they're talking about the temperature, wind, and moisture of a specific place at a specific time. Weather is localized, it changes, and (as your local meteorologist often has to admit), it can be unpredictable.

Climate refers more to the long term, or to average weather conditions over an extended period for large regions of the earth. The world can be divided into many different climatic zones that correspond roughly to their distance from the equator. (Keep in mind, though, that many other factors, such as altitude and proximity to a large body of water, also play a role in determining climate.) The most commonly used climate classification system is the Köppen system, named after its author, German climatologist Waldimir Köppen (1846–1940). Köppen divided the world into five climate zones, each with many subdivisions. A sixth zone—mountain—was added in a subsequent modification. The following map shows the location of Köppen's six main zones:

- ◆ **Tropical, rainy:** Near the equator

- ◆ **Dry, desert:** In bands just north and south of the equator

- ◆ **Humid, temperate:** Farther to the north and south than the dry, desert areas (the Mediterranean is an important subdivision of this climate zone)

- ◆ **Humid, cold:** Mainly north of the northern temperate zone, plus the southern tip of South America

- ◆ **Polar:** Toward each pole

- ◆ **Mountain:** The upper reaches of the world's great mountain chains: the Rockies, Andes, Alps, Caucasus, Himalayas

Why Is It Hot at the Equator and Cold at the Poles?

Even during the middle of winter at the equator, it's hot. Conversely, even during the balmiest summer day at one of the poles, it's cold. The primary reason is that the earth is hottest at the point where the sun's rays strike it most directly. At the equator, the sun is always close to being directly overhead. At the North and South poles, the sun never is directly overhead. During the winter months, the sun does not even appear. Even at the height of a polar summer, the sun remains close to the horizon.

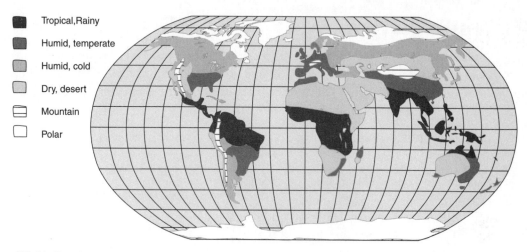

- ■ Tropical, Rainy
- ■ Humid, temperate
- ■ Humid, cold
- ■ Dry, desert
- ▤ Mountain
- ▢ Polar

World climatic zones.

The air temperature you feel when you step outside is caused only indirectly by the sun's rays. The process of heating the air is called *reradiation*. As the sun's rays penetrate the earth's atmosphere and strike the earth's surface, the surface absorbs the light energy from the sun and reradiates it as heat energy that warms the air.

Some surfaces are more effective than others at absorbing and radiating heat. The earth's vast oceans effectively absorb heat energy, as do dark surfaces, such as asphalt, soil, and tree leaves. Smooth and light-colored surfaces, such as ice and snow, are less effective in absorbing energy from the sun.

Because of this process, you probably don't want to sell air conditioners in Antarctica. There, the sun's rays are never direct, the surface does a poor job of absorbing what little energy it receives, it's always cold, and you always crave hot cocoa. Conversely, at the equator, the sun's rays are always relatively direct, the surface does a good job of absorbing the sun's energy, it's always hot, and you always need sunscreen.

Terra-Trivia

The average high temperature in July in Singapore—located almost directly on the equator—is 88 degrees F. Even in January, it's 86 degrees F! The average high temperature in July in Barrow, Alaska—located well within the Arctic Circle—is 45 degrees F. In January, its average high temperature is –7 degrees F.

GeoRecord

The earth's lowest temperature was recorded at Vostok Station, Antarctica. On July 21, 1983, the temperature dropped to a remarkable –129 degrees F. Although this is still the official world record cold snap, an unconfirmed reading of –132 degrees F was reported at Vostok during the winter of 1997.

Why Does the Wind Blow?

To understand this question, you have to understand air pressure. As you stroll along the beach, you feel some pressure. Although your walk might seem relaxing, you support a column of the earth's atmosphere that surrounds the earth and extends upward about 6 miles, as shown in the following figure.

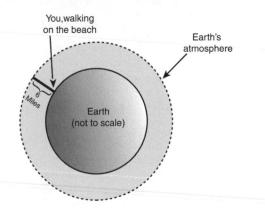

The weight of the atmosphere is about 14.7 pounds on every square inch of your body. If this pressure is wearing on you, head for the mountains. The higher the elevation, the smaller the column of air you're supporting and the lower the pressure. (The downside is that you have to follow those special baking instructions on the cake-mix box.)

Temperature also affects air pressure. As you've witnessed if you've ever watched a hot-air balloon drift gracefully overhead, the warmer the air, the lighter it becomes (and the higher the balloon rises). Warm air means lighter air, which means lower air pressure; conversely, cold air means heavy air, which means higher air pressure.

Nature follows predictable patterns. Water seeks its own level, mountains rise and valleys fall, and lawyers eventually move to the D.C. beltway. So it is with air pressure. The air in high-pressure areas moves toward areas of low pressure, resulting in wind. The greater the difference between the high pressure and low pressure, the faster the wind blows. Extreme differences in air pressure result in tornadoes and hurricanes.

They Call the Wind Many Things

People often use the terms cyclones, hurricanes, monsoons, tornadoes, twisters, and typhoons interchangeably to refer to any blast of windy, wet weather. Although some of these terms do refer to similar phenomena, they're not all synonymous. Here are the distinctions:

Twisters, tornadoes, and *whirlwinds* all refer to isolated points of extreme low pressure (warm, moist air) surrounded by rapidly spinning columns of wind. Although these roaring, funnel-shaped vortexes are not generally more than a couple hundred yards wide, they leave terrible destruction in their path when they touch down on the ground surface (often at a wind speed of as much as 300 m.p.h.). The most tornado-prone area in the world is the central plains states of the United States, or "Tornado Alley."

Waterspouts and *dust devils* are miniature and often short-lived versions of tornadoes. Waterspouts, as their name implies, are misty columns of spray over bodies of water. Dust devils are similar to waterspouts, except that they occur over land. Dust devils are often so small that they would be unnoticeable if not for the tiny column of sand and dust they swirl into the air. (Don't confuse dust devils with dust bunnies, the reclusive lint puffs found under beds and in the corners of rooms with hardwood floors.)

Hurricanes and *typhoons* are severe tropical storms with sustained winds of 74 m.p.h. or higher. These massive, circular- or oval-shaped storms can measure more than 300 miles across. A peculiar characteristic of these storms is their *eye*, an area of calm (often complete with blue sky) directly in the center of the storm. As with other types of storms, hurricanes and typhoons are associated with extreme low pressure. Hurricanes are storms that occur in North or Central America; typhoons occur in the western portions of the Pacific Ocean. Wherever they occur, these massive storms are associated with heavy rain, coastal flooding, and destruction of life and property.

Cyclone is an umbrella term (you should pardon the expression) for all the storms, of varying intensities, mentioned in this list (except for monsoons). Even low-pressure centers that do not develop into one of these types of severe storms are considered cyclones. The term *tropical cyclone* is also used more specifically to describe the Indian Ocean equivalent of a hurricane or typhoon. Interestingly, the circulation of winds associated with a cyclone varies in direction depending on the hemisphere in which it occurs. In the Northern Hemisphere, cyclonic winds spin counterclockwise. In the Southern Hemisphere, they circulate in a clockwise direction.

Terra-Trivia

On November 13, 1970, a tropical cyclone moved north through the Bay of Bengal and struck the low-lying delta that comprises the country of Bangladesh. The high winds, flooding, and resulting exposure and disease killed 300,000 people.

Monsoons: Monsoons differ from the other weather phenomena in this list in that they're not a particular type of storm. Instead, the term refers to a seasonal shifting of winds. Although generally associated with southern Asia, monsoonal climates are also found in northern Australia, West Africa, and other parts of the world. Monsoons may be "wet" or "dry." The "wet summer" phase of the Indian Ocean monsoon lasts from June to September and often brings torrential rains.

The Earth's Moving Surface

When geographers shift their attention from the sky above their heads to the ground beneath their feet, they encroach on the territory of geologists, but not too much. While geologists focus on the stuff the earth is made of and how it was formed, geographers are interested primarily in the uppermost layer of the earth's crust, where humans walk and live.

Solid Ground—or Not?

The earth's crust is good old terra firma—except that it's not so firma. What people like to think of as their anchor, as the bedrock on which they build, is really the equivalent of many giant rafts moving and bumping around on an orb floating in space. So much for stability.

If you were to assemble a jigsaw puzzle in which the pieces represent the seven continents, you might notice something that's crucial to explaining the movement of the earth's surface. If you jockey the pieces around a little, you would see that the shapes of the continents, especially South America and Africa, all fit together in one huge, interlocking piece—and for good reason: 225 million years ago, all the continents were once part of a huge supercontinent called Pangaea (Greek for "all earth"). About 180 million years ago, Pangaea began to split up. Its component parts slowly drifted away from (and in some cases, toward) each other, eventually resulting in the continental layout we know today. This process, called *continental drift*, still continues and is explained by the theory of *plate tectonics*.

Floating and Colliding Continents

According to the theory of plate tectonics, the earth's outer shell, called the lithosphere, is not continuous, but instead is divided into irregularly shaped rigid plates that float on an underlying molten layer called the *asthenosphere*. The plates move relative to each other at rates of as much as several inches a year, and they meet in a number of different ways.

At *divergent boundaries*, adjacent plates are actually forced apart by molten matter rising up from the asthenosphere to form ocean floor ridges. The Mid-Atlantic Ridge, which splits nearly the entire Atlantic Ocean floor in half north to south, is a good example of a divergent plate boundary.

When plates collide at what are called *convergent boundaries*, the older, heavier plate is forced to sink below the lighter, younger one, often forming deep ocean trenches. This process is known as *subduction*. If the colliding plates are carrying continents that also bang up against each, the result is new mountain formation. The Himalayas, for example, are the result of the collision between the Eurasian Plate and the Indo-Australian Plate.

Tectonic plates also meet at *transform boundaries*, or *faults*. In this case the plates don't collide head-to-head, but rather grind sideways against each other, creating lateral faults, such as California's famous San Andreas fault, and earthquakes.

> **GeoRecord**
>
> Some of the world's deepest subduction trenches are in the Pacific. The Marianas Trench in the western Pacific is the deepest spot on earth. There, in an abyss called Challenger Deep, the earth's surface plunges to 35,826 feet below sea level.

However it happens, the meeting of plates generates tremendous amounts of energy, which is released in part through volcanoes and earthquakes. In fact most of the world's volcanic and seismic (earthquake) activity occurs along plate boundaries and subduction zones. The most volatile of these zones is the perimeter of the Pacific Ocean, the infamous Ring of Fire, where the Pacific Plate is disappearing into deep subduction trenches. More than 75 percent of the world's volcanoes are located here, among them such celebrities as Mt. Fuji (Japan), Mt. Pinatubo (the Philippines), Kilauea (Hawaii), Mt. St. Helens (Washington State), and Popocatépetl (Mexico).

Tectonic plates and the Ring of Fire.

The Ring of Fire, as shown on the preceding map, stretches from New Zealand north through Indonesia, Japan, and the Pacific coast of Russia. Continuing across the north Pacific, the ring crosses Alaska's Aleutian Islands and continues south along the west coast of North and South America. Some of the world's largest population clusters are located along this treacherous ring.

The Great Shake: Earthquakes and Faults

If you were to draw a map with dots placed on the world's earthquake *epicenters* and then connected the dots, you would have almost an exact map of the earth's tectonic plates. When the massive pressure of two or more plates crunching against one another reaches a certain point—bam!—an earthquake occurs.

GeoJargon

The underground starting point of an earthquake is called its *focus* or *hypocenter*. The spot on the surface right above the focus is known as the earthquake's **epicenter**.

Earthquakes happen along faults, or cracks, in the earth's surface. As you've seen, transform faults, such as the San Andreas Fault, are caused by the lateral grinding or shearing of adjacent plates. Other types of faults are created when earth tension stretches or compresses the crust, causing it to crack. In an earthquake, a sudden release of energy causes the rock on one side of the fault to slip rapidly past the other, with either an up-and-down (vertical) or sideways (lateral) motion.

The wave vibrations generated by an earthquake are measured by a device called a seismograph. There are several ways to express the severity of an earthquake. The most common is the *Richter scale*, a logarithmic scale in which every whole-number increase represents a tenfold increase in earthquake magnitude. A seven on the Richter scale, for example, represents an earthquake ten times as powerful as a six. An earthquake's destructiveness, in terms of loss of life and property damage, depends on its magnitude and the degree to which the area around the earthquake epicenter is populated. By the time an earthquake hits seven or eight on the Richter scale, the damage can be catastrophic. A nine on the scale has never been recorded—an earthquake of that magnitude would result in virtually total destruction.

Tsunami or Tidal Wave?

Earthquakes that occur undersea can generate destructive ocean waves called *tsunamis*. You might also hear tsunamis referred to as *tidal waves*, but this term is inaccurate and misleading. Tsunamis, or tidal waves if you insist, have nothing to do with ocean tides.

Out at sea, a tsunami is not the huge wall of water you may remember from *The Poseidon Adventure*. A tsunami is barely noticeable, in fact, on the ocean's surface. When these fast-moving waves reach the shallow water of the shoreline, however, they build up to enormous heights and can cause extensive coastal damage. Seismic monitoring stations strung around the Pacific Ocean help warn vulnerable areas, such as Hawaii, of potential tsunamis.

Building Up the Earth's Surface

The movement of the tectonic plates is one of the earth's land-building mechanisms. As the continental plates press against one another, the earth's crust folds, warps, and faults. The process of building up the major features of the earth's surface through the movement and deformation of the crustal plates is called *diastrophism*.

Another primary land-building force is *volcanism*, which occurs most commonly along the fractured edges of continental plates. *Magma*, or molten rock, is forced up through cracks in the earth's crust to emerge at the earth's surface as *lava*. The buildup of lava and ash forms the familiar volcanic cones that soar majestically on the landscape.

Wearing Down the Earth's Surface

Nature's penchant for equilibrium is both minutely complex and beautifully simple. Just as nature's forces build up, they inevitably also tear down. Weathering, erosion, and other *gradational forces* combine to wear away in time even the highest mountains. Prying plant roots and the freeze–thaw cycle eventually reduce huge rocks to small stones. Acids and oxidation (rust) chemically break down the earth's minerals. The wind sculpts sand dunes, while windborne abrasives rasp away at exposed rock surfaces. But of all agents of erosion at work in the world, the most powerful by far is water.

Follow the Flowing Water

As a gentle rain falls to the earth's surface, the water runs off into countless tiny creeks that in turn feed into streams and eventually into rivers. Ultimately, the rush of water makes its way to the sea. At every stage of this cycle, water is eroding, wearing, and carving away at the earth's surface.

Sometimes the action of water on the landscape is dramatic and sudden as when raging floods shift the course of rivers. The magnificent Grand Canyon is a testament to the power of the Colorado River to cut through the earth's crust. Rocky coastal cliffs slowly yield to the ceaseless pounding of the ocean's waves.

In addition to cutting away at the earth's surface, water also moves around great quantities of its material. Streams typically originate in steep highlands and mountain areas, where they drop quickly, often forming rapids and waterfalls. This quickly moving water is capable of moving stones, sand, and silt (the *load* of the stream) as it descends through sharply cut gorges and ravines.

When fast-moving streams join to form a river in the lower flatlands, the flow begins to slow. The slower-moving water then spreads out to meander through *floodplains*, the wide, level valleys that store excess water when the river floods. No longer capable of moving large stones, the river now carries mainly sand and silt in its path to the sea.

As the widening river finally reaches the ocean, its flow has slowed to a crawl. Tremendous amounts of sand and silt are deposited at the mouth of the stream, often forming deltas. The deltas grow outward from the mouth of the stream, stretching their fingers into the sea. Deltas are often characterized by rich alluvial (water-deposited) soils renewed by the continual deposit of silt carried by the river from the land's interior.

The Crushing Effects of Ice

Water can also carve away effectively at the earth's crust in another form: ice. Many landforms in the Northern Hemisphere are the result of *glaciation:* During the ice ages, the last of which occurred between 8,000 and 15,000 years ago, huge sheets of ice spread southward from the North Pole. These massive 10,000-foot-thick glaciers scoured the land surface and left glacial lakes and deposits in their wake.

The remnants of these *continental glaciers* are still found today on the island of Greenland and on Antarctica. Smaller glaciers, called alpine glaciers, are also found in high mountain areas. *Alpine glaciers* advance and retreat seasonally, grinding the mountain rock into a fine mineral dust called *glacial till* that causes the spectacular aqua blue hues of many high mountain lakes.

The Least You Need to Know

- The earth's rotation on its axis once every 24 hours causes day and night.
- The earth's revolution around the sun once a year causes the changing seasons.
- The earth is hottest at the equator because that's where the sun's rays strike the earth most directly. It is coldest at the earth's poles because these areas receive the most indirect light from the sun.
- The earth's crust is composed of slowly moving plates that grind and bump against one another.
- The earth's surface is built up primarily by folding, warping, and volcanism; the surface is worn down primarily by running water and ice.

Chapter 3

Cultural Geography: The Hand of Humankind

In This Chapter

- Pick a language—any language!
- Religious persuasions
- The distinction between race, ancestry, ethnicity, culture, and nationality
- The political landscape
- The business of business

Cultural geography, or perhaps more accurately, human geography, is a fairly self-descriptive term. This chapter focuses on the people of the earth. If that statement sounds somewhat general, just compare this chapter to a winery tour: You get at least a taste of the wide selection offered.

The human experience incorporates language, belief systems, racial groups, political systems, customs, livelihoods, and interaction with the environment. Throw in a dash of history, and you're looking at just a few facets of this broad enterprise. This chapter briefly discusses political, economic, and urban geography as it relates to culture; these areas are covered in more detail in Parts 2, 3, and 4 of this book. First, you must understand a little of the terminology and classifications that cultural geographers use.

Parlez-Vous ... More than 3,000 Languages?

Thousands of languages are spoken in the world, grouped by linguists into a few *language families* consisting of languages believed to have had a common historic origin. The largest of these families is Indo-European, which includes languages spoken by about half the world's people.

Within language families are *language branches* (or *subfamilies*). The Indo-European family, for example, includes the Germanic branch and the Romance branch. Although languages within each branch might still be mutually unintelligible, they share common origins.

Branches are further broken down into *language groups*. Languages within a group share common origins in more recent history as well as a good deal of vocabulary and grammar. English, for example, is a language in the West Germanic group of the Germanic branch of the Indo-European language family. Complicated? You don't know the half of it. The relationship between English and German is evident in the similarity between, say, the English word *white* and the German equivalent *witte*. But if you remember your high school history, you'll know that England was conquered in 1066 by French-speaking Normans, who introduced a host of French words into the English language. Today's English is a far richer language thanks to it borrowings from French, a Romance language descended from the Italic/Latin branch of the Indo-European family.

It's the Same but Different: Dialects, Pidgins, Creoles, and Lingua Francas

You might think, after whittling this language jumble to a single language, that you would be pretty much home free. Not really. Although a resident of Chicago and a resident of Glasgow might both speak English, they can have a hard time understanding each other because they speak different dialects. *Dialects* reflect regional differences in vocabulary, pronunciation, and syntax that sometimes hinder fluid communication. The dialectic differences between a lobsterman in Maine and a catfish farmer in Mississippi, for example, might prove to be conversationally challenging.

A *pidgin* is a blend of different languages. Not the primary language of any of its speakers, a pidgin evolves so that speakers of different languages can converse. If, over time, the primary languages become lost and the pidgin language becomes the primary language, it's called a *creole*.

Another means by which people of different tongues converse is in a *lingua franca*, a language used by peoples whose primary languages are unintelligible to each other. As an example, ambassadors from Iran and Germany might converse in the language of diplomacy—which is now English. For ages, Roman Catholic priests could converse in Latin, no matter how different their native languages.

Terra-Trivia

The United States has no official language. Although the majority of the population is English-speaking, immigration continues to swell the number of people in the United States who speak other languages, especially Spanish. At the other end of the official-language scale, South Africa has 11 designated official languages.

Languages, Officially

An official language is one that has been designated by law as the tongue of the country. This type of decree might specify that the language (or languages) be used in schools, courts, businesses, and government. Mastery of the official language might even be a requirement for immigration into the country.

Some countries have not designated an official language, and others have several official languages. Areas within countries might promote languages different from the official language of the country.

A Look at the World's Religions: The Biggies

Because more than 6 billion people live on the earth and because religion is so intensely personal, you probably could make the argument that 6 billion religions exist on the earth. As with other geographic topics discussed in this book, some classifying and lumping together are in order here. This book defines a *religion* as any value system that incorporates worship and faith in a divine creative force.

A description of all the world's religions wouldn't fit in this book, much less in this chapter, nor would the intricacies of the ones briefly described, but the following list does include the religions with the most influence and the greatest number of adherents. (Although the list is incomplete, it does cover the religions practiced by more than 75 percent of the world's population.) Today more than ever it's important to have a basic understanding of these religions not only because of their huge impact on the world's geography, but because religion is increasingly a factor in so much of the strife afflicting the world today.

Geographers classify religions into two primary types: A *universalizing religion* is open to all human beings and attempts to spread its faith through missionary activities. (The three largest are Buddhism, Christianity, and Islam.) An *ethnic religion* generally encompasses specific groups of people in a particular location on earth. (Hinduism is the largest.) Because they are largely specific to a group and/or a place, such religions tend not to engage in missionary activity.

Christianity (2 billion adherents): This universalizing religion springs from Judaic roots. Centered on the life and teachings of Jesus, who lived in the first century C.E., Christianity's principal religious text is the Bible. Although this religion originated in the Middle East, today it's widely dispersed throughout much of the world.

Islam (1.3 billion adherents): Also a universalizing religion, Islam descends from the Judaic–Christian faiths and is centered on the life and teachings of the prophet Mohammed, who lived in the seventh century C.E. Its principal religious text is the Koran. Islam is the second largest religion and the fastest growing. It is practiced worldwide, but is most prevalent in the Middle East, North Africa, and Southeast Asia.

Hinduism (900 million adherents): Perhaps the world's oldest religion, the origins of Hinduism date back more than 5,000 years. Linked closely with the people and culture of India, this major religion is the world's largest ethnic religion. Not tied to any one prophet or theological system, it incorporates many doctrines and manifestations of the deity.

Buddhism (350 million adherents): A universalizing religion in the sense that it is open to all, Buddhism is highly tolerant of other beliefs and does not promote preaching or conversion. Buddhism originated as a Hindu reform movement in the sixth century B.C.E., based on the teachings and philosophy of Siddhartha Gautama, known as the Buddha, or "Enlightened One." Although Buddhism arose in India and has adherents around the world, today it is most prevalent in Southeast Asia, East Asia, and Japan.

Terra-Trivia

Although this chapter has identified the world's major religions, another group is worth mentioning: the *non-religious*. The number of people throughout the world who don't adhere to any specific religion, including atheists, who deny the existence of God, is thought to total nearly a billion.

Chinese traditional religion (225 million adherents): The major Chinese religion is a blending of Mahayana Buddhism, Confucianism, and Taoism, along with local beliefs and practices. More of an ethical code than a religion, *Confucianism* was founded in the sixth century B.C.E. by the philosopher Confucius (K'ung Futzu). *Taoism* also arose in China in the sixth century B.C.E. Its teacher, Lao Tsu, based his philosophy on a simple life in harmony with the way of nature. Chinese traditional religion is practiced primarily in China, Korea, and Southeast Asia. The communist campaign to eradicate religious practice in China, as well as the recent trend toward more openness and tolerance, make it difficult to arrive at an accurate estimate of the number of religious adherents in China today.

Judaism (14 million adherents): Rooted in the teachings of Abraham and the Torah of Moses, Judaism is thought to be between 3,000 and 4,000 years old. Although it's not one of the world's largest religions, it has global importance as the first major monotheistic religion and as the wellspring of both Christianity and Islam. Most of the world's Jews today live in the United States, Israel, and Europe, but there are also large Jewish communities in Canada, Latin America, and South Africa.

Race, Ancestry, Ethnicity, Culture, or Nationality— What Does It All Mean?

Cultural diversity, ethnic cleansing, and racial tension are in the news on almost a daily basis. Although the terms *culture*, *ethnicity*, and *race* are often used somewhat interchangeably, they have specific meanings to geographers. This section sorts through them and provides a few geographic examples.

The Diversity of Race

For better or worse, the peoples of the world have traditionally been sorted by race. A race is an identifiably distinct group of people, whose distinctiveness is based on inherited biological characteristics. These distinguishing characteristics include, among others, hair texture, height, facial characteristics, skin pigmentation, and even variations in blood type, each of which typically represents a physical response to the environment evolving over long periods of time.

The distinctions between racial groups can be extremely subtle, however, and variations within a racial group can be extremely wide. Nonetheless, for purposes of classification and study, geographers do categorize people by race. Although other geographer's lists might vary somewhat, these are the nine generally recognized racial groups:

- African
- Asian
- Australian (aboriginal, not *"Crocodile" Dundee*)
- Caucasian
- Indian (from South Asia, not American Indian)

- ◆ Indigenous American
- ◆ Melanesian
- ◆ Polynesian
- ◆ Micronesian (not always included in this list)

Ancestry: Climbing the Family Tree

The term *ancestry* simply refers to one's line of descent, or lineage. More and more, people attempt to piece together "family trees" and spend a great deal of time probing back through the generations. The term is cloudy from a geographical standpoint because it can mean different things to different people, including race, nationality, social status, and economic status. In census tables, ancestry typically refers to a connection to a particular country. In the U.S. census, a leading response to the ancestry question was not a foreign country, but rather simply "American."

Terra-Trivia

In the fiscal year ending on September 30, 2001, the United States admitted 1,064,318 immigrants.

The top five countries of origin were (in descending order): Mexico, India, China, the Philippines, and Vietnam.

The Nuances of Nationality

Although nationality is often confused with ancestry or even race, it is actually more of a legal concept, referring to a person's membership in a particular nation, either by birth or acquired citizenship. A nation can be thought of as a unified group of people sharing customs and ethnicity and usually occupying a territory, but not necessarily a state, as you'll see later in this chapter. Unlike race or ancestry, it's possible to change nationality by becoming a citizen of another country. The process of acquiring citizenship in a country other than the one in which you were born is called *naturalization*.

Coping with Changing Culture

Culture is a broad term that can be used to describe a society's entire way of life—its pursuits, behaviors, ideals, values, pleasures, and dreams. Technology, art, food, fashion, symbols, music, belief systems, and laws are all aspects of culture. As these elements are passed down from generation to generation and absorb outside influences, a culture becomes richer and more complex.

Los Angeles, California, offers an interesting example of how a culture develops and spreads. L.A. has emerged as the capital of U.S. pop culture. Trends, fashions, and fads emanate from Los Angeles (with the Hollywood film industry serving as transmitter and amplifier) and are picked up by the rest of the country and eventually the world.

An Eye on Ethnicity

A person's *ethnicity* can be based on many different criteria, such as racial characteristics, language (or dialects), religion, nationality, and culture. Any of these factors can identify an ethnic group as distinct from a larger population.

An example of the difficulties that can arise in defining ethnicity is the frequent misuse of the term *Hispanic* in the United States. The term is often listed in population tables as a race, although people of Hispanic origin can be of any race. Nor does it refer to any one nationality. Hispanics might be from Mexico, Puerto Rico, Cuba, the Dominican Republic, Central America, or South America, or a number of other countries. The common characteristic shared by people who describe themselves as Hispanic is either the use of Spanish as a primary or secondary language or family origins in a Spanish-speaking country.

Terms of Settlement: The Political Landscape

Geographers often use the term *political* to describe maps that feature human-made boundary lines and divisions, as opposed to *physical* maps that show mountains, rivers, and other features. Countries, states, territories, counties, and cities have defined (although sometimes contested) outlines that appear on a map but not on the actual physical landscape. Later chapters will address the world's political systems, forms of government, and national and regional divisions. The focus for now is on the variety of human constructs that define the political landscape.

Not Bound by Boundaries

Boundary lines surround us—from national borders to fences separating people's yards. The situation wasn't always like this, though. In some cultures (many Native American groups, for example), lines on maps were never used to demarcate one group's land from another. "No man's lands" were often established in which both parties could jointly use the middle ground without fear of conflict. With today's laser surveying equipment, those fuzzy boundaries are now often razor-sharp.

Terra-Trivia

Not all boundaries are razor-sharp. The Karakoram and Himalaya mountains, for example, are a boundary nightmare, with jagged peaks, violent storms, shifting glaciers, and avalanches. Unfortunately, Pakistan, India, and China share boundaries there, and regular disputes flare up as one country claims an area and another "occupies" it.

Here are some of the ways geographers define boundaries:

◆ **Natural versus artificial boundaries:** Boundaries based on landscape features (such as mountains, lakes, and rivers) are called *natural,* or *physical, boundaries.* The following map shows how easily they can be recognized (even on a map that doesn't show physical features) because they're rarely straight lines. The boundary between Texas and Mexico, for example, follows the Rio Grande and is therefore a physical boundary. If you see straight lines dividing places on a map, you're probably looking at *geometric,* or *artificial, boundaries,* which are usually surveyed lines often based on a parallel or meridian (a line of latitude or longitude). The straight boundary between Arizona and New Mexico is an example of an artificial boundary.

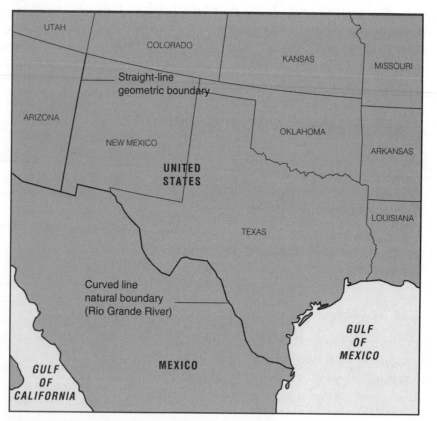

Natural versus geometric boundaries.

◆ **Antecedent versus subsequent boundaries:** Boundary lines drawn before an area has been well populated are *antecedent boundaries.* The western border between the United States and Canada is an antecedent border, drawn before "the West was won." (Native Americans were not factored into the governments' decision.) Borders drawn after an area has developed cultural characteristics are called *subsequent borders*, such as new boundaries drawn after the breakup of the former Yugoslavia.

◆ **Consequent versus superimposed subsequent boundaries:** Boundaries drawn after an area has well-established cultural characteristics can either respond to those characteristics or ignore them. *Consequent*, or *ethnographic, borders* are drawn in an attempt to respond to cultural dictates, as occurred when the Czechs and the Slovaks decided to go their own ways as separate nations in 1993. *Superimposed boundaries* are borders drawn rather arrogantly that "do their own thing" regardless of any preexisting cultural patterns. Superimposed boundaries placed by colonizing European powers, for example, typically divided indigenous peoples. Africa is laced with such boundaries. Tribal areas were simply cut in half as European countries divvied up the land of Africa by imposing boundary lines on a map.

Terra-Trivia

Geographers call the forces that tend to pull a country apart *centrifugal forces*. Differences of religion, language, or ethnicity all might act as centrifugal forces that cause a country to fracture. Forces that help to bind a country together are called *centripetal forces*. The sharing of religion, language, and ethnicity, the powerful force of nationalism, and even a common transportation system, can all act as centripetal forces.

Countries, Nations, and States

The distinction between countries, nations, and states can be somewhat confusing. First, forget about the use of the word *state* to define the political units that form a federal government (such as New York state or the state of California). What concerns you here is the use of the word *state* to describe a country or nation:

◆ **State:** An independent political unit with an established territory, boundaries, population, and government controlling its own internal and foreign affairs

◆ **Country:** A state, or the territory of a state (the word *state* has more of a constitutional and political connotation; *country* has more of a territorial connotation)

◆ **Nation:** A closely linked group of people occupying a territorial area

Stateless nations do exist. The most visible examples are the Palestinians of the Middle East and the Kurds, who live on lands straddling Turkey, Syria, and Iraq. A distinct nation of people can be the primary occupants of a state (Italy, France, and Japan, for example). Many examples exist of multinational states in which two or more nations of people occupy one state (the Flemish and Walloons in Belgium, for example). And, after reading this paragraph, it's entirely possible to be in a state of confusion.

Sorting Through Settlements

The people of the earth are becoming more urbanized. As global economies shift away from an agricultural base, more and more people are flocking to urban areas. Many of these areas are growing at an alarming rate, especially in the developing world. About 40 percent of the population in developing countries is now urban; that figure is expected to increase to 56 percent by 2030. In this chapter, *urban* refers to a centralized area regardless of size.

The following terms describe urban areas, in order from smallest to largest:

- **Hamlet:** Tiny; maybe just one or two dozen buildings (this one sounds quaint!)
- **Village:** Larger than a hamlet but not as large as a town
- **Town:** Bigger than a village and generally centered on a nucleated business area
- **City:** Same concept as a town, only larger and more complex
- **Conurbation:** A large, built-up area in which two or more cities have coalesced into one huge unit
- **Megalopolis:** A huge, built-up area in which several conurbations have coalesced; the Boston-to-Washington corridor on the East Coast of the United States is a prime example

Getting Down to Business

As you surely already know, vast economic differences mark the countries of the world. What might be surprising however, is just how vast these differences are. One gauge of economic status is per capita *gross domestic product* (or GDP), which is how much money a country produces per person living in the country. Countries such as Somalia (in Africa) have a per capita GDP of $550. At the other end of the scale are countries such as Switzerland, with a *per capita GDP* of $31,700.

GeoJargon

The term **gross domestic product** (GDP) refers to the total market value of all the goods and services produced in a country in a given year. It includes all consumer spending, investments, government spending, as well the value of all exports minus the value of imports. GDP, or more important, the annual growth in GDP is a key indicator of the health of an economy. **Per capita GDP** is the total GDP divided by the number of people in the country in question.

Why do such huge disparities exist? The answer, as with most in geography, is often complex. History, resources, education, and economic systems all factor into the mix. Although the economies of most countries are blends of two or more types, there are still only three basic types of economic systems:

A *subsistence economy* is based in rural areas and around family units. In this survival economy, people produce goods and services for their own consumption and that of their immediate family. The exchange of currency and goods is minimal, and people are often just one natural disaster away from famine. Typical of this economic type is *slash-and-burn* agriculture, nomadic herding, and intensive small-plot agriculture. Subsistence economies are prevalent in the world's least developed countries.

In a *commercial economy*, the most widespread in the world, people freely produce goods and services for sale in competitive markets, where prices are determined by the law of supply and demand. This is the dominant type of economy in developed countries and has been making rapid inroads in developing countries as well.

GeoJargon

Slash-and-burn agriculture involves the cutting and burning of forested areas as a preparation for planting crops. Because the soil in many tropical forests is poor, soil nutrients decline after just a few years of planting. Typically, the farming family simply moves to another area of forest, and the cycle is repeated. This type of agriculture is also called *shifting cultivation,* or *swidden.*

In a *planned economy*, the central government controls the production and distribution of goods and services and sets prices. Mainly associated with communist governments, planned economies are notoriously inefficient. With the dissolution of the Soviet Union and the collapse of Eastern European communist regimes in the early 1990s, the number of planned economies around the world has declined dramatically.

The Global Economy

You will be seeing the word *globalization* often throughout this book, for the simple reason that it is one of the major forces shaping the world in the early twenty-first century. But what is it, exactly, and why should you care about it? When most people hear the word *globalization* they think of multinational corporations and businesses that wheel and deal across far-flung borders. And to an extent, they are right. Globalization has to do with the freer, more rapid movement of goods, services, labor, technology, and capital around the world. This phenomenon is not new, but advances in technology and telecommunications, especially the Internet, have greatly accelerated it.

Globalization and the lowering of trade barriers have helped raise growth rates and living standards in many developing countries. China, still a communist country politically, has embraced market capitalism with gusto and is now the world's second largest economy. Economic liberalization in India continues to fuel robust economic growth. In Latin America, on the other hand, globalization is being blamed at least in part for the renewed political instability and financial woes plaguing much of the region.

There's more to globalization, however, than its purely economic aspect. The globalized world is a much smaller and more interconnected place, not only economically, but also culturally, socially, politically, and environmentally.

The Least You Need to Know

- Although thousands of languages are spoken on the earth, they are classified into just a few language families.

- Christianity, Islam, Hinduism, Buddhism, Chinese traditional religion, and Judaism are practiced by more than 75 percent of the world's population.

- *State* and *country* both refer to a political entity with an established territory, population, and government. *Nation* refers more to a people than to a specific area.

- The three major types of economies are subsistence, commercial, and planned.

- Globalization is bringing down barriers between nations and making the world a much smaller and more interconnected place.

Chapter 4

Maps: A Geographic Love Affair

In This Chapter

- ◆ Identify different types of maps
- ◆ Know why a map can never be perfect
- ◆ Let symbols on the legend help you read a map
- ◆ Use a map to locate positions and measure distances
- ◆ Read the ups and downs of the earth's surface
- ◆ Develop a sense of direction

Maps are the basic tools of geography. Even people who don't know much about geography are drawn to the beauty and usefulness of a map. This chapter provides all the basics you need in order to own and operate your map. Maps provide an enormous amount and variety of information. This book uses maps throughout to help you locate places around the world. Maps can also give you an idea of the relative size of places and the distances between them, and enable you to determine directions. The map skills described in this chapter will prove to be useful tools in your quest for geographic mastery.

Millions of Maps

Maps exist for virtually every type of information under (and including) the sun. Maps show the ocean bottom, temperatures in India, and the best route to get to a baseball game. They come in all sorts of sizes and formats: wall maps, folding maps, miniature maps, and globes. They are found in newspapers, atlases, and on the Internet.

One way to categorize maps is by the type of information they convey. Here are a few common map types:

Terra-Trivia

The oldest surviving terrestrial globe was created by the German Martin Behaim, in 1492. It represents the world as Columbus knew it before he set off on his voyage of discovery.

Terra-Trivia

Ancient Polynesians created maps by tying sticks together to make ocean-wave pattern charts. Seashells tied to the sticks represented the Pacific islands.

- ◆ **Weather:** A common feature in newspapers and TV news programs

- ◆ **Road:** Impossible to refold into its original form while you're in the car

- ◆ **Political:** Shows the boundaries of nations and their political subdivisions (states or provinces, counties, cites, townships, and so on)

- ◆ **Physical:** Presents the earth's physical features: mountains, valleys, deserts, rivers, lakes, and so on

- ◆ **Coastal chart:** Familiar primarily to old salts and weekend sailors

- ◆ **Star chart:** A map of the heavens (not to be confused with map of the stars)

- ◆ **Map of the stars:** Points out the homes of Hollywood luminaries

Pondering the Projection Dilemma

A map represents the ins and outs of the earth's surface on a flat piece of paper. If you've ever given a friend directions by scrawling a map on the back of an envelope, you know the basic process. Mapmakers (cartographers) attempt to do the same thing for the entire earth. Although this process might not seem difficult, there's a fly in the proverbial cartographic ointment.

When you peel an orange and attempt to lay the peel flat on a table, you begin to see the problem cartographers face: The earth is not flat. The spherical earth—an oblate spheroid, to be exact—poses a fundamental dilemma to cartographers: No matter how hard a mapmaker tries, the rounded surface can't be represented 100 percent accurately on a flat map.

Cartographers call their attempts to create flat representations of the spherical earth *map projections.* Creating a perfect projection is impossible. If you're a doubting Thomas, try to draw every side of a baseball accurately on a piece of paper. You might end up with a pile of sketches that look vaguely like torn-off baseball covers, a pile of eraser shards, and a bad headache.

GeoJargon

The term **map projection** comes from the old practice of placing a light source inside a transparent globe and projecting the image onto a piece of paper.

The only truly accurate representation of the earth is a globe because it preserves all the map characteristics of interest to a cartographer: area, shape, direction, and distance. A map projection can preserve only one or two of these characteristics. If you focus on getting an area's size exact, for example, its shape might be distorted, and vice versa. Cartographers have names for projections that preserve each of these characteristics:

- **Equal area:** Area
- **Conformal:** Shape
- **Azimuthal:** Direction
- **Equidistant:** Distance

Mercator's Famous Projection

One of the most important names in the history of mapmaking is Gerhardus Mercator, a sixteenth-century Flemish (Belgian) geographer, mathematician, and cartographer. He is best known today for the map projection that bears his name, which he introduced in a famous 18-sheet world map published in 1569.

What made Mercator's projection so influential was that it preserved shape and direction, allowing sailors for the first time to use a map to plot "true" direction. The problem with the Mercator projection is that it badly distorts area, especially toward the poles. Eurasia and North America loom very large in contrast to South America and Africa. Despite this basic flaw, the Mercator projection stuck for centuries, and although it has been largely superceded today, the mental map of the world you (hopefully) carry in your head is most likely a Mercator projection map.

Mercator's Projection.

Terra-Trivia

One of Mercator's other claims to fame was that he published the first bound collection of maps to be called an Atlas. Mercator named his book after a mythical king of Mauretania (ancient Libya), who was also a wise philosopher, mathematician, and astronomer. The Titan Atlas, a god of Greek mythology, was also associated with map compilations and was often depicted on title pages bearing the heavens on his shoulders.

Sorting Through the Symbols

Because maps represent a reduced part of the earth's surface on a small piece of paper, certain symbols (points, lines, and patterns on a map) are often used to represent landscape features. Cartographers also use color to help map users "read" a map:

- ◆ **Blue:** Water features (rivers, lakes, and ponds, for example)

- ◆ **Brown:** Height and depth (mountains and valleys, for example)

- ◆ **Green:** Vegetation features (woods, orchards, and scrub brush, for example)

- ◆ **Red:** Roads and urban development (important roads and built-up urban areas, for example)

When a map has many symbols or its symbols are tricky, it usually includes a *legend* (sometimes referred to as a *map key*) for explanation. See the following example of a legend.

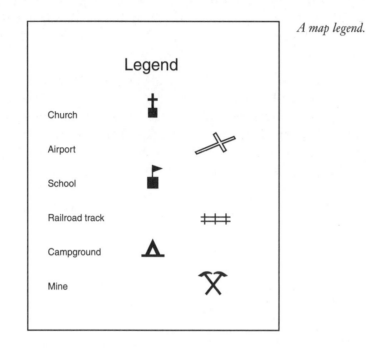

A map legend.

Now, Where Was I?

After you have your map in hand, you can locate a position on the map in two basic ways: *relative location* and *absolute location*.

Relative Location

You generally use relative location when you're giving someone directions. To direct a person to the local convenience store, you might say something like this: "Go to the airport, and continue through three lights; the Trendy Mart is on your right." These directions give the position of the Trendy Mart *relative* to the airport.

GeoJargon

Relative location specifies a particular position or place by relating it to another known position or place. **Absolute location** indicates position based on the coordinates of a grid system (such as latitude and longitude).

Absolute Location

Absolute location identifies the position of a place by means of a grid system. Although absolute location is a useful way to locate something in an atlas, it's not often used in everyday speech. The person who wants to find Trendy Mart might give you a strange look if you say, "Proceed on a 232-degree azimuth to coordinates 41.45 degrees north latitude and 72.40 degrees west longitude."

The old geographic standbys of latitude and longitude, often used to determine absolute location, can be confusing and difficult to work with. Many atlas publishers have given up hope that readers will ever master that system and no longer use latitude and longitude to locate positions on their maps. The National Geographic Society's *Atlas of the World*, for example, uses letters down the sides and numbers across the top and bottom of each map to help readers locate places. To locate Ulan Bator, Mongolia, for example, you would first look in the *index* of the atlas and see that it's on page 83, grid square D14. (The grid square system is similar to playing Bingo, but without the potential cash rewards.)

Despite the difficulties of working with latitude and longitude, no discussion of absolute location would be complete without an explanation of these terms.

Learning to Live with Latitude and Longitude

Lines of latitude and longitude form an imaginary grid on the surface of the earth that allows you to specify the exact location of any given place. Latitude indicates a location's north or south position on a map or globe; longitude indicates its east or west position. Both are stated in degrees and subdivisions of degrees. (Like a circle, a globe contains 360 degrees. Each degree is divided into 60 minutes of arc; each minute, into 60 seconds of arc.) When you specify a location using this grid system, latitude is given first and then longitude. Using both latitude and longitude, you can pinpoint any location on the earth.

Terra-Trivia

The earliest known system of latitude and longitude dates back to Hipparchus in the year 150 B.C.E.

Lines of latitude, or *parallels*, run east–west; lines of longitude, or *meridians*, run north–south. If you follow a meridian, you eventually run into the North or South Pole. If you follow a parallel, you head east or west and travel parallel to the equator.

Here's an overview of latitude basics:

- Latitude measures position north or south of the equator.
- The equator is 0 degrees latitude.

- The North Pole is 90 degrees north latitude.

- The South Pole is 90 degrees south latitude.

- Latitude ranges from 0 degrees to 90 degrees north or south.

- Always state direction (north or south) when you're specifying latitude.

And here's an overview of longitude basics:

- Longitude measures position east or west of the prime meridian, a north–south line that runs through Greenwich, England.

- The prime meridian is 0 degrees longitude.

- Longitude ranges from 0 degrees to 180 degrees east or west. Unlike latitude lines, longitude lines are not parallel because they come together at the poles.

- Always state direction (east or west) when you're specifying longitude.

Prime Meridian
(0 degrees
longitude)

North Pole (90 degrees
north latitude)

Lines of latitude and longitude.

Equator
(0 degrees
latitude)

Parallels

Meridians

Geographically Speaking

Longitude is not only essential in determining location, it is central to the world's method of keeping time. Because it takes the earth 24 hours to rotate once and the earth is a globe with 360 degrees of longitude, it's easy to figure out that it takes the earth an hour to rotate 15 degrees. Although this fact was known for centuries, it wasn't until the late nineteenth century that it was used to design a universal time standard that would end the confusion caused by each community's use of its own local solar time.

By international agreement in 1884, the prime meridian, or 0 degrees longitude, was declared to be the line running through the Royal Observatory at Greenwich, England. Starting at the prime meridian, 24 standard meridians spaced at 15-degree intervals were each designated as the center line of a one-hour time zone. For each time zone traveled east of Greenwich, the clock advances an hour. For each zone west of Greenwich, it loses an hour. When it's noon at Greenwich, it's midnight on the other side of the world at 180 degrees longitude. This line, where one day turns into the next, was designated the International Date Line. Although still roughly equivalent to the 180-degree meridian, the Date Line twists and turns to avoid splitting countries into different days. Although the time standard adopted in 1884, Greenwich Mean Time (GMT), has been superceded by the more precise Universal Coordinated Time (UTC), the principle is still the same.

Sizing Up Scale and Distance

Maps by necessity are reductions of the surfaces they represent; otherwise, in most cases, they'd be too huge to be of any use. So maps generally are drawn to a reduced scale. As you've already seen, scale is the ratio between a distance on a map and the corresponding distance on the surface of the earth.

If you have ever drawn up a plan for remodeling your kitchen, you know how scale works. Suppose that you draw your plan on ¹/₄-inch grid paper and decide that each ¹/₄-inch grid square on the paper equals a one-foot square in your kitchen. When you plot a refrigerator on your plan, you draw it as three grid squares across by three grid squares deep (or ³/₄-inch × ³/₄-inch). This *scale* drawing of the refrigerator in your kitchen measures three feet across by three feet deep.

The scale used on a map is generally much smaller than the scale used on an architectural plan because the geographic distances are so much larger. (Remember: The greater the ground distance per

Terra-Trivia

The circumference of a globe in a typical classroom is only 48 inches; the equatorial circumference of the earth is 24,902 miles.

map unit, the smaller-scaled the map is.) In contrast to a typical house-plan scale, in which ¹/₄-inch equals 1 foot, a typical map scale inch equals one mile. (Because 63,360 inches are in a mile, the scale could be written as 1:63,360 or ¹/₆₃,₃₆₀). One inch of map, therefore, is equal to one mile of the earth's surface.

What Are Those Numbers at the Bottom of the Map?

A map's scale can be expressed as a ratio (1:24,000) or as a *representative fraction* (¹/₂₄,₀₀₀). The first number in the ratio (the numerator in the fraction) represents a map distance; the second number (the denominator in the fraction) represents a ground distance. As long as you use the same unit for both the map and the ground distance, you can plug in any unit: 1 inch on the map equals 24,000 inches on the ground, 1 centimeter on the map equals 24,000 centimeters on the ground, or one pencil length on the map equals 24,000 pencils on the ground.

Bellying Up to the Bar Scale

Any discussion of scale is going to involve a lot of fractions, ratios, and proportions—the kind of stuff many people like to avoid! If you're one of those people, you might still be in luck. On some maps, an aid called a *bar scale* is printed in the margin.

Here's how to use a map's bar scale (sometimes called a graphic scale) to calculate the distance between two features on the earth's surface:

1. Locate two features on the map.

2. Lay a piece of paper along the two features.

3. Use a pencil to place a tick mark on the paper at each feature.

4. Move the paper to the bar scale, and compare the distance between the tick marks to the bar scale.

5. Notice that the distance on the scale represents the distance between the features on the ground.

6. Be sure to note the measurement used in the bar scale (feet, miles, or kilometers, for example).

Remember that a map is a *reduced* picture of the surface it represents. To determine actual ground distances, you can use either a map's numerical scale (ratio or representative fraction) or its bar scale. Knowing how to calculate distances on a map comes in very handy when you're planning a backpacking adventure, canoe trip, or bike route.

Be sure to measure along the actual route you intend to follow—don't just measure the straight-line distance between your starting and ending points.

Using a bar scale.

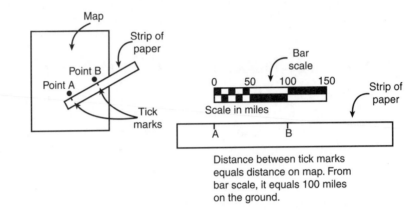

Distance between tick marks equals distance on map. From bar scale, it equals 100 miles on the ground.

If you're estimating distances for a trip you plan to make by car, you have some simpler options. Many road maps specify mileage along the major roads that connect towns and road junctions. Road maps and road atlases also often provide mileage charts that enable you to determine quickly the driving distances between large cities.

The Rise and Fall of Elevation and Relief

This chapter has already examined the problems mapmakers face when they try to map the spherical earth on a flat piece of paper. The dilemma has another component because, as everyone knows, the earth's surface is far from smooth. Soaring mountain peaks reach more than five miles above the surface of the sea, and plunging ocean abysses dip almost seven miles deep. The earth is spherical and its surface irregular—a cartographer's nightmare.

Two terms are often confused in describing the ups and downs of the earth's surface. *Elevation* refers to the height (or depth) of the earth's surface above or below a baseline (usually sea level). *Relief* refers to the variation in elevation from one place to another. An area of extreme relief is an area with dramatic fluctuations in elevation.

Cartographers indicate elevation and relief on maps in many ways. Usually, shading and hatching (small parallel lines) are used to depict major mountain ranges or river valleys, while different colored tints indicate different elevational zones. Sometimes the actual height of a hill or mountain is printed next to its top.

Contour Lines and Topo Maps

Perhaps the most accurate—and difficult to understand—method of indicating elevation and relief is with *contour lines*. A contour line on a map represents an imaginary, usually curvy line on the ground that follows a single elevation. This means that every point on the line is at the same elevation. To understand how a contour line works, picture a map of a mountain lake. If you trace on your map the point at which the water meets the shore around the lake, you trace a line of equal elevation (because water always serves as a level surface). If you trace the shoreline again after the lake floods and the water level rises 10 feet, the contour line of equal elevation is 10 feet higher than the original line.

If you continue this process, you eventually create a map that shows in great detail the elevation and relief (a.k.a. the topography) of the area around the mountain lake. Maps that depict relief by means of contour lines are called *topographic maps*. Hikers, developers, planners, and civil engineers find topo maps especially useful because they show surface relief with such a high degree of accuracy. The United States Geological Survey produces topographic maps in a variety of scales for most parts of the United States. You can obtain them from many public and private agencies or order them directly from the U.S.G.S. (call 1-888-ASK-USGA).

A topographic map.

Section of 1:24000 scale, New Britain Quadrangle, Connecticut-Hartford Co., 7.5 Minute Series (topographic) United States Geological Survey, Reston, VA 22092, in cooperation with Connecticut Highway Department.

> **Terra-Trivia**
>
> The spacing of contour lines on a topo map tells you a great deal about the lay of the land. If the lines are evenly spaced and wide apart, the terrain is flat or gently sloped. Lines that are evenly spaced but close together indicate a steep rise. You can even tell whether a slope is concave or convex by the configuration of contour lines. If the lines are close together at the top but widely spaced at the bottom, the slope is concave. Lines that are widely spaced at the top and close at the bottom indicate a convex slope.

Setting Off in the Right Direction

Everyone deals with direction all the time. Whether you're heading toward North Main Street or watching the sun slowly setting in the west, direction is key to orienting yourself in the world and describing the location of one place in relation to another. The most common way of expressing direction is still *north*, *south*, *east*, and *west*. These are the *cardinal* (meaning "most important") points of the compass. Most early maps from the fifteenth through the seventeenth centuries featured an elaborate compass rose like the one shown in the following figure. The rose indicated not only the four cardinal directions, but also four intercardinal directions (northeast, southeast, southwest, and northwest), and as many as twenty-four further subdivisions.

Compass rose.

North refers to the direction you follow to get to the North Pole. No matter where you are on earth, if you set off toward the North Pole, you are heading due north. If you're directionally savvy, you might have already guessed that to head toward the

South Pole is to travel due south, or directly opposite of due north. At a right angle to this north–south line lie east and west. On the face of a clock (nondigital, of course), if north is at 12 o'clock, then east is at 3 o'clock, south is at 6 o'clock, and west is at 9 o'clock.

Terra-Trivia

If you're standing at the North Pole, the only direction in which you can travel is south.

Northeast is located directly between north and east, at 1:30 on the clock face. I'll leave it to you to figure out southeast, southwest, and northwest, and all of the many other possible subdivisions of the compass rose.

Magnetic Attraction

In addition to looking at lines on a map, gazing at the stars, or observing the sun rise or set, people rely on another indicator of north: the magnetic compass. Perhaps you have mounted on your car's dashboard a compass orb that spins wildly as you careen around city streets. Or maybe you remember that 10-mile hike you made with the scout troop, relying on only your trusty compass (and the knowledge that your scout leader was trailing 50 yards behind). In either case, you rely on the magnetic needle in your compass to point north.

This method has just one minor problem: In most places on earth, the magnetic compass needle points to a different north than the geographic North Pole. A magnetic compass points to the earth's *magnetic* north pole, as shown in the following figure. The North Pole through which the earth's rotational axis passes (where meridians converge and where Santa Claus lives) is a different place from the magnetic north pole.

In many cases, the difference between the true north of the North Pole and magnetic north (called *magnetic declination*) is not significant. If you were to set off on a long hike, embark on a transoceanic voyage, or orient the photovoltaic cells on your solar home, however, you would have to consider this deviation. You can make the correction by using the *declination diagram* provided on most detailed (topographic) maps or with declination charts that show the magnetic variance over time. (These charts can be important because, in addition to not coinciding with true north, magnetic north moves around slightly over time.)

Terra-Trivia

The magnetic compass was in general use by the thirteenth century.

Magnetic north.

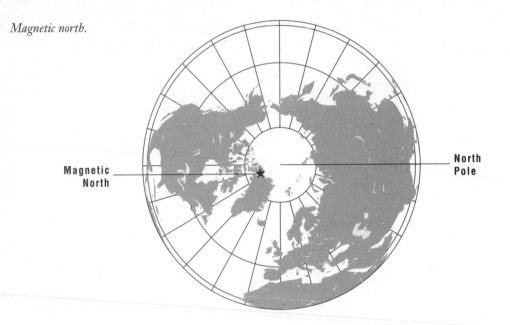

Technical Directional

Before finishing the subject of direction, you should be familiar with another term. *Azimuth* is a more precise way to express direction than by using the points of a compass rose. Topographic maps generally show directions as azimuths. Again, you need to visualize a clock face, but one that's divided into 360 degrees, as shown in the following figure.

An azimuth is measured clockwise from a north baseline (0 degrees or 12 o'clock on the clock face) to 360 degrees. Three o'clock, or due east, is a 90-degree azimuth; six o'clock, or due south, is a 180-degree azimuth.

Terra-Trivia

Few people realize that when you sail from the Caribbean Sea to the Pacific Ocean through the Panama Canal, you travel in a *southeasterly* direction. To see for yourself, check out the Central America map in Chapter 17.

New Directions in Mapping

In the second century C.E., the Roman geographer Ptolemy's atlas was the cutting edge of technology. Since that time, the field of geography has continued to evolve, and today technology is breaking new ground at a blistering pace. You can access maps and atlases from a computer with the click of a mouse and transform data into maps with the press of a key.

Azimuth.

Not that long ago, the idea of a video map display on the dashboard of your car, tracking your position and helping you navigate, seemed like something out of the future. Well, the future is now. The technology that makes it possible is the *global positioning system* (GPS), a worldwide radio-navigation system that uses 24 satellites and their ground stations to calculate positions on earth accurate to within yards now, to within inches in the not-too-distant future. Originally developed for the military, GPS has revolutionized marine navigation and is being used today to navigate and track vehicles, locate missing people, guide tractors, manage emergency responses, and increase the accuracy and versatility of maps. With GPS receivers getting smaller and cheaper and finding their way into laptops, who knows, you might even be able someday to locate your dream date via GPS.

Another technology that is transforming modern mapmaking is *geographic information systems* (GIS), a computer software/hardware system that links maps to databases and facilitates the manipulation, analysis, and display of spatial data. This technology has been a boon for environmental scientists, city and town planners, and marketing consultants—the list continues to grow. Geographers can use this digital information to determine, for example, the impact of local development on wetlands, track delivery of an Acme widget, or choose a prime location for the next Trendy Mart.

The Least You Need to Know

♦ No map can represent the earth's surface without distorting size, shape, distance, or direction.

♦ Symbols are used to represent landscape features on a map; a symbol key (legend) explains their meaning.

♦ Relative location is useful for locating one position with respect to another. Absolute location identifies the position of a place by means of a grid system, such as latitude and longitude.

♦ Elevation refers to a feature's height above or below sea level. Relief is the variation in elevation.

♦ The cardinal directions north, south, east, and west correspond to the directions of meridians and parallels.

Chapter 5

Here We Go! A World of Many Regions

In This Chapter

- ◆ Identifying hemispheres, continents, and regions
- ◆ Counting the continents
- ◆ Understanding the next three parts of this book
- ◆ Covering the essentials

You're about to go on a geographic journey around the globe, to places you know well and places that might seem strange and exotic. Prepare to witness a world filled with a mind-boggling diversity of environments and peoples. As you encounter indigenous populations living on the land and city dwellers who rarely step on anything other than a paved surface, how do you make sense of all the variety?

One way to get a handle on a confusing world is to divide it into parts called regions or subregions (also known as areas, territories, realms, ranges, fields, and domains). Although the terms are not important, you should understand each region's similarities and differences. This chapter identifies the major regions of the world and discusses the shared characteristics that qualify them as regions.

The Hemispheres: Half a World Away

If you choose the equator as the earth's dividing line (midway between the North Pole and South Pole), you can effectively divide the earth into two halves. A hemisphere is simply half a sphere. The northern half of the earth is called the Northern Hemisphere; the southern half, the Southern Hemisphere.

Northern Hemisphere.   *Southern Hemisphere.*

Terra-Trivia

A dozen countries around the world straddle the equator and are thus in both Northern and Southern Hemispheres. The divided dozen are: Ecuador, Colombia, Brazil, Gabon, Republic of the Congo, Democratic Republic of the Congo, Uganda, Kenya, Somalia, the Maldives (in the Indian Ocean), Indonesia, and Kiribati (in the Pacific Ocean).

The earth can also be divided into an eastern half and a western half. But this is a more arbitrary division because any meridian can be used to divide the earth into two equal halves (the equator, on the other hand, is the only line of latitude that divides the globe equally). The usual line of demarcation between Eastern and Western Hemispheres is either 0 degrees longitude (through the United Kingdom) and 180 degrees longitude (through the Pacific), or 20 degrees West longitude (through Iceland) and 160 degrees East longitude (through the Pacific). The advantage of the latter divider is that it leaves all of Europe in the Eastern Hemisphere and just the Americas in Western Hemisphere. Using 0 and 180 degrees to divide east from west, as in the following maps, leaves part of the western fringe of Europe and Africa in the Western Hemisphere.

Western Hemisphere.   *Eastern Hemisphere.*

The Seven Continents—or Maybe It's Five? or Four?

One of the first geography "facts" we all learn in school is that the world is divided into seven continents, but that division is just as arbitrary in many ways as the distinction between Eastern and Western Hemispheres. No matter which definition of "continent" you accept, you wind up with problems. A continent is generally considered to be a huge landmass surrounded entirely by water, a definition that works well for only about three of the seven continents.

The traditional seven continents (as shown on the following map) are, of course: North America, South America, Europe, Asia, Africa, Australia, and Antarctica. Some experts argue, however, that North and South America comprise only one continent because they're connected by the Middle American land bridge and that Europe and Asia comprise one continent because they're a single landmass with no intervening body of water separating them. If you accept these arguments, the world has only five continents.

Or maybe just four. Even Africa doesn't fit exactly this book's definition of a continent. In the northeast, Africa is separated from Asia by the Gulf of Aden and the Red Sea. The Red Sea tapers into the Gulf of Aqaba and the Gulf of Suez. Only the thin, man-made Suez Canal connecting the Gulf of Suez with the Mediterranean Sea severs Africa from continental Asia. Some geographers go so far as to consider Europe, Asia, and Africa a single supercontinent sometimes (rarely) referred to as Eurafrasia.

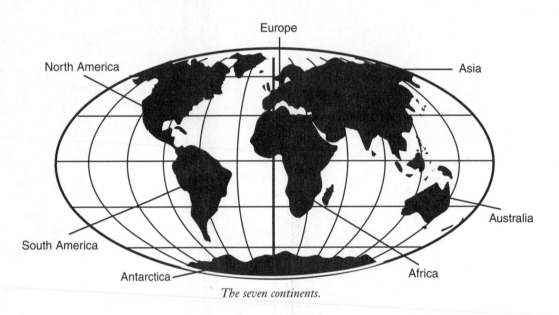

The seven continents.

The Europe/Asia/Eurasia Dilemma

Look at any world map and you see why referring to Europe and Asia as distinct continents is problematic. For the most part, they are a single continuous landmass, with neither side "entirely surrounded by water." You can throw that part of the *continent* definition out the window. Most of the boundary between Europe and Asia is not water but dry land. The following map shows the traditional dividing line between Europe and Asia (as you can see, it's a tortured distinction).

The line between Europe and Asia runs along or through a combination of mountains, seas, and straits: the Ural Mountains, the Caspian Sea, the Caucasus Mountains, the Black Sea, the Bosporus Strait, the Sea of Marmara, and the Dardanelles Strait. Because the distinction between Europe and Asia is difficult to justify using any definition of the word *continent*, many geographers simply refer to the continent of Eurasia.

GeoRecord

The largest continent in land area is Asia, with 17.3 million square miles. It represents about 30 percent of the earth's total land surface, followed in size by Africa, North America, South America, Antarctica, Europe, and Australia.

Australia: Island or Continent?

Earlier in this chapter, a continent was defined as a huge landmass surrounded by water. Although islands are also surrounded by water, most are not huge landmasses. (Unfortunately, "huge" in this context has no clear definition.)

Greenland is relatively huge (more than 840,000 square miles), but not quite huge enough for continent status. To be considered a continent, a landmass must be at least as large as Australia (nearly 3 million square miles). If it's any smaller than Australia, it's just an island.

Continents are tidy ways for geographers to divide the world into large chunks based on (mostly) obvious physical characteristics, but they usually are simply too big and too diverse to be studied as one unit, and they're not necessarily cohesive regions. To really begin to get handle on the world, geographers break the world up into smaller chunks called regions.

The Europe–Asia boundary.

Okay, So What's a Region?

Regions are hard to define exactly—five geographers might identify the world's regions in five different ways. When you're identifying a region, you look for some degree of sameness. What physical, cultural, economic, or historical attributes make one area distinct from another? Geographic proximity is a factor, as political boundaries often are. Depending on the weight you assign to any particular factor, definitions of regions can and do vary.

Although a region may coincide with an entire continent, that's not always the case. Regions usually encompass a much smaller area than an entire continent. Remember that the definition of a continent, as you read earlier in this chapter, has nothing to do with culture, history, economics, climate, or vegetation—all elements that factor strongly in the determination of a region.

Geographically Speaking

Don't confuse the division of the world into continents with the *Continental Divide* (or *Great Divide*). This term refers to the elevated "backbone" of a continent that divides rivers flowing in one direction from those flowing in the opposite direction, In North America, the Continental Divide runs north–south roughly along the Rocky Mountain chain. Rivers to the east of the divide flow into the Gulf of Mexico and ultimately drain into the Atlantic Ocean; rivers to the west of the divide drain into the Pacific.

The Idiot's Approach

In this book, I've used the points of similarity mentioned in the preceding section to identify the major geographic regions of the earth. No other book outlines these regions in exactly the same way as this one does. In fact, rarely do any two geography books agree precisely. Nonetheless, the regions outlined here roughly parallel regional breakdowns you might see elsewhere.

For the sake of clarity, this book does not split up countries between regions. Country lines are used as regional limits, though many countries are quite culturally and physically diverse. It's also not uncommon for two countries bordering one another to fall into different regions.

The larger the region you attempt to describe, the more general your description must be. As you've already seen, it's almost impossible to make any universal geographic assertions about an area as large and diverse as Asia. So the goal in this book has been to divide the earth into regions small enough and similar enough to allow you to make sense of their geography and understand what makes each a cohesive region.

On the following map, the earth is divided into 20 world regions:

A. Northern America	K. Sub-Saharan Africa
B. Middle America	L. Middle East
C. South America	M. Central Asia
D. British Isles	N. South Asia
E. West Europe	O. East Asia
F. Northern Europe	P. Japan
G. Southern Europe	Q. Southeast Asia
H. Eastern Europe	R. Australia and New Zealand
I. Russia	S. The Pacific
J. Saharan Africa	T. The Poles

The size of the regions varies greatly: Some are as small as the British Isles, others are as large as Russia. The populations of different regions also vary widely: East Asia and South Asia have more than a billion people each; at the North and South poles, only a few scientists are hunkered down in Quonset huts.

Similarly, the physical attributes of each region are quite diverse, from the tropical rain forests of Southeast Asia to the icy mountains and fjords of Northern Europe to the desert expanses of Saharan Africa. By discussing each region in detail, this book gradually paints a portrait of the entire earth.

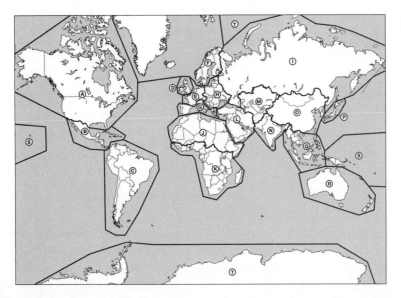

The 20 regions of the world.

Highlighting the Essentials

Parts 2, 3, and 4 of this book describe the essential geography of each of the world's regions. Although the topics covered will be similar from chapter to chapter, the focus sometimes varies, depending on the region. In some regions, the cities are the defining factor; elsewhere, a physical feature, such as a major river, is the primary focus. In other places, such as at the poles, the climate is the most striking geographic element.

Each regional chapter provides a bit of historical background, an overview of the countries that make up the region, and a description of its physical geography. Also touched upon are the region's major cities, its cultural and human characteristics, and recent newsworthy trends and events.

The Least You Need to Know

- A hemisphere is simply one half of the earth.

- Geographers traditionally have recognized seven continents: North America, South America, Europe, Asia, Africa, Australia, and Antarctica.

- Regions are areas of the earth that share a variety of geographic characteristics that make them distinct.

Part 2

A Regional Look: The Developed World

This part of the book examines the world's developed regions. The fact that you're reading this book means that you're probably from one of those regions, so at least some of the places described in Part 2 will be very familiar to you. Because these places are also the world's primary economic players, they're typically not far from the headlines.

Distinguishing between the "developed" and "developing" regions of the world isn't always easy. Although the distinction is primarily an economic one, many other factors such as demographics, quality of life, and social infrastructure also come into play. The point to remember is that a region's development level is neither fixed forever nor is it a value judgment.

Chapter 6

Northern America: Land of Opportunity

In This Chapter

- ◆ Recognizing that the New World isn't so new
- ◆ Focusing on physical features
- ◆ Poring over population
- ◆ Examining ancestry, religion, and language
- ◆ The ongoing repercussions of September 11, 2001

Northern America is a place of immense variety, as reflected in its peoples, environments, and geographic features. Within this single region, you can see palm trees in southern Florida and polar bears in Canada's Northwest Territories. You can dine on enchiladas rancheros in Albuquerque or pâté de foie gras in Quebec. Brokers in pinstripe suits hail taxis in the canyons of Wall Street, and ranchers drive their pickups down the dusty roads of Eagle Butte.

Part of the job of a geographer is to generalize. In this chapter, you'll step back and look at the big picture to find the similarities in the sea of differences that is Northern America. But before we begin, a word of explanation about that name, "Northern America."

As you saw in the previous chapter, most continents are too vast to make them worthwhile subjects of geographic study, and North America is no exception. The continent of North America consists of Canada, the United States, Mexico, the countries of Central America (which serve as the land bridge to South America), the huge island of Greenland, and the islands of the Caribbean. Although there are several ways to slice the Americas, North and South, for geographic study, it is common to divide North America into two regions: Northern America, consisting of Canada and the United States, and Middle America, consisting of Mexico, Central America, and the Caribbean islands. Confusing? Yes, but after you've examined both regions, it will all fall into place.

The New World That Isn't

The Americas, North, Central, and South, are often referred to as the New World, but this is far from correct, because the American continents were new only to the Europeans who stumbled upon them on their way to Asia. The so-called New World certainly wasn't so new to the indigenous peoples who had lived there for thousands of years. To lay claim to inhabited lands, whose people walk to the beach to greet you as you arrive, is a classic example of European ethnocentricity (the belief that one's own culture is superior to all others). Still, to the Old World, the Americas were certainly new.

Terra-Trivia

"What about Greenland?" you might ask. The world's largest island, 84 percent of which is covered in ice, is geographically part of North America, but politically it is European. The sparsely populated island is a self-governing administrative division of the Kingdom of Denmark, on which it depends heavily for economic support.

There is still much debate over who the first "discoverer" of the Americas was. The Vikings almost certainly beat Christopher Columbus to the New World by some 500 years, but their settlements in what is now Newfoundland, Canada, were short-lived. Far less solid claims have been put forth for the great Chinese admiral, Zheng He, whose fleet, according to some theories, circumnavigated the globe in 1421 and established settlements on the west coast of North America. What is beyond dispute is that Columbus was the first European navigator to establish sustained contact between the New World and the Old. For better or worse, it was Columbus who began the European colonization of the Americas.

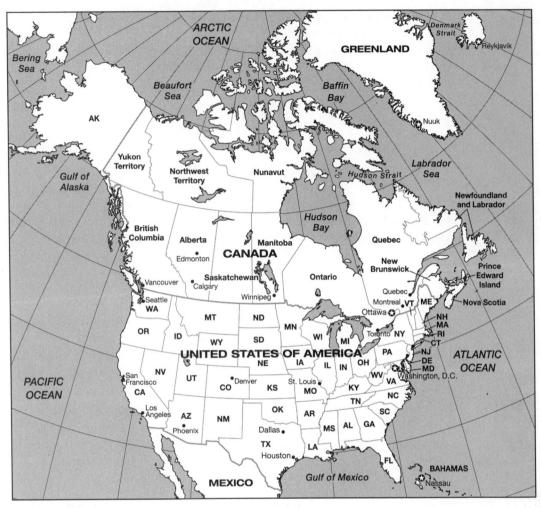

Northern America.

Two Countries, One Region

The United States and Canada share many cultural and economic features as well as the world's largest trading relationship and longest international open border (a political border that doesn't inhibit crossing). They share a lot of history, too. Both began as British New World possessions. Canada retains its ties to the British crown; the United States severed its connection in 1776. Both nations are highly industrialized and their populations enjoy very high standards of living. Because of their geographic proximity, intertwined economies, and cultural similarities, the United States and Canada constitute a natural and cohesive geographic region.

Canada: A Look at the Provinces and Territories of the Great White North

The Great White North, as Canada is popularly called, is a vast country with relatively few people. (Its entire population is less than that of California.) It is divided into ten provinces and three territories, and although Canada shares many ties and a very long border with the United States (not to mention a host of professional sports teams), its geography and history are pretty well-kept secrets south of the border.

In the spirit of promoting geographic understanding between neighbors, we'll name these divisions (shown on the preceding map). The four small eastern provinces bordering the Atlantic Ocean are the Maritime Provinces: New Brunswick, Prince Edward Island (the smallest province), Nova Scotia, and Newfoundland and Labrador (which before 2001 was known as just Newfoundland). Moving west, you can see Canada's two largest provinces (in land area and in population): Quebec and Ontario. Farther west are the provinces known as the Prairie Provinces: Manitoba, Saskatchewan, and Alberta. All the way west, bordering the Pacific Ocean, is the tenth province, British Columbia.

Canada's territories lie to the north of the provinces and are large in land area and short on people. To the far northwest is the Yukon, which shares a long border with Alaska. East of the Yukon are the huge Northwest Territories and the newest territory, Nunavut, which was carved out of the Northwest Territories in 1999 as a homeland for the indigenous Inuit. (Nunavut is an Inuit word meaning "our land.")

Geographically Speaking

The United States consists of 50 states, a federal district (Washington, D.C.), an associated commonwealth (Puerto Rico), and a number of dependent areas including the U.S. Virgin Islands in the Caribbean and American Samoa, Guam, Midway Islands, the Northern Mariana Islands, Wake Island, and other islands and atolls in the Pacific.

Some of these areas are tiny; others are much more substantial. With a population of more than 3.5 million, Puerto Rico is larger than some states. American Samoa, Guam, the Virgin Islands, and the District of Columbia are represented in the House of Representatives by a delegate; Puerto Rico by a resident commissioner. Although these representatives cannot vote on the floor of the House, they can vote on legislation in the committees to which they've been assigned.

The United States: Fifty States Plus

If you received any geographic training during your school years, you likely had to take a place-name quiz covering the states and capital cities of the United States. Because there are so many of them (50) and they're on the map at the beginning of this chapter, we don't list them in this section. (We don't want to stir up dark fears of blank maps and panicked memorization, either.)

What Are the "Lower 48"?

Of the 50 United States, 48 form a single, uninterrupted landmass. These are the so-called "lower 48," also known as the 48 contiguous or conterminous states. (*Contiguous* means "touching or being in contact with each other"; *conterminous* means "sharing a boundary.") The remaining two states, however, are more far flung. Alaska is separated from the lower 48 by Canada; Hawaii, by the Pacific Ocean. Geographers call countries like the United States that are not contiguous or conterminus *fragmented* countries.

> **Geographically Speaking**
>
> Try not to use the term "continental United States" to refer to the lower 48 states because it can lead to confusion and is basically incorrect. Strictly speaking, the continental United States comprises the lower 48 as well as Alaska, which is also part of the continent of North America.

Purple Mountains' Majesty and Fruited Plains

Extensive plains and great mountain chains dominate the Northern American landscape. The mountain chains, shown on the following map, all run north–south, which can present an obstacle to anyone traveling east–west. (Early settlers undoubtedly wished that North America's mountains ran east–west.)

In eastern North America, the dominant mountain chain is the Appalachian. Although these old, eroded mountains no longer reach great heights, they still stretch from Alabama to the Maritime Provinces in Canada. In the west, the higher Rocky Mountains and the Pacific ranges (the Cascades, Sierra Nevadas, and Pacific Coastal Range) run the length of the United States and Canada.

Along the Atlantic Ocean and the Gulf of Mexico are the coastal plains on which the largest concentrations of the region's people live. Stretching between the great mountain chains of the east and west are the vast central plains that extend northward past the Great Plains of the Midwest to the icy tundra of the far north. The central plains of Northern America are one of the world's most important breadbaskets, producing year in and year out vast yields of grain for domestic consumption and export.

> **GeoRecord**
>
> The highest mountain in North America is Mount McKinley (also called Denali) in Alaska. At 20,320 feet, it's the only North American mountain higher than 20,000 feet. The lowest point in North America is located in California's Death Valley. The bottom of Death Valley drops to 282 feet below sea level.

The Mighty Miss

If you can remember only one river in this region, make it the Mississippi, by far the largest river in all of North America. Feeding the Mississippi River are its two largest *tributaries* (feeders), the Missouri and Ohio rivers. The tributaries of the Mississippi reach to the Rocky Mountains in the west, the Appalachian Mountains in the east, and just below the Canadian border in the north. They all drain south to the Mississippi's great delta and the Gulf of Mexico. The Mississippi has nurtured westward expansion, the riverboat era, historic expeditions to the interior (such as the great Lewis and Clark expedition to the Pacific in 1804), and *The Adventures of Huckleberry Finn*.

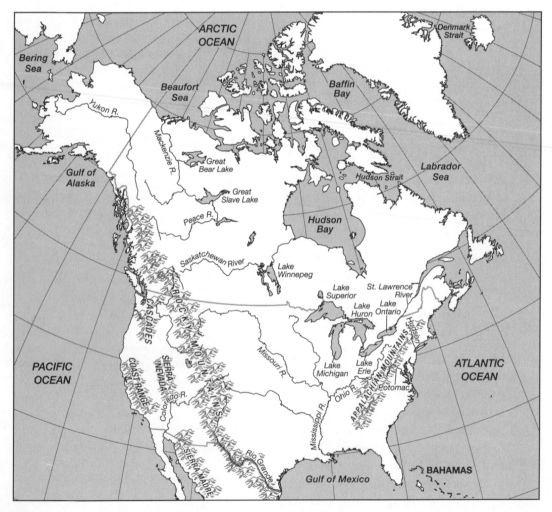

Northern America—physical features.

In addition to the Mississippi, noteworthy rivers in Northern America include the Columbia, Colorado (creator of the Grand Canyon), Mackenzie-Peace, Rio Grande, Saint Lawrence, and Yukon. Many smaller, well-known rivers lace the eastern part of the United States, including the Connecticut, Delaware, Hudson, James, Potomac, Savannah, and Susquehanna. (If you feel overwhelmed by this list of names, just pick two from the preceding map of North America: the Mississippi and the river closest to your hometown.)

> **GeoRecord**
>
> The Mississippi–Missouri river (the combination of the Mississippi River and its largest feeder, the Missouri River) is the fourth-longest river in the world. At 3,740 miles long, it's exceeded only by the Nile, the Amazon, and the Chang Jiang (Yangtze) rivers.

Water, Water, Everywhere

The Northern American region is surrounded by oceans, seas, gulfs, and bays. Within the landmass of North America are several of the world's largest lakes. (Look at the preceding map to get yourself hydrographically oriented.) The three oceans in this region are the Pacific to the west, the Atlantic to the east, and the icy Arctic to the north. The Bering Sea is notable because it and the narrow Bering Strait were all that separated the world's two greatest superpowers, the United States and the Soviet Union, during the Cold War. Hudson Bay is memorable because of its vast size (it's the world's largest bay) and the historical importance of the great fur-trading company named after it. The Hudson's Bay Company is Canada's oldest corporation (founded in 1670) and still one of its largest retailers. Many smaller bays in Northern America, including the Chesapeake Bay, have great local importance.

This region's primary gulf, the Gulf of Mexico, is formed by the Caribbean Sea and the Atlantic Ocean. The gulf borders five states in the southern United States as well as Mexico, for which it is named. As mentioned previously, the Gulf of Mexico is the exit point for the mighty Mississippi River. It is also important for its oil reserves.

Northern America also has 8 of the world's 12 largest freshwater lakes. Three of them—Great Bear Lake, Great Slave Lake, and Lake

> **Terra-Trivia**
>
> One way to keep track of the names of the five Great Lakes is to remember the acronym HOMES: Huron, Ontario, Michigan, Erie, and Superior. Lake Superior is the largest freshwater lake on earth. (The only lake larger than Lake Superior is the saltwater Caspian Sea.) All the lakes straddle the United States–Canada border except Lake Michigan, which is located entirely within the United States.

Winnipeg—are located entirely within Canada. The other five lakes all comprise the Great Lakes, which are located along the border of the United States and Canada and drain primarily into the Atlantic Ocean via the Saint Lawrence River.

Weathering the Climate

Northern America stretches from near the tropics in the south to well within the Arctic Circle in the north, and these latitudes define the region's varied climate. The region is surrounded on three sides by oceans, which restrain climate extremes along coastal areas. Northern America's location in the northern hemisphere results in June–September summers and December–March winters.

Northern Canada and Alaska have cold polar climates. Summers are typically short and cool; winter, long and cold (not the ideal destinations for sunbathing). Most of the vast Northern American landmass experiences a range of humid climates, colder in the north and warmer in the south, with year-round precipitation. Southern Florida has a humid tropical climate, with typically dry winters. The southwestern United States and the south–central plains of Canada have a dry climate, with arid conditions prevailing in the far southwest.

From Tundra to Mangrove

Northern America's vegetation generally parallels its climatic areas (as happens worldwide). Tundra (treeless Arctic plains) in the extreme northern reaches of the region gives way to pine forests as you move south. As you move farther south, the pine forests yield to deciduous forests of broadleaf trees that seasonally lose their foliage, while in the southeast pine forests again predominate.

Terra-Trivia

The number of bison (or buffalo) roaming the prairies in the early 1800s exceeded 40 million. The westward migration of people of European descent into the Great Plains in the nineteenth century resulted in a wholesale buffalo slaughter. By 1890 the number of bison had been reduced to fewer than 200. Today buffalo herds, raised largely on private ranches, are rebounding.

Tallgrass prairies cover most of the drier central areas. The prairies, which once supported vast herds of bison and the Plains Indian culture that depended on them, are now primarily used for ranching. In Northern America's arid southwest, desert vegetation with cacti, shrubs, sage, and chaparral growth is the norm.

Another type of vegetation, unique to Northern America, is found in the Pacific Northwest. There, in old-growth temperate rain forests, are the largest trees on earth. Redwoods, Douglas firs, sitka spruces, and giant sequoias grow to be more than 300 feet tall in this area. The tallest tree on record is the National Geographic Society Tree, a California redwood that has been measured at 365 feet, 6 inches tall. California also boasts the largest tree on record, the General Sherman Sequoia, with a trunk circumference of 101.6 feet. In California's Inyo National Forest, you can also see the world's oldest living tree, a 4,600-year-old bristlecone pine.

Black Gold

Because fossil fuels are so important to modern economies, you should know about Northern America's reserves. The United States has larger coal reserves than any other country on earth. In Northern America, coal is abundant, accessible, and relatively cheap, but the environmental cost of extracting and burning coal—the "dirtiest" of the fossil fuels—is high. Despite large oil and gas deposits in the south central and southwestern United States and in northern Alaska and Alberta, Canada, the United States is the world's largest petroleum importer and not coincidentally the largest producer of air pollutants and greenhouse gases. For more on the energy situation in Northern America and around the world, see Chapter 28.

Eye on the Environment

With petroleum reserves declining in both Canada and the United States, the U.S. economy is increasingly dependent on oil imports from some of the world's most volatile regions, particularly the Middle East. In an effort to lessen that dependence, the administration of President George W. Bush has embarked on an ambitious campaign to increase domestic oil production. A focus of that campaign is ANWR, the Arctic National Wildlife Refuge, an ecologically critical and pristine area on Alaska's northern coastal plain, which the administration proposes to open to oil drilling. Environmentalists argue that the ANWR reserves would hardly put a dent in U.S. oil needs and that its benefits would be far outweighed by the potential cost—to the area's indigenous peoples, wildlife, and ecology—of a major oil spill or other environmental disaster. As of presstime, the matter is still unresolved and ANWR remains a rallying cry in the struggle between pro-development and pro-conservation forces.

Mother Nature's Dark Side

Although Northern America is blessed with natural resources, it also has its share of natural disasters.

Because the west coast of North America lies along the Pacific Ring of Fire, it is subject to volcanic eruptions and frequent earthquakes. A notable eruption occurred in

Washington state in 1980. In May of that year, Mount St. Helens exploded with the force of 500 atomic bombs, blowing away its entire north face. California, infamous for its San Andreas Fault, is subject to the continual threat of earthquakes. The 1989 "World Series quake" in San Francisco dramatically illustrated this threat to a worldwide television audience.

If you're ready to pack up your possessions and move east, you might want to consider a few more natural disasters before you leave. Droughts ravage the southwest, devastating tornadoes frequently plague the central plains states, the Mississippi River and its tributaries are subject to severe flooding, and the Gulf of Mexico and eastern coast of the United States are vulnerable to tropical storms and hurricanes. You might try Montana, where heavy snows and occasional forest fires are all you have to worry about.

Terra-Trivia

The two countries of Northern America are both among the world's largest in total land area (Russia is largest). Canada is the world's second largest country, with 3,849,674 square miles of land area. The United States is the world's fourth largest country, with 3,618,770 square miles of land area.

Profiling the People

For hundreds of years, people from all across the globe have made their way to Northern America. They brought with them their ethnic characteristics, languages, customs, and religions. Because America is indeed a melting pot, any generalizations about its culture must necessarily be broad. After years of "melting," Northern American culture is unique.

GeoJargon

The **northeastern megalopolis** is an area of large coalescing cities in which most of the people can be called "city folk." Encompassing some of the largest cities in the United States, the megalopolis is anchored in the north by Boston and in the south by Baltimore/Washington D.C., but extends even farther—up to Maine and into North Carolina.

Where the People Are

Population distribution in Canada is easy to visualize: Think south. Almost all of Canada's relatively small population resides in the southernmost part of the country in a small east–west strip next to the United States. More than half of Canada's population lives in the area called Main Street, which runs from the north shore of Lake Erie northeast along the Saint Lawrence River.

In the United States, the largest population cluster is the *northeastern megalopolis*. A second major population cluster extends west from the megalopolis and encompasses the manufacturing areas around the southern fringes of Lakes Ontario, Erie, and Michigan. Two more major population clusters are located in Florida and southern California. Except for urban areas such as Denver and Salt Lake City, the interior of the United States remains relatively sparsely populated and is becoming even more so in some places. Rural areas of the Great Plains states, from Texas to Montana and the Dakotas, have been losing population for decades, a trend driven in part by the shift from small family farming to large-scale corporate agriculture.

The Urban Landscape

Northern America is a region of many large and well-known cities. Selecting important cities to include in this book was a treacherous business: Most people think that the most important city is their own. Nonetheless, here's a list of the major cities in Northern America (in no particular order): Canada's most important cities are all located in the south, close to its border with the United States:

◆ **Toronto:** Largest city, financial center

◆ **Montreal:** Second largest city, the business and cultural center of French Canada

◆ **Ottawa:** Capital, seat of Parliamentary government

◆ **Vancouver:** Canada's link to trade with the Pacific Rim

◆ **Other important Canadian cities:** Quebec, Quebec; Winnipeg, Manitoba; Calgary, Alberta; Edmonton, Alberta

At one time all major U.S. cities were located east of the Mississippi, most of them on the Atlantic Seaboard. The following list reflects a population shift to the west:

◆ **New York, New York:** The largest city in the Northern American region (Mexico City is the largest city in the continent of North America), New York is the financial and cultural capital of the United States

◆ **Washington, D.C.:** The nation's capital

◆ **Los Angeles, California:** Entertainment capital, pop culture center

◆ **Chicago, Illinois:** Home of some of the world's tallest buildings (the Sears Tower is North America's tallest building, at 1,454 feet)

◆ **San Francisco, California:** Tourist center, hub for high-level technology

◆ **Other important U.S. cities:** Boston, Massachusetts; Philadelphia, Pennsylvania; Atlanta, Georgia; St. Louis, Missouri; Houston, Texas; Dallas,

Texas; Detroit, Michigan; Denver, Colorado; Seattle, Washington; New Orleans, Louisiana; Phoenix, Arizona; and Miami, Florida

Roots of the People

In both the United States and Canada, a large percentage of the population is of European ancestry. But both countries have also been magnets for immigrants from around the world, and today both have large ethnic minorities. Canada has a substantial Asian population, concentrated mainly in the cities of Ontario and British Columbia. In northern Canada, indigenous Indian and Eskimo ancestry is common. The United States has a large African American minority concentrated primarily in urban areas and throughout the south. As of 2003, however, Hispanics, or people of Latin American heritage, outnumbered African Americans to become the largest minority group in the United States. Hispanic communities are located in the cities and suburbs of the Northeast and Midwest, in south Florida, and in the states that border Mexico.

Indigenous Peoples

The native populations of the United States and Canada have shown remarkable resiliency in the face of overwhelming economic and educational disadvantages. Despite predictions at the beginning of the twentieth century that Native Americans would melt into the majority population and disappear, Indians have revived their tribal governments and cultural traditions and are asserting their legal rights. In 1999, Canada took a dramatic step toward reconciliation with its indigenous people by creating the new Inuit territory of Nunavut. In 2002, the Cree Nation reached an historic agreement with the government of Quebec, ending a decades-long struggle over the development of hydroelectric power projects that threatened Cree homelands.

Terra-Trivia

In an ironic twist, the western world refers to the indigenous people of the northern plains as the *Sioux*, a French distortion of a Chippewa word meaning "snake" or "enemy." The Sioux have always called themselves Lakota (or Dakota, depending on dialect), which means "ally" or "friend."

In the United States, people have become more aware and concerned about the squalid conditions that prevail on many of the nation's reservation lands. Although gambling facilities have provided an economic stimulus for some groups, casino culture often undermines traditional culture. In an interesting twist, however, the Indian population of the Great Plains is on the rise, even though the general population is declining. Native Americans are leaving the cities in growing numbers and repopulating their ancestral homelands.

Finding the Faith

Christianity is the dominant religion throughout much of Northern America. In the southeast and central portions of the region, Protestant faiths are most prevalent. Roman Catholicism is the dominant faith in the northeastern and southwestern sections of the region. Mormonism is common in the west central part of the United States (in and around Utah).

Judaism is practiced primarily in urban areas across Northern America. On reservation lands and in the far north, traditional Native American religions are being practiced with renewed vigor. There are also sizable Muslim populations throughout the region, as well as many followers of Eastern faiths.

Speaking of Language

For generations, immigrants have come to the United States speaking little or no English. Although the United States does not have an official language, English is the de facto official language and is spoken by the vast majority of its citizens. Although areas with growing Spanish-speaking populations are becoming increasingly bilingual, English is the norm for most of the United States.

Canada's language story is more complicated because it has two official languages: English and French. Francophones (French speakers) are concentrated primarily in the province of Quebec, where they constitute about 85 percent of the population. Quebec's struggle to preserve its French language and culture and keep it from being overwhelmed by English has fueled strong nationalist sentiment, and the province has on several occasions in the past seriously considered breaking away from Canada and forming a separate nation. Since the last referendum on separation was narrowly rejected in 1995, separatist feeling in Quebec is apparently on the decline.

In the News: September 11

With two mighty oceans on either side of them, the United States and Canada have long felt secure from the threat of foreign attack. That precious sense of security was shattered, perhaps forever, by the Al-Qaeda terrorist assaults on New York City and Washington, D.C., on September 11, 2001. Although the attacks were directed at the United States—specifically at its centers of economic and military power—Canada suffered, too, with at least 24 of its citizens killed at the World Trade Center. For the first time, the rage and violence of "foreign" religious strife swept onto American shores, and the region has not been the same since. With the threat of another terrorist attack still in the air, even that long open border between two old friends and neighbors is not quite as open and free as it once was.

The Least You Need to Know

- ◆ The region of Northern America contains only two countries: Canada and the United States.

- ◆ Canada is politically divided into ten provinces and three territories; the United States has 50 states, plus a number of outlying possessions and dependencies.

- ◆ Mountain chains in North America run north–south. The largest are the Appalachians and the Rockies.

- ◆ The Mississippi River is the largest and most important river in North America. The major water bodies surrounding North America are the Pacific, Atlantic, and Arctic oceans and the Gulf of Mexico.

- ◆ Canada is officially bilingual; the United States has no official language.

The British Isles

In This Chapter

- A brief look at the rich history of the British Isles
- Distinguishing between Great Britain, the United Kingdom, and England
- Why Ireland is known as the "green land"
- Why an umbrella is useful in this region
- Why everything revolves around London
- Why all Celtics are not basketball players
- Whether the UK will give up its Pounds for Euros

The British Isles are located on the western fringe of Europe. Although their story parallels and meshes with much of Europe's history, their island status caused them to develop quite distinctly. Protected by water from every angle, Britain has withstood for nine centuries. The islands also dominated the surrounding seas for hundreds of years and achieved a power and greatness well out of proportion to their size.

Once Upon a Time

The British Isles have also had a disproportionate influence on world literature and science and played a leading role in the development of parliamentary democracy. British concepts of individual liberties and the legal system to ensure them are reflected in the constitutions and governments of many countries around the world. Until recently, this small grouping of islands ruled more than a quarter of the earth. Although the empire is gone, the region remains an industrial power. In the early twenty-first century, the British Isles are still a major global player.

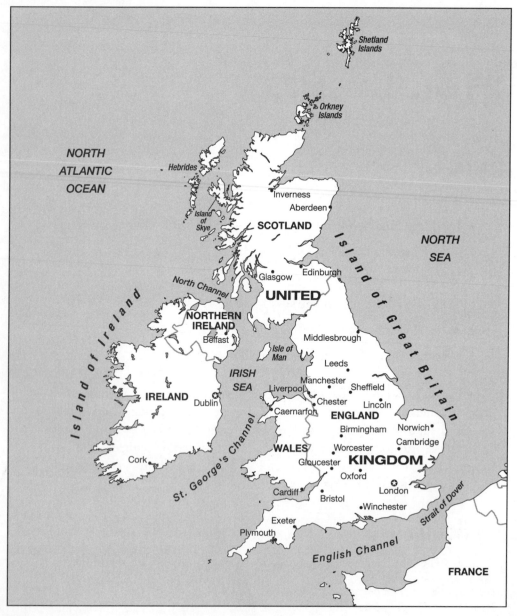

The British Isles.

The Early Years: Celts, Romans, Angles, Saxons, Vikings, and Normans

People from the continent of Europe have had their eyes on the British Isles from time immemorial. The Celts began arriving around 800 B.C.E. and became the dominant population of the islands. They brought with them their language, metalworking skills, religion, and religious leaders, the Druids. Roman armies under Julius Caesar invaded Britain in 55 B.C.E., but it wasn't until nearly 100 years later that the main island of Great Britain was conquered by the Romans, and then only in part. What are now Scotland and Wales were never fully subdued. In what is now England, however, the Romans settled and prospered. They farmed the land, built impressive villas and towns, and constructed excellent roads as they had throughout the empire. Even though the Romans stayed in Britain for some 400 years, the native population never became fully Romanized, nor did Latin replace Celtic.

The last Roman troops left Britain in 410 C.E. and the resulting power vacuum was filled by Germanic tribes, among them the Angles (for whom England is named), Saxons, and Jutes. The new immigrant groups gradually pushed the existing Celtic residents back to the extremities of Great Britain—Wales, Cornwall, and Scotland— and to Ireland. Waves of Viking invaders brought Danish and Norwegian influences to the British Isles during the ninth and tenth centuries. The last successful invaders of the British Isles were the Normans under William the Conqueror in 1066. Descended themselves from Vikings who had settled in northwestern France, the Normans brought with them their castle-building style (the Tower of London is a prime example) and their French dialect. The language of the Norman elites mingled with the Anglo-Saxon (Old English) spoken by their new subjects to form the basis of the English we speak today.

Terra-Trivia

On Salisbury Plains in southwestern England stand the ancient stones of Stonehenge. The huge, 21-foot-high stones date back 3,000 to 5,000 years. Its original purpose is unknown, although it is thought to have been used for astronomical calculations or for important long-ago burials. How these huge stones were moved to the site in the first place is also a mystery.

The Norman Conquest turned the focus of the British Isles away from Viking Scandinavia and onto France. Throughout much of the Middle Ages, England and France were locked in a deadly struggle for power and supremacy. In the late fifteenth century, a bloody civil war, known as the War of the Roses, placed the Tudor family on the throne of England, initiating a period of great religious and political turmoil, but also of expanding prosperity and power. The second Tudor monarch, King Henry VIII (of the ill-fated wives) broke with the Roman Catholic Church in the beginning of the sixteenth century, and established himself as the head of the Church of England (the Anglican Church). The last of the Tudor monarchs was Henry's second daughter, Elizabeth I. "Good Queen Bess" presided over a remarkable flowering of English literature, commerce, and sea-power. This was the age of Shakespeare and the beginning of England's overseas empire. After Elizabeth's navy under Sir Francis Drake defeated the Spanish Armada in 1588, England was on her way to becoming a global maritime power.

The Empire

Despite losing the American colonies in 1781, Britain retained a vast empire that would soon grow even more vast. Although the Thirteen Colonies had gone their own way, there was still immense Canada and the rich sugar colonies of the Caribbean. India, Australia, and large chunks of Africa would also become parts of the British Empire. The combination of a growing empire and an Industrial Revolution at home turned nineteenth-century Britain into the economic engine of the world. A processing economy developed, with raw materials pouring in from the colonies and flowing out again as manufactured goods.

During the long reign of Queen Victoria, from 1837 to 1901, the United Kingdom reached the height of its power and influence. The British empire extended around the globe and controlled a fourth of the world's land and one-fifth of its people. The phrase "the sun never sets on the British Empire" was literally true.

Geographically Speaking

If the nineteenth century was a high point for England, it was the lowest of low points for Ireland. Long oppressed by British colonial rule, Ireland suffered a tremendous blow in 1846 when the Irish Potato Famine began to destroy the island's agricultural base. In the four years that the famine lasted, millions died or emigrated to the United States and other English-speaking countries. Those who survived and stayed became convinced of the need for political change.

With the exception of six northeastern counties, which remained British, Ireland finally gained its independence in 1921, in the aftermath of World War I.

World War II marked the beginning of the end of the rest of the British Empire. Under Churchill's "blood, sweat, and tears" leadership, Britain held off the threat from Nazi Germany and prevented its leader Adolf Hitler from becoming the first successful invader of the British Isles since William the Conqueror. Although Britain emerged victorious from World War II, the empire did not long survive the independence movements that swept the world in the years following it.

The Recent Past

The two world wars of the twentieth century seriously depleted Britain's economic strength. But since the end of World War II and the dissolution of the empire, Britain has reinvented itself as a modern industrial, hi-tech European society. Its economy is still one of the world's biggest, and as one of five permanent members of the United Nations Security Council and a founding member of NATO, it is a major player in Europe and the world. In 1973, the United Kingdom and Ireland belatedly joined the European Union, then known as the European Community. Britain has always been ambivalent about its integration with Europe. Even now that it is a key member of the European Union, the United Kingdom, unlike Ireland, has held off joining the European Monetary Union, as this would entail giving up the mighty Pound Sterling in favor of the upstart pan-European currency, the Euro.

The United Kingdom—It Doesn't Equal Great Britain

The region of the British Isles is composed of two large islands and a collection of smaller, peripheral islands. Although the entire region is known as the British Isles, only one island is named Britain. (The map of the United Kingdom and Ireland at the beginning of this chapter can help you sort this all out.) The two large islands are …

- **Great Britain:** The larger, more easterly island is sometimes called just *Britain*. People often get confused because almost any island, administrative division, and country in the entire area is referred to as British. (Calling the region the British Isles is an example.) The thing to remember is that Great Britain is an island and not a country.

- **Ireland:** Although Ireland is the name of the smaller, more westerly island, it's also the name of a country that occupies part of the island—so don't get confused.

Skip the smaller islands for now, and stay focused on the two large islands and the two countries that occupy them:

- **United Kingdom:** Also called the United Kingdom of Great Britain and Northern Ireland, it's the larger of the two countries occupying the islands in both land area and population. The United Kingdom occupies the entire island of Great Britain and the northeast portion of the island of Ireland.

- **Ireland:** The Republic of Ireland, or Eire, is the smaller of the two countries the make up the region. Ireland, independent since 1921, occupies all of the island of the same name, except the northeast corner which, as noted, is part of the United Kingdom.

Though these distinctions are confusing enough, things get even more bewildering when you factor in the four component parts, or constituent countries, as they are called, that make up the United Kingdom. At various times in history, each has been a separate political entity and two have been independent kingdoms in their own right. Although the four are joined together as the United Kingdom of Great Britain and Northern Ireland, letting go of the rivalries and grievances of past is not always so simple:

- **England:** The largest and historically the most dominant of the constituent countries that form the United Kingdom, England occupies the bulk of the southern and eastern portions of the island of Great Britain.

- **Scotland:** Like England, Scotland was once an independent kingdom. King James VI of Scotland also became James I of England upon the death of Elizabeth I in 1603, and the two countries were ruled by the same monarch from then on. In 1707, they agreed to became a single constitutional entity, called Great Britain. Scotland occupies the northernmost third of the island of Great Britain.

- **Wales:** Situated on the southwestern portion of the island of Great Britain, Wales never developed into a fully integrated kingdom as did Scotland and England. The Celtic principality was joined by force to England back in 1284, but has retained its distinctive language, culture, and sense of nationhood through the centuries.

- **Northern Ireland:** All of Ireland and Great Britain were united constitutionally in 1801 as the United Kingdom of Great Britain and Ireland. This became the United Kingdom of Great Britain and Northern Ireland in 1928 after the island of Ireland was divided into the independent Republic of Ireland and British Northern Ireland, which in a nutshell is how Northern Ireland came to be the only constituent country of the United Kingdom not on the island of Great Britain. Although you might hear Northern Ireland referred to as Ulster, the two are not identical. Ulster is one of the four historic provinces of Ireland, six of whose nine counties constitute British Northern Ireland; the remaining three are part of the Irish Republic.

Although England, Scotland, Wales, and Northern Ireland are called constituent countries, none of them are separate, independent nations. In recent years, however, Scotland, Wales, and Northern Ireland have acquired a considerable amount of autonomy from the central government in London, especially since the establishment of a Scottish Parliament and Assemblies for Wales and Northern Ireland in 1999.

Although the coverage of the United Kingdom in this chapter focuses on the four constituent countries, the UK still includes a number of holdovers from the empire. These include: Bermuda, the British Virgin Islands, the Cayman Islands, the Falkland Islands, Gibraltar, Pitcairn Island (settled by mutineers from the ship *Bounty*), and St. Helena.

Terra-Trivia

Each of these constituent countries contains smaller political subdivisions. England, Wales, and Ireland have counties; Scotland has regions; and Northern Ireland has districts.

Cool Britannia

Throughout the United Kingdom, smokestack economies have largely given way to a booming service economy, focused on banking, insurance, and business services. England, especially London and its surrounding counties, was a prime beneficiary of the tech explosion of the 1990s.

The UK has large reserves of coal, natural gas, and petroleum. Its agriculture is highly mechanized and relatively efficient, compared to that of Europe. Although the economic repercussions of the September 11, 2001, terrorist attacks in the United States and the bursting of the tech bubble in the early '90s have had a dampening effect on the UK economy, it remains one of the largest and strongest in Europe.

Eye on the Environment

The Industrial Revolution of the nineteenth century was fueled by dirty-burning coal, and lots of it. London in particular was notorious for the lethal quality of its polluted air. Although coal emissions were the primary air pollutants in Victorian England, in the twentieth century, the combustion engine added a new element to the noxious brew that often passed for air in London and other heavily industrialized cities. The worst of London's killer fogs, or more accurately smogs, killed an estimated 4,000 people in December 1952; another 1,000 died in the killer fog of 1956. Fortunately, these were the last of the great killer fogs. In 1956, the British passed the nation's first Clean Air Act. Today, London's air might not be pristine, but killer fogs are a thing of the past.

The Bonnie Land of Scots and Lochs

Though sharing a country with the English, the Scots have remained distinct. Helped by their highland location and tight clan system (which is the basis of family life), they have maintained a rugged character and an independent approach to life. Teased about their presumed frugality and production of haggis (a concoction of oatmeal and sheep's liver, heart, and lungs), the Scots maintain a strong sense of nationalism.

The coal and iron once mined in Scotland have been depleted, and manufacturing in general continues to decline. Scotland has rebounded, however, with research and development, electronics, and North Sea oil and gas. Tourism is also a mainstay, and playing golf at St. Andrews has become sort of a "Holey Grail" for golf players.

The Slate and Coal of Wales

Wales, from the Anglo-Saxon word that means "foreigner," is the home of a tough people pushed into the infertile, wild, and mountainous western edge of Great Britain. The Welsh traditionally eked out a living by mining coal and slate or working the machinery of the industrial revolution. Aside from strip-mine scars on the landscape, there are few reminders left of the principality's recent industrial past. In the 1980s, overseas investment by multinational firms in the electronics, automotive, and chemical industries began to transform the Welsh economy. The services sector is also playing an increasingly important role here. Still, Wales lags behind the other regions of the UK in terms of economic development and prosperity.

Both Wales and Scotland have seen a resurgence in nationalism in recent years. Road signs bear both English and Celtic names, and the traditional languages are increasingly being taught in school. The new Welsh National Assembly and Scottish Parliament are both an expression and a fulfillment of these national aspirations.

The Troubles of Northern Ireland

Northern Ireland was settled primarily by Scottish and English immigrants who crossed the Irish Sea to occupy one-sixth of the territory of the island of Ireland. Unlike the almost totally Catholic Republic of Ireland (Eire), only a third of Northern Ireland is Catholic; the rest of the population is primarily Protestant.

And therein lies the conflict: Catholics decry discrimination from the Protestant-controlled government and yearn for union with the Republic of Ireland, while the Protestants are determined to remain part of the United Kingdom. For more than 30 years, the words Northern Ireland have been synonymous with violence and terrorism, perpetrated by sectors of the Irish Republican Army (IRA), which is out-lawed in the Republic of Ireland; Protestant paramilitary groups; and local police and British Army forces.

In 1997, however, a cease-fire among the main paramilitary groups somehow managed to hold, and in the following year a real glimmer of hope broke through with the signing of the Good Friday Agreement of April 10, 1998. The agreement called for the establishment of a National Assembly and an executive body in which power would be shared by both sides. Since then, the British military presence in Northern Ireland has been scaled back, the IRA began to decommission weapons, and the cease-fire more or less held. But the peace process has been fraught with setbacks, the most serious of which was the suspension of the Assembly in October 2002 and its dissolution in 2003. What happens next is anyone's guess, but both Britain and Ireland, and many in Northern Ireland itself, are committed to making the peace process stick. And in the meantime, life does go on in Northern Ireland. Its economy, though hindered by the "troubles," was the fastest growing in the UK through the 1990s.

Ireland: The Green Land

After resisting English domination for centuries, Ireland finally fell during the seventeenth century. Not until 1921 did the Republic of Ireland (Eire) again emerge as a separate country. Ireland never became as industrialized as the United Kingdom, and until recently the economy of the appropriately named "green land" (or "emerald isle") was based on agriculture.

Today, however, agriculture has been eclipsed by industry and Ireland has transformed itself into a modern, trade-dependent economy with one of Europe's fastest growth rates. Tourism is another important sector of the economy, as Ireland's green hills and lush countryside draw travelers from around the world. In a small village pub, you can still sample the famous Irish gift of gab and an equally famous pint of stout.

Natural Diversity

The British Isles offer a great deal in the way of natural diversity. During the ice age, much of the region was enveloped by massive ice sheets, which left their mark on the northern landscape. Because the region consists of islands, the surrounding waters play a significant role in modifying its climate.

Landforms: From Scottish Highlands to Peat Bogs

The Northwestern Uplands, or Highlands, of Scotland are as high and spectacular as those in Norway, but are too rocky and wet for all except the hardiest settlers. The Highlands continue south from Scotland, turning into the Pennines Mountains in north central England and the Cambrian Mountains in Wales. The north country is also sprinkled with lakes that form the magnificent Lake District, celebrated by the romantic lake poets William Wordsworth and Samuel Coleridge. The crags of the mountains here loom over stone-walled, sheep-filled dales below.

Terra-Trivia

Separating Scotland from England is the famous 75-mile-long Hadrian's Wall. Built in England by the Romans in 200 C.E., it was designed to keep northern barbarians at bay. It symbolically separates the Scots and English even now.

Glacial remnants endure in the jagged, fjordlike terrain of northern Scotland. Bogs and moors abound as glaciers from the Grampian Mountains poke their way past lochs and through heather coastward, where the fishing is rich and the people more numerous. There, you can find Loch Ness and maybe glimpse its fabled monster, Nessie.

Aside from the Highlands of Scotland and Wales, most of the rest of the island of Great Britain is lowland, aptly called "downs." The lowlands are low (not surprisingly) and arable (easily cultivated) and have adequate rainfall. Southern England's lowlands form rolling hedgerowed farm country. The thick hedges were used instead of fences to separate farmlands. The poet William Blake called this area "England's green and pleasant land." In this agricultural area, every inch is cultivated intensively to produce such famed English delicacies as roast beef and Yorkshire pudding, topped off with tangy local strawberries and clotted Devonshire cream.

Terra-Trivia

Mosses, heather, and various grasses compress over time to create peat, a relative of coal. When the chunks of peat are cut, they're 95 percent water; when dried, they can be used as fuel. Peat-burning produces one-quarter of Ireland's electricity. Unfortunately, peat pollutes as it heats thatched cottages and factories alike.

Marking the westernmost edge of Europe, the Emerald Isle of Ireland is lush and green and rimmed with low hills spaced around the fertile central plain. Half of Ireland is made up of meadows; the other half is filled with lakes, mountains, and the peat bogs that cover one-seventh of Ireland.

Although this section has focused on the two larger islands, the British Isles include many smaller islands, most of which you can see on the maps in this chapter. To Scotland's northeast lie the Orkney Islands, replete with race tides, skerries (small, rocky islands), and rocky reefs; only 20 of the 70 are inhabited. Farther to the northeast are the Shetland Islands, the home of every child's pony dreams. The wool of Shetland sheep is especially soft, and sweaters knitted from it make fine souvenirs.

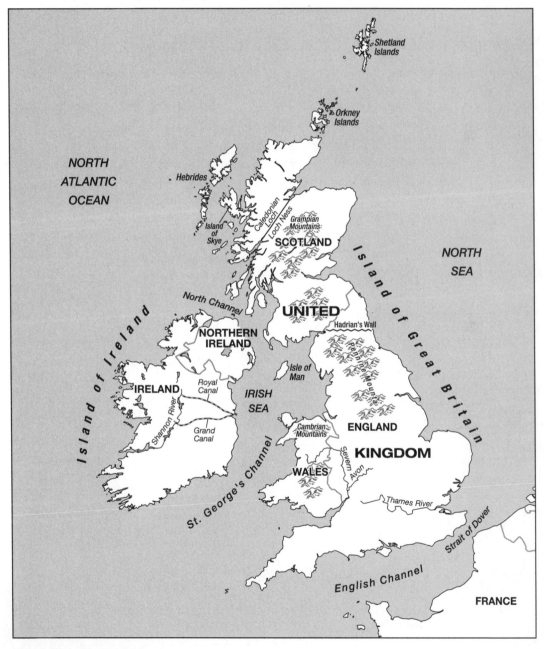

British Isles—physical features.

On Scotland's northwest coast are the more than 500 Hebrides Islands. In the Irish Sea between Great Britain and Ireland is the Isle of Man. The Channel Islands are in the English Channel off the coast of France, and off Land's End, in extreme southwest England, are the tiny Isles of Scilly.

Waterways: From the Gulf Stream to the English Channel

As you can see from the preceding map, an important part of the geography of the British Isles is the water that surrounds them. The primary bodies of water surrounding the islands are the Atlantic Ocean to the west and the North Sea in the northeast. Separating the British Isles from continental Europe are the English Channel and the Strait of Dover. Southeast England's cornerstone, the famous white cliffs of Dover, has challenged every invader from Caesar to Hitler. Ireland is separated from Great Britain by the North Channel in the north, the Irish Sea in the middle, and St. Georges Channel in the south. Great Britain and Ireland are also home to many well-known rivers, such as the Severn, Thames, and Avon in Great Britain and the Shannon in Ire-land. Canals abound, including the 60-mile-long Caledonian Loch that slices across Scotland. England is laced with hundreds of miles of canals that connect virtually every corner. Ireland features two famous canals, the Royal Canal and the Grand Canal.

A Balmy Climate at 55 Degrees North?

England is known for its fog and rainy skies, and Ireland is even rainier. What's surprising is that the region isn't known for its heavy snows. Great Britain stretches from about 50 degrees to 59 degrees north latitude. Carried over to North America, the latitude would approximate that of Hudson Bay, with a climate that's ideal for polar bears. How is it that North America is adrift in ice and snow and the British Isles are relatively balmy?

The answer lies in the Gulf Stream, a vast, warming ocean current. Originating in the tropical waters of the Gulf of Mexico, it becomes the North Atlantic current that warms the British Isles. This huge flow of warm water tempers the region's climate and helps to make it milder and wetter than its latitude would otherwise dictate. Although umbrellas are requisite in the British Isles, snow shovels can stay largely in storage.

The Cities: London and Beyond

London is the dynamic hub of the British Isles. It's the capital of the United Kingdom and one of the world's largest and most cosmopolitan cities. Straddling the Thames

River, which gives it access to the North Sea, London is also a large and bustling port. It is a global center of finance, publishing, communications, and retailing, and one of the world's great tourist attractions and cultural centers.

The region has more cities than just London, however. Here are some of the more notable:

England

- ♦ **Liverpool–Manchester:** This conurbation (network of cities) was a leader during the industrial revolution. Manchester was known for its textiles, and Liverpool served as a major port. Manchester, one of England's largest cities, has become a major air transportation hub and commercial center; Liverpool, forever famous as the home of the Beatles, remains a great shipping port and industrial center.

- ♦ **Birmingham:** England's second largest city was badly damaged by World War II bombing and has not yet fully erased the scars of its post-industrial decline. But Birmingham, now England's biggest convention center, is sprucing up and becoming a greener and more enjoyable city than in the recent past.

Scotland

- ♦ **Glasgow:** This old industrial city benefits from oil and gas exploration, industrial research, and electronics.

- ♦ **Edinburgh:** Presbyterianism began in this capital and cultural center of Scotland. Lovely parks and great shopping contrast with tall, narrow houses made of granite. Nearby is Leith, where Scotch whisky, its most famous export, is blended.

Northern Ireland

- ♦ **Belfast:** Despite the sectarian violence that has plagued Belfast, the Northern Ireland capital is a surprisingly happening place, with a lively economy and vibrant nightlife. Even at the height of the "troubles," Belfast wasn't a particularly dangerous place for visitors.

Wales

- ♦ **Cardiff:** Largely rebuilt after World War II bombing, the Welsh capital on the Severn River is rapidly developing into one of Europe's finest maritime cities, famous for its parks, choirs, and ancient Cardiff Castle.

Republic of Ireland

- **Dublin:** The once dowdy Irish capital has in recent years become the Celtic "Silicon Valley" and the hub of Ireland's recent economic boom. It is home to Trinity College (and its 1200-year-old *Book of Kells*), world-class museums, and Europe's largest urban park, Phoenix Park.

The People and Their Languages

One legacy of the British Empire is the large number of ethnic minorities who live in the United Kingdom, mostly Chinese, West Indian, and South Asian. Although church attendance is down dramatically here as in most of Western Europe, most people are at least nominally Protestant. The largest denomination is Anglicanism, followed by Presbyterianism and Methodism. There is a large Catholic minority, and sizable numbers of Muslims, Hindus, Sikhs, and Jews. The Republic of Ireland is overwhelmingly Roman Catholic. Both islands are densely populated, except for the northern reaches of Great Britain.

Most people in the United Kingdom and Ireland speak English, a Germanic-based language. As Anglo-Saxon peoples advanced into the British Isles, the existing Celtic populations were pushed to the peripheries of the region, where they never completely adopted the invaders' speech. Three remnant Celtic tongues are still spoken on the islands. Wales has the complex Welsh language. In northwestern Scotland, highlanders still speak the Gaelic language. Irish is spoken throughout Ireland, especially in remote clusters on the western coast.

In the News: The Euro Future

The integration of the United Kingdom into the European Union was late in coming and has not been altogether smooth. A sizable percentage of the British population was against joining the EU in the first place back in 1973, and as of 2003 a majority is against joining the European Monetary Union and adapting the Euro. One argument against the Euro is that the UK economy is stronger than the Eurozone economies and would be dragged down by monetary union. Proponents of the Euro argue that the Eurozone wields too much economic clout and is too important to British trade for the UK to remain out of it for much longer.

The Least You Need to Know

- The British Isles are made up of many small islands and two large ones, Great Britain and Ireland. The region consists of two countries, the United Kingdom and the Republic of Ireland.

- Despite the high latitude of the British Isles, the influence of the Gulf Stream makes its climate fairly mild.

- London is by far the largest city in the British Isles and a major world financial and cultural capital.

- The people of the United Kingdom are largely Protestant; Ireland's population is overwhelmingly Roman Catholic.

- English is the dominant language in the region, along with remnants of ancient Celtic tongues.

Chapter 8

West Europe: The Heart of the Continent

In This Chapter

- ◆ The mixed blessing of this region's location
- ◆ The heart of Europe
- ◆ The natural side of West Europe
- ◆ The region's spectacular cities
- ◆ The declining population

The West Europe region has been the mother of great philosophers, scholars, writers, artists, traders, and explorers. Though it's home to only a fraction of the world's population, it's a dominant player in trade, manufacturing, finance, art, fashion, and technology.

Small but complex, West Europe is a collage of separate nations characterized by different economies, politics, and physical features. Although old animosities, religious differences, and language barriers have not completely gone away, West Europe is at the core of the increasingly cohesive and expanding European Union.

Blood and Glory: West Europe's Turbulent Past

West Europe's location on the continent has proved to be both an asset and a curse. The region has served as a crossroads for millennia and has thrived on the trade and culture that such a strategic position fosters. The downside is that, as a crossroads, West Europe has for centuries also served as the battleground of the western world.

West Europe.

Apparently, if you're going to throw a war, this is the place to do it. Julius Caesar marched his Roman legions through the area in 55 B.C.E., crushing the Germanic tribes he encountered along the way. During the decline and fall of the Roman Empire, Franks and Visigoths returned the favor, ravaging the European countryside. Centuries of warfare later, Charlemagne's Holy Roman Empire briefly united the countries of the West European region. Dynastic competition and religious strife tore the region apart during The Thirty Years' War in the seventeenth century. The early nineteenth century brought more turmoil under another empire builder, Napoleon Bonaparte.

Things didn't improve in the twentieth century to say the least; in fact, the warring reached new heights. West Europe became the primary battlefield in both World Wars I and II. Flanders Fields (in the Belgian countryside) and the Normandy beaches on the western coast of France recall scenes of wartime carnage unprecedented in human history. The loss of life during these conflicts staggers the imagination. Between Germany and France, more than 3 million people died in the First World War. In the Second World War, more than 3 million German men died in battle.

The effect of this hideous loss of life and economic devastation on West Europe cannot be minimized. The heart of European culture reeled as priceless art and centuries-old architectural masterpieces were destroyed in these conflicts. Although West Europe holds a strategic geographic position, it has paid a high price for it.

Countries in the Crossroads of Europe

West Europe can be grouped into three primary geographic subregions:

- **France and Germany:** Core countries that form the largest subregion, in terms of land area and population

- **Switzerland, Austria, and micro Liechtenstein:** Landlocked Alpine countries

- **Netherlands, Belgium, and tiny Luxembourg:** The "Low Countries" that comprise the smallest subregion

The Biggies: France and Germany

The largest of West Europe's countries (it's not quite as large as Texas), France covers more than 200,000 square miles of varied terrain, ranging from the beaches of Normandy to the peaks of the Alps and the Pyrenees. It is the westernmost country in the region, and with nearly 60 million people, it is the second most populous. The French countryside is laced with vast networks of rivers and canals that unite the country and connect it with the rest of the continent.

France is West Europe's leading agricultural producer. Grains from the north (France is the only European country able to export surplus wheat) are the dominant crop.

France has more than 3,000 miles of canals. They link the country's rivers in such a way that grain grown in northern Normandy can float to Mediterranean Marseilles in the south.

Farms in the south benefit from long growing seasons to produce olives and oranges. And of course, France is the source of the world's finest wines.

From the casked brandies of Cognac to the red wines of Bordeaux and Burgundy to the white wines of the Loire valley and the sparkling "bubbly" of Champagne, each district of France has its vineyards and distinctive wine-making style. For France, wine is a very serious business indeed.

France also rules supreme in the design, production, and retailing of luxury goods and fashion. French brand names, such as Chanel, Dior, Givenchy, St. Laurent, Vuitton, and many other others, are international icons of chic. With arguably the world's finest cuisine, wines, and fashion, France is the global capital of good living.

Germany, the home of bratwurst and beer, Volkswagen automobiles, and the speed-limit-less autobahn, is called the "locomotive" of Europe. The continent's most populous nation and largest economy, Germany has one of the world's greatest concentrations of industrial centers and factories and dominates the European coal, iron, and steel industries. Every type of machine—military equipment, oil pipelines, and even roller coasters—is made by German industry, which in turn is dependent on coal and oil.

Geographically Speaking

After separating East and West Germany for 28 years, the Berlin Wall, the hated symbol of Communist oppression in the east, finally came down in 1989. West Germany and East Germany were united politically once again in 1990, but the unification of their two very different economies has been slow and difficult. The cost of bringing the infrastructure and economy of the former East Germany up to western standards has been far greater than anticipated and has weakened Germany's economic performance in recent years.

The industrial heart of Germany, and of Europe, is the Ruhr Valley in the northwestern part of the country. From such industrial centers as Essen, Duisburg, and Dortmund, manufactured goods move easily via Rhine waterways to Rotterdam in the Netherlands and out to sea. But as in other parts of the developed world, the industrial pulse of the Ruhr Valley is slowing as the economic focus of Germany economy shifts to commerce and services.

Germany relies heavily on a low grade of coal to fuel its industry, and emissions from its coal-burning utilities contribute to air pollution. Environmental problems are especially acute in the former East Germany, where acid rain and other industrial pollutants are to blame for what is know in German as *waldsterben*, or forest death.

The Alpine States

In the southeast of the West European region lie the Alpine states. This subregion consists of three countries: Switzerland, Austria, and the microstate of Liechtenstein. As the term *Alpine states* implies, these countries are dominated by West Europe's vast mountain chain, the Alps.

Switzerland, like its Austrian neighbor, is famous for its Alpine beauty and picturesque qualities. Although the name Switzerland conjures up images of superb skiing, fancy watches, luscious chocolates, Swiss cheese, and yodeling shepherds, this small landlocked mountain country is one of Europe's most prosperous with a highly skilled work force, a strong currency, and a secretive banking system that attracts foreign investment.

Terra-Trivia

The Swiss are known for their fierce independence and strict neutrality. They stayed out of both world wars in the last century, have not joined the European Union, and while they have long participated in United Nations organizations, the country did not officially join the UN until 2002.

Having few energy resources, the Swiss have harnessed hydropower and nuclear power to fuel the renowned Swiss precision industry. Machine instruments, watches, and diesel engines make up the bulk of Swiss exports.

Unlike Switzerland, Austria was at the center of both twentieth century world wars, and despite being on the losing side both times, it has emerged as a modern, prosperous nation with a thriving, mainly service, economy. Like Switzerland, Austria uses "white coal" (hydropower), of which it is a net exporter, to fire its machine and chemical industries. Long a leader in forest management, Austria has been careful in using its Alpine forests. Because the forests protect against erosion and avalanches, they have been harvested selectively to provide fuel and lumber—no clear-cutting here.

The microscopic principality of Liechtenstein is set high in the Alps astride the Rhine River and between Switzerland and Austria. Small though it may be, Liechtenstein has a big economy, focused on industry and financial services. Gorgeous mountains and fairy-tale castles make Liechtenstein a popular tourist destination.

The Low Countries

The third subregion of West Europe is composed of three small, low-lying countries located in the north central part of the West European region. The Netherlands, Belgium, and Luxembourg are known collectively as the Low Countries or the *Benelux* countries (from the combination of the first two or three letters of *Bel*gium, the *Ne*therlands, and *Lux*embourg).

> **Geographically Speaking** _____
>
> Low might be in the eye of the beholder, but the Netherlands is low by any standard. Its average elevation is less than 40 feet above sea level, and much of the country is below sea level.
>
> Although Belgium isn't as low as the Netherlands, most of this flat country can still slide in under that label. Luxembourg, however, is mostly a hilly plateau and hardly belongs in the low club. Because Luxembourg is not big like France or Germany nor Alpine like Switzerland or Austria *and* because it's close to Belgium and the Netherlands, it is forever known as one of the Low Countries.

The Netherlands, which literally means "low lands," is also known as Holland, although strictly speaking that's the name of two of the country's provinces (North and South Holland). The people of the Netherlands, or Holland, are called the Dutch in English. The Netherlands are traditionally associated with windmills, tulips, and wooden shoes, but what really defines the country is water. With two-fifths of the Netherlands lying below sea level, much of Dutch history has been about flood prevention. (For more on how the Dutch deal with their water problem, see the section "Creating Land from Sea," later in this chapter.)

Although water has played a destructive role in Dutch history, it has also made the country's fortune. Situated at the mouths of three major rivers, the Rhine, the Maas, and the Schelde, the Netherlands is one of the world's most important commercial and transportation hubs. Its main port, Rotterdam, is Europe's largest. The Netherlands is also a major producer and exporter of food and flowers, and of course, mountains of colorful Edam and Gouda cheese.

Belgium is situated below the Netherlands and is in the middle of the Low Countries in terms of land area and population. The country is divided between French-speaking Walloons in the south and Dutch-speaking Flemings in the more industrialized north. Although conflicts between the two groups once threatened to tear the country apart, increased regional autonomy has helped ease tensions considerably. Belgium today is the bureaucratic heart of Europe, with NATO and the European Union both headquartered in its capital, Brussels.

Belgium is famous for it hundreds of different types of beer and for a cuisine that rivals (and according to many surpasses) that of its neighbor to the south, France. Tourists often skip Belgium, which is a pity as it has much to offer, including two cultures for the price of one.

The last of the Low Countries in this list is the pseudo-low country of Luxembourg. At about the size of Rhode Island, Luxembourg is just a notch above being a micro-state. Despite its reputation as an outdoor paradise filled with lush forests and rushing streams, Luxembourg has a very modern, high-growth economy. Although steel was once king in Luxembourg, its economy has diversified, with the focus increasingly on financial services. Multilingualism and stability make Luxembourg attractive for for-eign business, and its scenery, pâtés, wild game, and Moselle wines don't hurt either.

The U.S. of E?—The European Union Today

The countries surveyed in this chapter form the core of what is today known as the European Union (EU), formerly known as the European Community (EC) and before that, the European Economic Community (EEC). Germany and France are its largest and most populous countries and are two of the four trillion-dollar economies in the EU (the other two are the United Kingdom and Italy).

The idea for a European Federation was first proposed in France in 1950, but the foundation for it had been established in 1948, when the three Benelux nations joined together in a customs union. The original six members of what would become known as the European Economic Community (the Benelux countries, plus Italy, France, and Germany) became nine in 1973 when the United Kingdom, Ireland, and Denmark joined the European Community. As of 2003, the European Union included 15 Member States (Greece, Spain, Portugal, Austria, Finland, and Sweden joined in subsequent expansions) and is preparing to take in another 10 in 2004 and possibly another three in 2007. After the 2004 expansion, the European Union will include for the first time former Soviet-bloc countries of Eastern Europe.

Despite rifts and occasional setbacks (see "In the News" at the end of this chapter), The European Union has brought about something that no one would have thought possible after seeing the rubble of post-World War II Europe—namely a prosperous, unified, border-free, economic and increasingly political entity stretching from the Atlantic to the borders of the old Soviet Union. In January 2002, 11 of the 15 Mem-ber States cemented their economic union by exchanging their historic currencies—francs, marks, liras, pesetas, escudos, shillings, drachmas, guilders, punts, and crowns—for the single European currency, the Euro.

The Expanding European Union.

West Europe's Natural Side

Pushing into the Atlantic Ocean on the west and the North Sea on the north, much of West Europe has a maritime feel. The moderating influence of the Atlantic Gulf Stream tempers the plains of the northwest. Interior and mountain areas face colder climes. Protected by mountains from the cold of the north, the French Riviera (the Mediterranean shore of France) enjoys sunny temperatures with only a little rain.

Physical features—West Europe.

Landforms: From Polders to Alps

Contrasting with the lowlands of the Netherlands is the most striking of Europe's physical features: The mighty Alps. This rugged chain is a 680-mile-long collection of snowy peaks and valleys filled with glaciers and dotted with lakes. The Alps form an arc that stretches from France's Mediterranean coast through Switzerland and Germany to Austria's Danube.

With an elevation of 15,781 feet, France's Mont Blanc is the tallest of the Alps. Switzerland's mountain landscape is dotted with romantic castles and charming chalets. Soaring over all are its famous mountains Jungfrau and Matterhorn on the Switzerland–Italy border. The Bavarian Alps in southeastern Germany offer some of the most beautiful scenery and the best skiing in the nation, as well as the nineteenth century fantasy castles of Bavaria's "Mad" King Ludwig.

Terra-Trivia

A characteristic weather pattern of Southern France is the *mistral*, an incredibly forceful wind that occurs when Mediterranean storms pull cold, north European air down the Rhône River valley. During these mistrals, the south receives a chill lasting several days.

Two great mountain chains separate West Europe from Southern Europe. The Alps run between West Europe and Italy, while the Pyrenees Mountains form a natural boundary between southern France and Spain. The Pyrenees are the home of the Basques, a people whose language is related to no other and whose homeland straddles the French and Spanish sides of the western Pyrenees. The Basque struggle for a separate state has periodically involved violence and terror attacks.

Despite the size of these huge and magnificent mountain chains, they have not completely barricaded nor isolated one region from another. Because of natural passes, valleys, and rivers, north-south trade, communications, and invasions have been a fact of life for millennia in this part of the world.

Creating Land from Water

The Netherlands is cradled around what was once a shallow inland sea called the Zuider Zee. At the mouth of this sea, the Dutch built a huge earthen dam, or *dike*, that cut off the Zuider Zee from the North Sea. The dike created a lake from the former Zuider Zee that the Dutch renamed the IJsselmeer. Over the years, the Dutch have drained portions of this shallow lake bed to form reclaimed lands called *polders*.

The poldering process that has enabled the Dutch to create land out of what was once the sea involves the construction of massive dikes around areas to be reclaimed. After the dikes have been constructed, the brackish water is pumped out and the area's famous canals are then used to drain the land. (Windmills once had the job of constantly pumping out the ever-seeping water. Today diesel pumps are used and the windmills are primarily scenic elements.)

North Sea storms have breached the dikes in the past, and the loss of life and property has been catastrophic. After a devastating flood disaster in 1953, the Netherlands began the construction of a cutting-edge system of dikes, dams, and surge barriers on the mouth of the eastern Scheldt river. The Delta Works, as the project is called, is one of the engineering wonders of the world.

Waterways: From the Danube to the Rhine

Snowmelt from the Alps gives rise to three of Europe's key rivers: the Rhine, the Rhône, and the Danube, as shown on the preceding map. Other notable rivers are the Seine, the Loire, and the Garonne, which together with the Rhône form the basis of France's great waterway system. The flat, level plains of Northern France and the Low Countries have made it easy to join the area's major rivers, the Rhine, Meuse, Seine, and Marne, through a complicated network of bridges, rail links, and canals.

West Europe's most significant river is the majestic Rhine, which starts in the Swiss Alps, flows through western Germany (where it forms part of the France-Germany border), and empties into the North Sea at the great delta in the Netherlands. The Rhine is the major lifeline of the West European region, carrying goods and people from the industrial heartland out to sea. It is also one of the most romantic and scenic rivers in the world, its banks dotted with castles and vineyards. (Germany's Rhineland is a major winemaking area.)

Cities of West Europe.

A Place of Wondrous Cities

Many of the world's most spectacular and storied cities are located in the West Europe region. There are so many of them in fact that covering them in depth is way beyond the scope of this book. Here are a few highlights:

France

- ◆ **Paris:** The center of just about everything in France, Paris is the seat of government and the primary expression of national pride and culture. Rich in historical and cultural treasures, including the great Gothic cathedral of Notre Dame and the magnificent Louvre museum (home of da Vinci's Mona Lisa), Paris is unquestionably one of the world's most beautiful cities.

- ◆ **Lyons:** With its prime location on the confluence of the Rhône and Saône rivers, Lyons is second only to Paris in population and economic clout. It is a historical center of silk and textile manufacturing and site of major annual trade fairs.

- ◆ **Bordeaux:** This historic port on the Garonne River is a major cultural center and the heart of one of the most important wine-producing regions in the world.

Germany

- ◆ **Berlin:** This vibrant city of 3 million people is once again the capital of a reunified Germany. From 1961 to 1989, Berlin was divided by the most notorious expression of the Cold War, the Berlin Wall.

- ◆ **Hamburg:** Germany's second largest city, Hamburg is a major port at the mouth of Elbe River and one of the world's greenest cities.

- ◆ **Munich:** The main city of southern Germany and Bavaria, Munich is the site of the world's most famous annual beer festival, its raucous Oktoberfest.

Switzerland

- ◆ **Zurich:** Switzerland's largest city is also a global center of banking and business.

- ◆ **Bern:** The capital and seat of Swiss government.

- ◆ **Geneva:** The home of the Red Cross and the League of Nations, Geneva also boasts the towering Reformation monument honoring Calvin, Knox, and others who made the city a center of Protestantism.

Austria

- **Vienna:** Famous for its musical heritage and rich desserts, Vienna is now the capital of small but prosperous Austria, but was once the center of the mighty Hapsburg Empire.

The Netherlands

- **Amsterdam:** The art-rich capital of the Netherlands is known as the "Venice of the North" for its many canals and magnificent town houses. Amsterdam shares capital honors with The Hague, the seat of the Dutch government and site of the World Court.

- **Rotterdam:** Almost completely destroyed during World War II, Rotterdam today is one of the world's largest and busiest ports. Rotterdam is known as the "gateway to Europe" because of its strategic location on the mouth of the Rhine.

Belgium

- **Brussels:** Beyond its magnificent medieval central square, the Belgian capital is a booming modern city that serves as headquarters for NATO and the European Union.

The People

Although it's not a large region in area, West Europe does have a very large population. Especially heavy concentrations of people live in northern France, the Low Countries, the Rhine River valley, and along the Rhône in southern France. West Europe is highly developed and industrialized, and most of its people live in cities.

The Netherlands, with more than 1,200 Dutch squashed into each square mile, is the most densely populated country Europe (with the exception of the microstates of Malta and Monaco); Belgium is a close second. Because Europe's birthrate is relatively low and life expectancy is high, however, you might hear mutterings about a "demographic winter" whose projections suggest a high old-to-young ratio. As this ratio becomes reality, West Europe faces a situation of having fewer workers and higher costs for social programs.

In Germany, the declining population trend is counterbalanced somewhat by the continuing influx of *gastarbeiter*, or guest workers, who have been filling jobs in Germany since the 1960s. Immigrant workers from Turkey, Greece, Spain, and more recently Russia and the former Yugoslavia have also kept the country's growth rate on the positive side (barely). Without this immigration, Germany's population would actually be shrinking.

Like Britain, France had an overseas empire as well. Immigrants from former French colonies in Southeast Asia and especially North Africa have settled in the cities and suburbs of France, where ethnic prejudices and rivalries have led to profound cultural conflicts.

Protestants and Catholics

Most of the West European region is nominally Christian, although church attendance is in decline here as in the United Kingdom. The north is predominantly Protestant; the south is mainly Catholic. The Dutch, Swiss, and Germans are fairly equally divided between Protestant and Catholic. The French, Austrians, Belgians, and "Luxembourgians" are primarily Catholic. Although the Jewish population of West Europe was decimated during the World War II holocaust, the world's third largest Jewish population (about 600,000) lives in France.

Languages: Germanic and Romance

West Europe has two primary language subfamilies: Germanic and Romance. Both subfamilies belong to the Indo-European language family. French, the primary representative of the Romance subfamily, is spoken primarily in France. French is also spoken in southern Belgium and western Switzerland.

The Breton Peninsula in northwestern France is considered a *refuge area*. It has an isolated pocket of people who speak Breton, a language in the Welsh group of the Celtic sub-family. Celtic languages are considered *relic languages*. Once spoken over a wide area, they were overwhelmed by other languages and now survive only in refuge areas.

The Germanic languages (including Dutch, Flemish, and German) are dominant in the remaining countries in the north and east. Almost all Europeans are at least bilingual, with much of the population proficient in three or more languages. Switzerland has four official languages: Italian, French, German, and Romansch (an umbrella term for several Romance dialects spoken on both sides of the Swiss-Italian border). By the way, many West Europeans also have mastered English.

In the News: Old vs. New Europe

One of the many repercussions of the United States-led Iraq War of 2003 was a bitter rift between, on the one hand, the United States and its main ally in the war, the

United Kingdom, and on the other, France and Germany, which strongly opposed the war. U.S. Defense Secretary Donald Rumsfeld made headlines and raised hackles around the world when he characterized France and Germany as "old Europe" and the pro-war governments of Eastern Europe, as "new Europe." In Secretary Rumsfeld's view the center of gravity in Europe was shifting east to the former Soviet-bloc.

The rift was not only a trans-Atlantic one, it also cut through the heart of the European Union on the eve of its historic expansion east. It underscored some of the great unanswered questions for the coming decades, namely how will the vast new European entity develop, and will it emerge as a political counterweight to the sole remaining post-Soviet superpower, the United States of America?

The Least You Need to Know

- West Europe has been the site of major wars throughout history, and was the focal point of the two twentieth-century world wars.

- The European Union emerged from the devastation of World War II. Although its core is in West Europe, it is about to expand to the borders of the former Soviet Union.

- The Alps are the dominant physical feature of the region. The Rhine is its principal river.

- The people of West Europe are primarily urban, affluent, and Christian. They speak either Romance or Germanic Indo-European languages.

Northern Europe: Land of the Midnight Sun

In This Chapter

- ◆ Weathering the wrath of the Norsemen
- ◆ Europe's rugged northern fringe
- ◆ Legacy of the Vikings
- ◆ Great cities of the north
- ◆ A bridge to the future

The region of Northern Europe stretches northward from the continental core of Europe through the Baltic Sea and the North Atlantic Ocean to the frozen Arctic Circle. Northern Europe is appropriately named because all of this region is far to the north (it's the northernmost group of countries on earth). The sea has provided a livelihood, albeit a difficult one, but has distanced people of the North from the rest of Europe. The result is a sense of neutrality and unity—and no small degree of isolation.

Democratic and literate, sturdy and industrious, Northern Europe's sparse population is homogeneous and mostly Protestant and enjoys a high standard of living. An egalitarian outlook on life is typical, and the "cradle to grave" social support system is an accepted part of life. Sports (especially winter sports) pervade the national scene, adding color and glamour to the incredible natural beauty of the northern lakes, vast wilderness areas, and rugged mountains. This is a region of modern, advanced societies still marked by the romance of its Viking past and the fantasy of Hans Christian Andersen's world-famous fairy tales.

Northern Raiders and Resisters

Beginning in the eighth century, Viking warriors from Northern Europe carried out raiding expeditions throughout a large part of the continent. As they sailed their open longboats (adorned with dragon heads), they raided, plundered, and generally terrorized coastal inhabitants, leaving a path of destruction in their wake. Despite their "wrath of the Norsemen" reputation, the Vikings were more than raiders: They explored and discovered as they traveled, and Viking settlements existed far beyond Scandinavia for centuries.

The seafaring Norsemen also made forays across the Atlantic. Settlements were established in Iceland and even in Greenland. Around the year 1,000 C.E., Leif Eriksson and his comrades abandoned Greenland and sailed west to become the first Europeans to land in North America. Remains of a Viking village from Eriksson's era were unearthed in Newfoundland, Canada, in the 1960s. By the eleventh century, the Vikings had more or less stopped raiding as they became more settled in their homelands. Although the Viking settlements in Iceland survived, the Greenland colonies were wiped out between the fourteenth and the fifteenth centuries by a combination of disease and an increasingly colder climate in the North Atlantic.

In the seventeenth century, Scandinavia was again on center stage. The Swedish kingdom was at the height of its power and extent, encompassing all of present-day Sweden, much of Finland, a large part of the Baltic region, and sections of Germany east and west of Denmark.

In the twentieth century, the Northern European countries remained neutral during World War I, but only Sweden managed to do so in World War II. Finland fought off attacks from both Germany and the Soviet Union. Norway and Denmark were both occupied by Nazi Germany. Norway's government was taken over by Nazis and run by a Norwegian Nazi named Vidkun Quisling, whose surname has become synonymous with "traitor." The Danes distinguished themselves by their tenacious resistance and determination to save their Jewish fellow-citizens from the Nazi death camps. When rumors of plans to arrest all Jews surfaced, the Danes acted immediately and managed to spirit away most of country's 8,000 Jews to neutral Sweden.

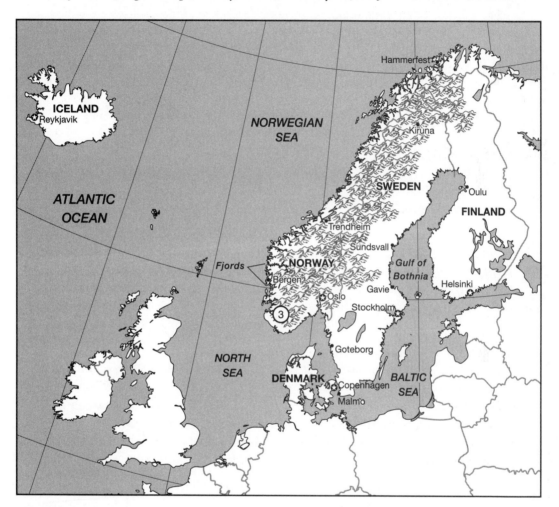

Northern Europe.

The Land of the North

Europe's northern tier includes five countries, which as a group are referred to as Norden. Scandinavia refers to the Scandinavian peninsula containing Norway and Sweden. Finland, Denmark, and far-off Iceland (900 miles away) complete the Nordic group. The five Nordic countries have a great deal in common: a shared heritage and culture, rugged landscapes (except for low-lying Denmark), closely related languages (except for Finnish), strong economies, and a long and close association with each other. The result is a very distinctive and cohesive geographic entity within Europe.

Geographically Speaking

Northern Europe has long been called the "land of the midnight sun." During the Arctic summer, the sun never dips below the horizon. The $23\frac{1}{4}$-degree tilt of the earth's axis causes the polar areas never to leave the illuminated half of the earth around the summer solstice. Summertime brings 24-hour sunshine, midsummer's Arctic bramble bush wine called "mesamarja," and long evening celebrations.

This phenomenon has a downside. Northern Europe can also be called land of the noon moon (it's not astronomically correct, but it has a nice rhyme). During the dead of winter, the sun never rises above the horizon.

Norway, Sweden, and Finland all lie in the Northwestern Uplands, Europe's oldest mountain range, which runs all the way to Great Britain. The countries of Northern Europe benefit from the moderate temperatures generated by the Gulf Stream (a warm current) as it pours into the North Atlantic. Southern coastal Norway does not have a frigid climate; harbors remain open and free of ice year-round. The same cannot be said, of course, of the interior or mountain regions, where the weather, like the terrain, is much more severe.

Despite some relief from the Gulf Stream, the region still encroaches on the Arctic Circle. Moisture and wind from Atlantic storms are—pardon the expression—nothing to sneeze at. Winter days in this region are short, dark, and cold. The poor, rocky soil clings to the steep terrain, and the growing season lasts about 15 minutes. The far north is wild and mountainous, with much of it above the Arctic Circle. Most of this area is blanketed with vast stands of coniferous forests, called *taiga*, and with extensive tracts of birches.

Terra-Trivia

Measuring 125,000 square miles and reaching 1,000 miles from tip to toe, Norway is huge compared to most European countries. Despite Norway's size, much of its territory is inhospitable. Only 4 million people live here, most of them on the coast. Of the Northern European countries, only Iceland has fewer residents.

The Fjords of Norway

Norway is famous for the magnificent *fjords* that define its jagged and indented western coastline. A fjord is a long, narrow inlet, or arm of the sea, created by glaciers and lined by steep cliffs. Some are as long as 100 miles. Fjords are fed by river waterfalls that cascade from the mountains to join the sea. As the fjords reach the sea, *skerries* (small, rocky islands) form a buffer between the Atlantic and the shore. Norway's fjords are a must-see tourist destination during the summer.

In addition to the tourist bonanza the fjords attract, thousands of rivers and streams from Norway's high mountain ranges produce more kilowatt hours of energy per person than anywhere else in the world. In addition to hydro-electricity, the country enjoys other resources in the form of oil and natural gas yielded by the North Sea. Half of Norway's exports are petroleum products, and a great deal of industry has developed around power sources, both hydro and petroleum.

The Swedish Core

On Sweden's west and north sides, rugged highlands mark the boundary with Norway. Coniferous forests grow in the northeast, and large lakes and low agricultural land are found in the south. The country is laced with rivers that originate in the highlands of the west and drain to the Gulf of Bothnia in the east. Although lacking Norway's oil and gas deposits, Sweden does possess substantial mineral wealth, primarily in the form of metal ores.

The Forests of Finland

Finland's terrain is flatter than its neighbors' to the west, but poor soils and cold temperatures make agriculture nearly impossible. Mineral resources are few. Although this situation sounds dismal, Finland does have trees, and lots of them, which account for its large and thriving paper, pulp, and lumber industries.

Although nearly a third of the country lies above the Arctic Circle (its neighbor to the east is Russia), Finland, like its neighbors, benefits from the warming influence of the Gulf current. Despite its northern location, Finland never has permanent snow cover or ground frost. The waters of the Baltic Sea and the Gulf of Bothnia also contribute to the relative mildness of Finland's maritime climate. Some 30,000 islands form micro-*archipelagos* (island clusters) along the Finnish coast.

Denmark

As the southernmost Nordic country, Denmark occupies yet another peninsula, Jutland. In the south, the Jutland peninsula shares a mere 42-mile border with Germany. The low flatlands of the Jutland peninsula strike a sharp contrast to the rugged highlands of the Scandinavian peninsula to the north.

Also notable are Denmark's 482 islands. They range from mite-size to medium-size to Greenland, the world's largest island. In the midsize category is Zealand, home to

Copenhagen, the capital. Denmark also has the 21 Faeroe Islands, off Scotland's north shore. Some of the Faeroes, meaning "sheep islands," have a language and flair of their own and even a separate set of laws. All make up the Kingdom of Denmark.

The country is more agriculturally blessed than its Nordic cousins. With productive soils, milder temperatures, level lands, and high-tech methods, Danish farmers are net exporters of food, with an emphasis on meat and dairy products. With farmland that covers three-quarters of the countryside, Denmark is indeed the Nordic breadbasket.

Iceland's Heat from Below

Of the five Nordic countries, Iceland is the worst off in terms of quality terrain. Three-quarters of the interior is filled with volcanoes, glaciers, lava formations, frozen lakes, and varied wastelands. Most of Iceland, in fact, is uninhabitable. All the island's settlements, and the few farms that make hay for the famous Icelandic ponies and hardy sheep, are coastal. Nonetheless, Icelanders have learned to make the best of their situation.

Perched on the spreading mid-Atlantic ridge, a volcanic hot spot, Iceland treats its 100 volcanoes as an additional energy resource. Geothermal heat from volcanic sources and hydroelectric power from Iceland's rivers and waterfalls are used to warm apartment buildings as well as greenhouses for vegetable production. Not bad for a barren land with no fossil fuels.

Eye on the Environment

Although most of Iceland's electricity and heating needs are now met by hydroelectric power and geothermal sources, its cars, buses, and all-important fishing fleet are still powered by conventional fossil fuels, of which it has none. To free itself from its dependence on imported fuel, Iceland is at the forefront of what might be the coming hydrogen revolution.

Hydrogen is an efficient, clean-burning fuel that is found wherever there is water, and there's certainly a lot of water in and around Iceland. The problem with hydrogen (H) is that extracting it from water (H_2O) requires electricity and lots of it, and that usually means burning (dirty) fossil fuels to produce (clean) hydrogen. And that's where Iceland has an edge: Its electricity is abundant, cheap, and clean. Using it to extract hydrogen to power vehicles and ships makes both environmental and economic sense. Iceland's first energy revolution involved hydroelectric power; its second, geothermal heat; its third will most likely be about hydrogen. For more about the promise of hydrogen fuel, see Chapter 28.

The Descendants of the Vikings

Considering Norden's location at the top of the world, you probably aren't surprised to hear that few people live there. The hardy souls who do reside there live on the coasts and as far south as possible. With just fewer than nine million inhabitants, Sweden boasts the largest population in the region. Population densities are fairly low throughout Norden (Denmark has the highest), and the bulk of the people are urban dwellers.

> **Terra-Trivia**
>
> With fewer than 280,000 residents, Iceland is one of the world's least populated countries (the population of the entire country is less than that of Birmingham, Alabama). Population density in Iceland is a paltry 7 people per square mile. To put this number in perspective, Iceland has about 18,160 fewer people per square mile than Hong Kong and only 7 more people per square mile than the moon.

The Lapps: Life on the Top of the World

Along the Arctic Circle across the top of Norway, Sweden, Finland, and part of Russia live the aboriginal people of Norden, the hardy, nomadic Lapps, or Sami, as they prefer to be called. The Sami traditionally have made their living hunting, fishing, and herding reindeer on the frozen, moss-covered treeless arctic plains known as tundra (a Sami word). Although Sami today engage in many different economic activities, they are still best known for their association with the reindeer, which provide them with food, clothing, tools, shelter, and transportation. In recent years, however, increased mining and logging activities are putting pressure on traditional Sami reindeer herding lands.

Almost Intelligible

Most of Norden shares a common linguistic heritage. Danish, Icelandic, Norwegian, and Swedish are all Germanic languages in the Indo-European language family, and all are more or less mutually intelligible. Finnish, however, is the Nordic oddball. It originated in central Asia and is part of a separate language family, the Uralic family, to which the Sami language and Hungarian also belong. Fortunately for the sake of Nordic mutual comprehension, Swedish is Finland's second official language.

Living the Good Life

The Nordic countries present a true *smorgasbord* of experiences and contrasts. Despite a reputation for being uncommunicative and introspective, Scandinavians know how to have a good time and how to live the good life. The homeland of Nordic skiing is mad about sports, especially of the winter variety, and they take great pleasure in their *saunas*—hot Finnish-style steam baths. The sophisticated capitals of the region, especially Copenhagen and Stockholm, offer world-class cuisine, rich cultural attractions, and a vibrant nightlife.

The Many Religions of Northern Europe?

Think Lutheran. Although the Nordic region today is largely secular, its main religion is Lutheranism.

Making Ends Meet on the Tundra

Northern Europe today has emerged as one of the continent's most prosperous regions. The economies in the north have thrived on fishing (especially Iceland), wood products (especially Finland), industry and manufacturing (especially Sweden), agriculture and food processing (especially Denmark), and petroleum revenues (especially in Norway). The people of Northern Europe generally are well educated and have high incomes.

 Geographically Speaking _____

Known as "the land of fire and ice," Iceland shares a maritime economy with its Nordic neighbors. It sits amid the richest fishing grounds in existence and annually delivers more than 1.5 million tons of fish to the rest of the world. Because the fish industry provides 70 percent of Iceland's export earnings, it has no interest in having foreign fleets encroaching on its domain and takes great pains to make that information public knowledge. It has also resisted joining the European Union for that very reason.

The Cities of the North

Norway

- **Oslo:** Norway's capital and center of commerce and culture. Nature-loving Oslo is one of the world's most heavily forested cities.

- ◆ **Bergen:** A favorite starting point for many tours of the fjords and the host of an international festival of arts.

Sweden

- ◆ **Stockholm:** The capital of the most industrialized of the Nordic nations, Stockholm was built on islands where the Baltic meets Lake Malar, hence its renown as "the queen of the waters."

Finland

- ◆ **Helsinki:** Finland's capital and largest city, Helsinki, known as the "white city of the north," is famous for its striking modern architecture.

Denmark

- ◆ **Copenhagen:** The capital of Denmark and the most cosmopolitan of Nordic cities, Copenhagen is home to the world-famous amusement park Tivoli Gardens, which features symphonies and ballets, glitzy rides, and tranquil gardens.

Iceland

- ◆ **Reykjavik:** The Icelandic capital is a small but booming city, home to more than one in three Icelanders. One of Europe's newest tourist hotspots, Reykjavik is also the world's northernmost national capital.

In the News: Bridging the Ore Sund

Although Denmark shares a border with Germany and is perhaps the most "European" of the Nordic countries, Scandinavians have always seen themselves as somewhat apart from Europe, much as the British do. But just as the Chunnel created a land—and psychological—link between England and the continent, the new Ore Sund bridge between Copenhagen and the Swedish port of Malmö provides the first fixed link between Sweden and Denmark (and by extension the European heartland) since receding glaciers split them apart 7,000 years ago.

Opened to traffic in July 2000 after nearly 10 years of planning and construction, the Ore Sund Bridge is the world's longest single road and railway bridge. It has shortened the travel time between Denmark's largest city and Sweden's third largest city to 20 minutes (an hour less than by the old ferry route) and has turned the combined

Copenhagen-Malmö area into Europe's eighth largest metropolitan area. In many ways, the Ore Sund Bridge is a fitting symbol of what the European Union is all about—eliminating boundaries and fostering integration and growth.

The Least You Need to Know

- Five countries are in the Northern European, or Nordic, region: Norway, Sweden, Finland, Denmark, and Iceland.

- The Nordic countries are the most northern countries in the world.

- With the major exception of North Sea oil and gas, the region is relatively poor in fuel and mineral resources, but is highly developed and affluent.

- The region's relatively small population is located in the south and along the coasts.

Southern Europe: Embracing the Mediterranean

In This Chapter

- ◆ The Greco-Roman heritage
- ◆ The Southern European quartet
- ◆ Glorious cities of art and history
- ◆ The peoples and cultures of Mediterranean Europe

The four countries of the Southern European region share intertwined pasts and common cultural roots. Greece is the fountainhead of Western civilization, and the imprint of ancient Rome is evident throughout Italy and the countries of the Iberian peninsula, Spain, and Portugal. The Southern European quartet also shares a Mediterranean climate and vegetation, mountainous landscapes, and generally poor soil. All either occupy or share peninsulas.

There are exceptions, of course, to all this sharing. Portugal, Spain, and Italy are predominantly Roman Catholic and speak Romance languages. Greece is Orthodox and speaks a non-Romance language. Greece and Italy are fully in the Mediterranean Sea, whereas Spain also faces the Atlantic, and Portugal faces only the Atlantic. Although Italy is one of Europe's four trillion-dollar economies, Spain, Portugal, and Greece are among the least developed of the current European Union member states, although their ranking will change after the smaller Eastern European economies join the EU in 2004.

The Legacy of the Greeks and Romans

The ancient Greeks established an extensive trading network that spanned the Mediterranean. By the fifth century B.C.E., classical Greece was at the height of its power and cultural influence. Athens in particular played a key role in laying the foundations of Western arts, science, architecture, mathematics, philosophy, literature, and democratic government. The achievements of the Greeks were later absorbed and expanded on by the Romans.

Southern Europe.

The Roman Empire eventually controlled all the Mediterranean and beyond. The territory under the empire shared one law, one language, and, later, one religion—Roman Catholicism. Rome dominated the Mediterranean for nearly 600 years, until Germanic tribes swept south in the early fifth century (420) C.E. The Lombards occupied northern Italy, the Visigoths swept through Spain, and the Vandals dominated Africa and (later) Spain. In the eighth century, Muslim invaders pushed north into Spain from Africa and established a Moorish dominion that lasted until the late fifteenth century. All these varied peoples contributed to the genetic mix in Southern Europe today.

With the rest of Europe to the north, Asia to the east, and Africa just below, the Mediterranean Sea and Southern Europe are at a crossroads of the world, and not surprisingly the region became a springboard for great explorations. The Venetian

Marco Polo traveled to China in the thirteenth century and wrote of his exploits. Portugal's great geographer prince, Henry the Navigator, sent forth expeditions to the southern coasts of Africa. Other great explorers and adventurers followed, including Christopher Columbus (Italian sailing for Spain), Vasco da Gama (Portuguese), Hernán Cortés (Spanish), Ferdinand Magellan (Portuguese sailing for Spain), and Francisco Pizarro (Spanish). Through these explorations and subsequent conquests, Mediterranean culture was spread (often brutally) across the entire globe.

The rebirth of classical Greek and Roman arts and learning known as the Renaissance began in the fourteenth and fifteenth centuries in Italy and later spread throughout Europe. The countless masterworks created by Italian poets, such as Dante and Petrarch, and artists, such as Giotto, Botticelli, Raphael, Michelangelo, da Vinci, and Bernini, to name only a tiny fraction, established Italy as one of the greatest artistic and cultural centers the world has ever seen.

Terra-Trivia

Although Portugal and Spain became fully integrated nation-states in the twelfth and fifteenth centuries respectively, the various states and entities on the Italian peninsula did not unify until 1861. Greece achieved its independence from the Ottoman Turks in 1829.

The Countries of Southern Europe

As shown on the preceding map, Southern Europe consists of four countries (and several microstates, mentioned later in this chapter). In the west are Portugal and Spain, the only countries in the region to share a border; in the center lies the famous boot of Italy; and to the east is Greece, with its many islands. This section looks more closely at each country's particular Mediterranean character.

Seafaring Portugal

Portugal is surrounded by Spain and the Atlantic Ocean. Occupying 35,550 square miles of the Iberian Peninsula, Portugal is Southern Europe's smallest and westernmost country. At one time, Portugal was the center of a vast overseas empire that at its peak was 23 times the size of the motherland. Its major colonies included Brazil in South America; Angola, Guinea-Bissau, and Mozambique in Africa; Goa in India; Macao on China's southeast coast; and East Timor in Southeast Asia. The last remnant of Portugal's empire was Macao, which it returned to China by agreement in 1999.

Terra-Trivia

Portugal is a world leader in the production of cork. Cork is extracted from the outer bark of the Mediterranean cork oak tree, which abounds in Portugal.

Income levels in Portugal are lower than in most other European Union countries. Although the Portuguese economy has expanded and diversified since it joined the European Union with Spain in 1985, agriculture is still a mainstay of its economy. Portugal's farmers raise sheep, cattle, and poultry, and grow typical Mediterranean crops: olives, fruit, grapes, and grain. Don't miss spring in Portugal—the wildflowers are unforgettable. Tourism in the southern Algarve area is a mainstay of the economy.

Spain—Land of Toros and Tradition

Spain is Southern Europe's largest country in land area and its second most populous (after Italy). It is more urbanized and industrialized than Portugal and enjoys a higher income level, although Spain, too, is in the lower economic tier of the European Union. As with the other countries in the region, agriculture figures prominently in the Spanish economy. Both Iberian nations are major wine producers: Portugal is famous for its Vinhos Verdes and Port wines; Spain is known for its Riojas and Sherrys.

Tourists from Northern Europe and beyond flock to the cosmopolitan cities, Moorish-flavored towns, and sun-drenched beaches of Spain. The local café is a perfect place to enjoy a breakfast of *churros con chocolate* (deep-fried dough eaten with dark chocolate) or sip a refreshing sangria made of red wine and fruit. You can top off an afternoon with a *corrida de toros* (bullfight) in the bull ring and spend your evening enjoying the flamenco (an Andulacian gypsy-based dance) and nibbling curls of *jamon serrano* (thinly sliced spirals of aged ham).

Italy: Europe's Cultural Magnet

Italy is the giant of the region. Although second to Spain in area, Italy has the largest population and the greatest population density in the Southern European region. It is also the Mediterranean's economic leader, with the bulk of the region's industry and its highest income levels. Its economic health, however, is clouded by the persistent disparity in income levels and standard of living between the prosperous north and the poorer, more agricultural south. Other hard-to-shake problems confronting Italy include government corruption and the influence of organized crime.

Italy, however, has so much to offer it's hard to know where to begin. Its cuisine is one of the world's two or three finest (in fact, many of the classics of French cuisine originated in Italy). Its horde of art and archaeological treasures is beyond compare. And if any city rivals Paris in the realm of fashion design it is surely Milan.

> **Geographically Speaking**
>
> Although many visitors are drawn to Italy's famous and exquisite cities (described in more detail later in this chapter), if you visit, you should take time to tour the tiny towns typically perched atop the rugged hills that dot Italy's landscape. Built from the same stone on which they stand, these towns seem to be simply extensions of the natural terrain. Some favorite hill towns are Orvieto, Assisi, Gubbio, Urbino, Siena, and San Gimignano.

Greece and the Aegean

Greece is a land of archeological wonders and shimmering islands set in the blue Aegean. Given these two resources, it is not surprising that tourism accounts for 15 percent of the country's gross domestic product. A whopping 20 percent of its work force is involved in agriculture, even though only a quarter of the land is arable, plots are small, and rainfall is unreliable. Nevertheless, Greek farmers produce tobacco, cotton, and wheat in addition to the typical Mediterranean mix of crops: olives, grapes, oranges, and vegetables. Although Greece has benefited enormously from membership in the European Union, its income levels still lag behind those of most other EU members. Greece, especially the Athens area, has recently undergone a major infrastructure overhaul as the country prepares to host the Summer Olympic Games in 2004.

Greece was part of the empire of the Ottoman Turks for centuries before it gained its independence in 1829. Bad feelings between Greece and its neighbor Turkey linger to this day, even though the two are members of the North Atlantic Treaty Organization (NATO) and Turkey seeks to join Greece in the European Union. Although confrontations between the two countries have diminished in recent years, a lingering bone of contention between Greece and Turkey is the island of Cyprus, which has been divided into Greek and Turkish sides since 1974. In that year, an unsuccessful Greek-sponsored coup led to an invasion of the island by Turkish troops and the formation of the so-called "Turkish Republic of Northern Cyprus," which is recognized only by Turkey. Unfortunately, UN-led talks on the reunification of the island collapsed in 2003, further clouding Greek-Turkish relations.

Directly and indirectly, many Greeks are tied to the sea. The population is clustered largely along the coasts, and even now many fish for their livelihood. Shipping is also an economic mainstay, and Greece boasts a substantial merchant marine fleet. The sea, fishing villages, and Greece's many islands provide a perfect backdrop for tourism. Dazzling white houses, quaint harbors, and delightful cafés and tavernas are the order of the day on these sun-kissed islands.

The Microstates

In addition to its quartet of large countries, the Southern European region includes a handful of tiny (but full-fledged) countries called microstates:

♦ **Monaco:** Located in the southwest corner of France, on the dazzling French Riviera right near Italy, this infinitesimally small constitutional monarchy is occupied almost entirely by the city of Monte Carlo. Monaco's reigning prince, Rainier III, was famously married to the late Hollywood superstar Grace Kelly (Princess Grace). In addition to its magnificent setting and celebrity royal family, Monaco is also famous for its casinos and its jet-set residents and visitors.

GeoRecord

With more than 41,000 people per square mile, Monaco is the world's most densely populated country. But because Monaco is such a speck of a country, that record should probably have an asterisk next to it. The most densely populated full-size state is teeming Bangladesh, with almost 2,600 people per square mile.

♦ **Andorra:** This miniscule microstate is sandwiched between Spain and France, where Catalan-speaking farmers tend to their goats deep in the Pyrenees.

♦ **Malta:** South of Sicily in the Mediterranean Sea, this tiny island did not become a republic until 1974, when the British departed after almost 200 years of military presence.

♦ **San Marino:** Perched atop Italy's Mount Titano, San Marino was a haven for refugees during World War II.

Southern Europe's Microstates.

 ◆ **Vatican City:** In the heart of Rome, this Lilliputian state, wrapped around Saint Peter's Basilica, is the home of the Pope and the spiritual center for more than 1 billion Roman Catholics worldwide.

Mother Nature's Chosen One

For centuries, writers have waxed poetic about the natural beauty of the Southern European region, with its sunshine and spectacular coasts. Although the Mediterranean was once cloaked in virgin forest, centuries of human occupation have left it largely deforested. Where forests once stood, vineyards and olive groves now grace the gentle slopes of the region.

The Lay of the Land: Peninsulas and Mountains

As you can see from the following map, Southern Europe can be likened to a series of peninsular fingers poking into the Atlantic, the Mediterranean, and the region's assortment of lesser seas. The largest of these peninsulas, the Iberian Peninsula, is located at the region's western edge. Occupied by Spain and Portugal, Iberia is bounded on the north by the Pyrenees Mountains, which separate the peninsula from France. To the south, only eight miles and the Strait of Gibraltar separate Iberia from Africa. The interior of the Iberian Peninsula consists largely of high, rugged plateaus.

Southern Europe—physical features.

To the east of Iberia is the boot-shaped peninsula of Italy. A mountain range also separates this country from Western Europe: This time, it's the Alps. Notable passes through the Alps include the Brenner from Austria, Maloja from Switzerland, and Little St. Bernard (yes, of canine fame) from France. Running the entire length of the

Terra-Trivia

Perhaps the most famous volcano in Europe is Mount Vesuvius, near Naples, Italy. In 79 C.E., Vesuvius erupted, spewing toxic gas, volcanic ash, and mud across the surrounding countryside. Although some 15,000 people were estimated to have perished in the cataclysm, the Roman cities of Pompeii and Herculaneum were buried in ash and preserved remarkably intact for archaeologists of our era to study.

"boot" is Italy's spine, the Apennine Mountains. To the south, on the island of Sicily, is the still-active volcano Mount Etna.

Greece is located on the southern tip of the Balkan Peninsula on Southern Europe's eastern edge. This is rugged, mountainous country. From the jagged southern end of the Dinaric Alps and the Balkan Mountains, the Pindus Mountains run south through the peninsula, dominating the Greek landscape.

South of the Balkan Peninsula is yet another peninsula, the Peloponnesus, site of the famous Peloponnesian Wars between ancient Sparta and Athens. The isthmus of Corinth connects, or rather connected, the Peloponnesus to the mainland. Since the Corinth Canal was cut through the isthmus in the late nineteenth century, the Peloponnesus, technically speaking, has become an island.

Islands Everywhere

In addition to its three main peninsulas, Southern Europe abounds in islands. Greece wins the prize for the most islands, with more than 2,000 of them making up the Greek archipelago (string of islands). The largest of the Greek islands are Rhodes, Lesvos, and Crete.

Italy also has its share of islands, including the large islands of Sicily and Sardinia. Another Italian island of note is a favorite tourist spot, the tiny Isle of Capri. Located off Sorrento (near Naples), Capri is home to the famous Blue Grotto, where limestone formations are visible underneath the iridescent waters.

Spain controls the Balearic Islands (Mallorca is the best known) on its Mediterranean east coast and also owns the Canaries off the west coast of Morocco. Portugal's islands include the Azores and the Madeiras (as in the wine), located to the west and southwest in the Atlantic Ocean.

Warm Waterways

The Mediterranean Sea dominates Southern Europe. Only Portugal has no coastline on this sea (its coastline is exclusively on the Atlantic Ocean). The Mediterranean is subdivided into several smaller seas.

Spain has coastlines on the Atlantic Ocean on the west and on the Mediterranean's Balearic Sea on the east. To the south is the Strait of Gibraltar, which connects the Atlantic and the Mediterranean. To the north of Spain lies the Bay of Biscay.

As Italy juts southward into the Mediterranean, it edges the Adriatic Sea on its northeast and the Tyrrhenian Sea on its southwest. Between the sole of the Italian "boot" and Greece is the Ionian Sea. The bulk of the Greek islands are located in the Aegean Sea between mainland Greece and Turkey.

Southern Europe has several rivers of note, as shown on the map earlier in this chapter. Italy's major river is the vital Po, whose basin is home to the bulk of the country's agriculture, industry, and population. Of great historical significance is the Tiber River in central Italy, on which Rome was founded.

Spain also has several significant rivers. The Ebro rises in the Cantabrian Mountains in north central Spain and cuts deep gorges in Catalonia before emptying through a wide delta into the Mediterranean. Because Northern Spain uses the Ebro (or Spanish Nile) for irrigation, the Ebro River basin has developed into a productive and populous area. The tiny district of Rioja, for example, produces some of the best wine in Spain.

The Guadalquivir River in the southwest passes through Córdoba before flowing to the Atlantic. The fertile lowlands of the Guadalquivir valley are one of Spain's primary agricultural areas. Originating deep in the heart of Spain is the Tagus River, which passes through Toledo before bisecting Portugal and emptying into the Atlantic Ocean at Lisbon.

Climate: The Mediterranean Sun

Almost all of Southern Europe enjoys a Mediterranean climate. Hot and dry summers prevail, and the brown land lies dormant until the winter brings cooler and wetter days. Northern mountainous areas throughout the region can be cold and snowy. On the Spanish interior plains, arid is the word of the day, and temperatures are extreme. Endlessly flat and treeless, this interior region called *secano* (unirrigated) is by no means barren. Fields of corn, rows of gnarled olive trees, and vineyards crawl over the prairies of La Mancha, the home of the legendary Don Quixote, complete with windmills to grind the corn.

The common vegetation of the entire Southern European region is what the Italians call *macchia*, the French, *maquis*, and the Spanish, *matorral*. Similar to what Californians call *chaparral*, this dense, scrubby growth is able to survive arid conditions thanks to water-holding thorns and leaves. The maquis has replaced the native forests of the region, which were long ago burned or cut for lumber. Remnants of these once prolific forests, however, survive in Italy's Po River valley.

The Historic Cities

The cities of Southern Europe are treasure houses of art and history. Although the capital cities of Lisbon, Madrid, Rome, and Athens each have storied pasts and were primary contributors to the course of Western history, many secondary cities in the region are also important:

Greece

GeoJargon

Athens is a classic example of what geographers call a **primate** city, a country's largest and the most important historically, culturally, and economically. A primate city is often the capital of its country.

♦ **Athens:** The glory of the Greek capital is its wealth of classical antiquities, beginning with the hilltop citadel and religious center, the ancient Acropolis. Around the antiquities is a sprawling, congested, smog-filled city that can be exasperating. In recent years, however, this *primate* city has made strides in controlling its pollution and congestion, especially as it prepares to greet the world as the host of the 2004 Summer Olympic Games. Piraeus, the port of Athens, handles most of the country's massive amount of shipping.

Cities of the Mediterranean.

Italy

♦ **Rome:** The Eternal City offers a mix of the ancient, medieval, Renaissance, and, of course, modern eras. Built on the famous seven hills, Rome embraces its ancient lifeline, the Tiber River. The city is Italy's capital and is home to spectacular piazzas, churches, palaces, and of course the awe-inspiring vestiges of its imperial past.

◆ **Milan:** Unlike cities farther south where the twentieth century is less in evidence, Milan is a fast-paced commercial city and financial center, complete with skyscrapers and an extensive subway system. One of the world's major capitals of fashion and contemporary design, Milan is also home to the world-renowned La Scala opera house, Leonardo da Vinci's *Last Supper*, and the ornate Gothic Duomo (cathedral).

◆ **Venice:** Built on more than 100 islands in a lagoon off Italy's northern Adriatic coast, Venice is one of the world's most beautiful and art-filled cities. Today, however, The Queen of the Adriatic faces multiple crises. The increasingly flood-prone city is sinking at an alarming rate, while damp and pollution are eating away at its monuments. A long-delayed anti-flood project was finally approved by the Italian government in 2003, but whether it will solve the problem or worsen it is still a matter of heated debate.

◆ **Florence:** The Renaissance and, in many respects, the modern age got their start here. Florence was the home of Dante, Donatello, Ghiberti, Michelangelo, and Galileo—the list goes on and on. The arts of Renaissance Florence flourished under the patronage of the ruling Medici family and today are on display in the city's magnificent museums and churches for all the world's art lovers to enjoy.

◆ **Naples:** Italy's third largest port and a former royal capital in its own right, Naples is one of the country's most fascinating cities, rich in art, culture, street life, and scenery. Its setting on the Bay of Naples under the shadow of Mount Vesuvius is one of the most spectacular anywhere.

Spain

◆ **Madrid:** Situated on Spain's high central plateau, the national capital is a political and financial center and industrial hub. Relatively new by Southern European standards, Madrid once had a reputation for stodginess and isolation. No more. Madrid has come out of its shell and morphed into a vibrant, fast-paced European capital.

◆ **Barcelona:** This is the wealthy, sophisticated capital of the Catalan region in northeastern Spain. Traditionally regarded as more "European" than insular Madrid (although that's debatable today), Barcelona is proud of its distinct culture and language, which had been suppressed for so long under the dictatorship of Francisco Franco. An important seaport and center of commerce, Barcelona hosted the 1992 Summer Olympics.

Portugal

- ◆ **Lisbon:** From its harbor at the mouth of the Tagus River, Portugal's capital once ruled a trading empire that spanned the globe. Today, Lisbon is reinventing it-self as a modern, vibrant European city, where soaring skyscrapers jostle with medieval churches and magnificent relics of Portugal's proud seafaring past.

The Peoples of the South

As might be expected in so mountainous a region, the people in this region have tended to cluster in river valleys and along the coast; this is especially true in Greece. More Southern Europeans derive their livelihoods from agriculture than is typical elsewhere in Europe; urbanization, not surprisingly, is also lower than the European norm. The linguistic and religious divide in Southern Europe is fairly straightforward, although there are a few twists, as you'll see.

Roman Catholic and Greek Orthodox

Although the Roman Catholic and Orthodox churches share common origins, the Great Schism of 1054 C.E. split them apart for good. Southern Europe is overwhelmingly Roman Catholic, with the notable exception of Greece, which is almost entirely Greek Orthodox.

Romance Languages—and a Few Others for Good Measure

Romance languages aren't the only tongues spoken in this region; in fact, not all the languages spoken there even belong to the Indo–European language family.

The Romance branch of the Indo–European family is dominant in three of the four countries in the region. In Portugal, virtually everyone speaks Portuguese. In Spain, most people speak either Castillian Spanish or Catalan. The latter is a distinct Romance language spoken by some 10 million people, primarily in northeastern Spain (Valencia, Catalonia, the Balearic Islands, and parts of Aragon), as well as in Andorra and in parts of southern France and Sardinia. In the far northwest portion of Spain, people of Celtic stock speak Galician, which is related to Portuguese. The Basque language, spoken in north central Spain, is an odd bird discussed below.

Italy is also in the Romance camp. Although dialects and regional differences exist throughout the country, Italian is the language of the land. The only exceptions are pockets of the far north, where Romansch (another Romance tongue) and German are spoken.

Greece is the only major country in the region not to speak a Romance language. Greek is in the Hellenic branch of the Indo–European language family, and is spoken across almost all of Greece and its islands. It might not be a Romance language, but at least it's in the Indo–European family.

The same thing can't be said of Basque. Unique to Southern Europe, Basque is not a Romance language and is unlike any other language known today. Iberia's oldest native inhabitants, some 50,000 Basques live in a beautiful pocket of the Pyrenees in northern Spain. With their unique language and Asian nomadic stock, the Basque people tend to be taller and fairer than most Spaniards. Dissatisfaction with centralized control from Madrid has spawned a powerful separatist movement among the Basques that has often spilled over into violence and terrorism. Remember that language differences mark cultural differences and often give rise to autonomy movements.

> ### CAUTION Eye on the Environment
>
> Enclosed by land on all sides, the Mediterranean Sea's only exits are by way of the Strait of Gibraltar and the tight Suez Canal. Most of the rivers from Southern Europe empty into the Mediterranean, as do rivers from Western Europe, Eastern Europe, the Middle East, and Saharan Africa.
>
> Pollution is especially acute in the Mediterranean Sea because with so narrow a link to the Atlantic Ocean, little flushing action occurs. Salinity (saltiness) and pollution prevail as pollutants from northern Italy's industry, rivers, and coastal cities and resorts pour into the Mediterranean. With so many people's livelihoods depending on these waters, it's no surprise that Southern Europeans are deeply concerned.

In the News: Illegal Immigration

The countries of Southern Europe have traditionally been net exporters of labor, sending emigrant workers to Northern America and West Europe in search of better jobs and better lives. As living standards in the Southern European region rose in the latter part of the twentieth century, and especially after Greece and the Iberian countries joined the European Union, migration patterns reversed dramatically. No longer net exporters of workers, each country in Southern Europe is now viewed as a destination as well as a transit point to West Europe by emigrants from Africa, Asia, East Europe, and South America.

The proximity of Spain and Portugal to Saharan Africa and their historical ties to Latin America make them magnets for African and Latin American emigrants seeking to escape joblessness and poverty in their own countries. The collapse of the Soviet Union and the civil wars and upheavals that accompanied the breakup of Yugoslavia in the 1990s sent streams of refugees to Italy and Greece in search of security and jobs. Greece's position at the eastern fringe of the European Union, coupled with its jagged coastline and thousands of islands, makes it particularly vulnerable to illegal entry.

The nations of Europe find themselves in a tricky bind with regard to illegal immigration. On the one hand, their native populations are growing slowly if at all and are increasingly reluctant to take on low-wage jobs. As current demographic trends play themselves out, Europe will become even more dependent on foreign labor, legal or otherwise, than it is now. On the other hand, unorganized immigration brings with it a wide range of problems, from organized crime and drug trafficking to illegal arms smuggling and terrorism. In the post 9/11 world, secure borders are much more than just a legal formality, they can be matters of life and death. How Europe copes with the problem of illegal immigration in the coming years will have widespread repercussions not only in Europe, but throughout the developing world.

The Least You Need to Know

- The Greeks and Romans left a lasting cultural imprint on the region of Southern Europe.

- All four countries in the region (Portugal, Spain, Italy, and Greece) are located on peninsulas and have many islands. Greece has more than 2,000 of them.

- All the Southern European countries except Portugal border the Mediterranean Sea, and all share a Mediterranean climate.

- The region is mountainous and tends to depend more on agriculture than does the rest of Europe.

- Higher living standards and membership in the European Union have made all four Southern European countries magnets for illegal immigration.

Japan: The Pacific Dragon

In This Chapter

◆ Tracing the sword of the samurai

◆ Touring Japan's islands, mountains, and plains

◆ Meeting the island-nation's people

◆ Sizing up the economic giant

◆ Assessing the future

The Japanese name for their country is Nihon (or Nippon) Koku, meaning "Land of the Rising Sun." It's an appropriate name because Japan is one of the first major countries to see the sun rise each day. By the time the sun finally rises in New York, it has set in Tokyo. Just as Japan leads the world into each new day, it has also been a world economic leader, although today it is a sputtering one.

Japan is an island country in the Pacific Ocean off the eastern coast of Asia. Geographically, the country is well positioned to access the markets both of Eastern Asia and across the sea in the Americas. Although Japan is the world's eighth-largest country in population with 126 million people, its 146,000 square miles make it only the fifty-ninth largest in area. The result is that Japan's population density (people per square mile) is among the highest on the earth.

Song of the Samurai

Japan's earliest inhabitants were the Ainu, an obscure tribe that probably populated all the islands by the first millennia B.C.E. As Asians began to migrate into Japan, the Ainu were pushed farther and farther north, where a few now remain. Legend has it that, around 660 B.C.E., the migrants united under the first emperor of Japan, Jimmu Tenno.

Japan.

The Chinese began to arrive on the Japanese islands about 400 C.E. They brought culture, writing, Buddhism, and, later, Confucianism. The Japanese readily absorbed all these elements and at first demonstrated a desire to learn from others. In the ninth century the influence of China was on the decline and power gradually shifted from the emperor to the great landowning families, most notably the Fujiwara. By the year 1000, the Fujiwaras had become the absolute power behind the throne, hereditary dictators able to dethrone emperors at will.

As the power of the Fujiwaras began to decline after 1016, the landowning families became warlords, hiring their own bands of professional soldiers, called samurai, to protect their properties and maintain order. The various samurai groups fought bitterly to achieve greater power for their warlord. Finally, in 1192 C.E., the Minamotos defeated the Tairas, and Minamoto Yoritoma became the first shogun (the military commander-in-chief under the emperor) of Japan. The shogun ruled over the daimyo (local feudal lords, each with his own samurai and fiefdom). The Shogun system, which lasted from the twelfth to the nineteenth centuries, seemed designed to promote constant strife. One warlord versus another, the daimyo versus the shogun—the possibilities for intrigue were endless.

Geographically Speaking

From 794 to 1868 C.E., the splendid city of Kyoto was the capital of Japan and the inland center of this isolated country. Kyoto also was the country's cultural center, where theater, music, and the arts all flourished. In nearby Nara, a huge Buddhist shrine encloses one of the world's largest bronze statues of the Buddha.

The city of Kyoto boasts more than 2,000 shrines, temples, and magnificent miniature gardens. It's also the home of the former Imperial Palace and the Nijo Castle. Today only small cottage industries are encouraged because the past is deliberately being preserved as a revered springboard to the future. Because Kyoto didn't have a war industry, the United States spared it from bombing in World War II.

In the thirteenth century, Kublai Khan's Mongols tried to invade Japan twice, but were repulsed each time. In the mid-sixteenth century, Europeans traders began to arrive in Japan. The Portuguese were first in 1543, followed by the Dutch in 1600. The Spanish Jesuit missionary St. Francis Xavier arrived in 1549 to convert the Japanese to Christianity. Despite disapproval and persecution, some 300,000 did convert before the shogun put a stop to the conversions and most things foreign in the seventeenth century. In 1639, the Portuguese traders were expelled from Japan, leaving the tiny Dutch outpost in Nagasaki as Japan's only link to the West.

Although a ban on foreign books was eased in 1720, Japan's self-imposed isolation lasted until 1854, when an American fleet under Commodore Perry sailed into Edo Bay (Tokyo) and essentially forced the opening of Japan. In 1867, the last shogun resigned, and in 1868 the Emperor Mutsuhito reestablished himself as supreme ruler. He took the name Meiji (meaning "enlightened government"), moved the imperial capital to Edo and renamed it Tokyo (meaning "eastern capital"), and by 1871 abolished all fiefs. The Meiji Restoration had begun.

Emulating Colonialism

For Japan, the modern era truly began with the Meiji Restoration, in 1868. The planners of that time took stock of what they had and what they needed in order to become a modern country, capable of exerting world leadership.

Japanese leaders looked to the West, where they saw that type of capability. They looked first to Britain, whose industrial revolution in the preceding century had transformed the country from a primarily agrarian island society to the world's leading manufacturing and trade center. The Japanese sought technology, economic and educational reform, and advice on city planning, communication, and transportation. The West served as the model on which Japan's leaders built their modern nation.

Unfortunately, Japan lacked coal and iron and other raw materials necessary for industrial production, and it was also short on money for investment in plants and machinery. Japan realized that it could earn some money from the export of the goods it was already producing (textiles, porcelain, wood, and small metal products) and could raise taxes on its own farmers. But how to secure a reliable supply of coal, iron, and other raw materials? Although it could purchase the materials, that would take even more money. Then again, it could use the old tried-and-true European system: They could colonize and exploit.

Forging the Japanese Empire

First, Japan seized the Ryukyu Islands in 1879. China objected, but so what? Why the Ryukyus? Their people were ethnic Japanese who spoke Japanese and produced excess food. In addition, the islands, which offered homeland protection (a buffer zone to protect the main Japanese islands), were an easy place to begin building an empire. They were also a strategic jumping-off place for Japan's next move. With its new navy and new army, Japan took Taiwan from China in 1895, getting coal, petroleum, natural gas, and a market for its finished goods.

Japan defeated the Russians in 1904, annexed Korea in 1910 (gaining iron, more coal, and another market), and grabbed Manchuria in 1932 (more iron and coal and yet another market). The empire then expanded to several Pacific archipelagos by conquest or forced mandate (homeland protection). In 1937, the Japanese invaded China and within a year controlled almost all of eastern China—a huge market.

To fully establish itself as a superpower, Japan had to eliminate Britain and the United States from its sphere of influence. The outbreak of World War II in Europe presented Japan with a window of opportunity. (See Chapter 24 for more information about Japan's role in World War II.) The American naval fleet in Pearl Harbor, Hawaii, was a vulnerable target. By destroying it, the Japanese could go on to seize the Philippines, Malaysia, Singapore, and the East Indies, plus a few more Pacific islands for use as an extra buffer for the homeland.

Japan almost pulled it off, and probably would have if it had completely destroyed the American fleet at Pearl Harbor (it didn't because key ships were at sea at the time of the bombing). The Imperial navy went on to lose two critical sea battles with the United States at Coral and Midway, and Japanese leaders completely underestimated the will of the United States and its ability to fight and win on two fronts. After undergoing months of homeland bombings and eventually two atomic bombs on Hiroshima and Nagasaki, Japan—morally and physically devastated—surrendered in 1945 and was occupied by the United States until 1952.

> **Terra-Trivia**
>
> Considering the extent of the devastation that Japan suffered at the end of World War II, its recovery and rise to economic dominance fully deserves the name "economic miracle." The Japanese people pulled themselves up by dint of their strength, cohesiveness, and determination, aided by massive financial aid from the United States that by 1949 had reached $1 million a day.

An Economic Giant Reborn from the Ashes

With massive financial aid from the United States and a willingness to learn and plan ahead, Japan began its economic miracle all over again. Again, it looked to the West, this time to industrial strategies employed by Germany and Italy in their buildups for World War II. A key notion seized on by Japan was the cohesion of government, business, and labor. This would become the guiding principle of Japan, Inc.

Japan adapted a series of policies aimed at protecting domestic industries and encouraging export. Earnings were invested in research, and expansion was financed through low-interest, government-sponsored bank loans (made possible by high personal savings due to a deliberate scarcity of consumer goods). It took Japan until the mid-1950s to get back to pre-war production levels. By the 1960s, its economy had skyrocketed. Growth rates throughout the decade averaged a phenomenal 10 percent a year. By the 1980s growth had "slowed" to a more sustainable 5 percent a year, but Japan was now number one or two in the world in shipbuilding and in producing steel, automobiles, trucks and buses, watches, VCRs, TVs, cameras, microwave ovens, refrigerators, computers, and copying machines.

Terra-Trivia

Until recently, Japan was the second-largest consumer of electricity on the planet. With only 2 percent of the world's population, it consumes more than 7.5 percent of the world's total electricity. China has now jumped ahead of Japan. With 20 percent of the world's population, China consumes more than 9 percent of its electricity. The number-one consumer of electricity remains the United States, which, with 4.5 percent of the world's population, consumes more than 25 percent of the world's generated electricity.

In 1991, the Japanese economic juggernaut began to falter, as many countries to which Japan exported its goods were hard hit by recession. In addition, Japan's Asian neighbors and export partners had begun to produce higher-quality and lower-priced goods of their own. Japan felt increased pressure to open its own markets to lessen its imbalance of trade with other nations. All of these factors took a toll on Japan's economy. Throughout the 1990s, real-estate and stock-market values plummeted, thousands of small companies went bankrupt, and a series of financial and political scandals rocked the nation. A country that once seemed poised to rule the world economically now faces serious and troubling questions about its economic future.

Eye on the Environment

One of the defining factors of Japan's economy and recent history is its almost complete lack of natural resources. A whopping 61 percent of Japan's electricity is generated by imported fossil fuels (home-grown nuclear power accounts for 29 percent and hydropower for another 9 percent.)

But Japan is not sitting idly by. In a concerted effort to break its dependence on imported fossil fuels and meet its Kyoto Protocol commitments to reducing greenhouse gas emissions, Japan has become a leader in the solar revolution. The Japanese government has invested heavily in solar photovoltaic technology, and some 400,000 Japanese homes are expected to be fitted with solar roof panels by 2005. Plans are now underway in Japan to launch a satellite equipped with solar panels that would beam energy back to earth. The first power transmission test is expected to take place by 2025. For more on the promise of solar power, see Chapter 28.

Land of the Rising Sun

Millions of years ago, tectonic action caused the Pacific plate to smash into the Asian plate, pushing up great mountains on the edge of the Asian continental shelf. A deep gorge formed where the Pacific plate and Philippine plate folded under the Asian plate. Today, the mountains are Japan, and the gorge is the Japanese Deep (28,000 feet deep, off Japan's southeast coast). Japan's most beautiful mountain, Mount Fuji-san, is also its highest mountain (12,389 feet). This snow-capped volcanic cone was last active in 1707.

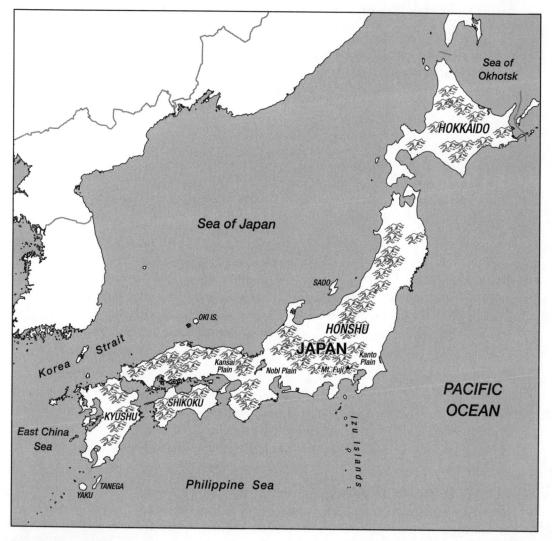

Japan—physical features.

The Waters and the Islands

Japan is an archipelago, which means that it's made up of many islands. Although the country has about 1,000 islands, you have to keep track of only the "big four," which loosely form the shape of a huge dragon. To the north is the dragon's head, the island of Hokkaido. Its northern latitude makes it a chilly place in the winter. Although Hokkaido is large, it's not as heavily populated as the islands to the south.

The next island south of the dragon's head (the body of the dragon) is Japan's largest island, Honshu. Considered to be Japan's mainland, it's home to Japan's largest cities and the bulk of its population and industry. Farther south, the other two large islands, Shikoku and Kyushu, form the tail of the dragon.

The Kuril Islands, which link northern Japan with Russia, are an issue of great concern. Although Japan claims the islands as part of its national territory, Russia says "nyet"—it got them in a last-minute grab at the end of World War II and it occupies them. Even without the Kurils, Japan is almost 1,200 miles long and, at most, only 200 miles wide. No location in Japan is more than 100 miles from the sea.

Mountains and Plains

Japan has a major north–south spine of mountains called the Japanese Alps. The Japanese islands are 80 percent mountains and hills, and only 25 percent of Japan's land slopes less than 15 percent. The islands are dotted with more than 200 volcanoes, 60 of which have been active in recent history.

Japan is on the Pacific's Ring of Fire (see Chapter 2) and is subject to volcanic eruptions and at least 15 earthquakes a year. The country also experiences a similar number of typhoons and at least one or two tsunamis (seismic sea waves) each year. All this happens right on top of the spot where the earth's Eurasian plate, Pacific plate, and Philippine plate all collide! (Plates are also discussed in Chapter 2.)

Geologically, Japan is one of the world's most dangerous places to live. Tokyo sits largely on fill (land created artificially by filling in parts of Tokyo Bay) that could liquefy in the event of a major quake. Tokyo has narrow streets dotted with many skyscrapers built before stringent regulations were in force, and it's part of the world's largest population concentration. If the next "big one" hits in or near Tokyo, the results could be catastrophic.

Japan's largest island, Honshu, has three major plains: Kanto, site of much of Japan's agriculture, industry, and population, as well as the world's largest urban conurbation—Tokyo, Kawasaki, and Yokohama; Nobi, centered around the major port of Nagoya toward the southwest; and Kansai, further to the southwest and home to Kyoto, Kobe, and Osaka. The Honshu plains and their mega-cities all are connected by super-fast trains known as "bullet trains."

Japan's most dominant body of water is its Inland Sea (240 miles long and from 8 to 40 miles wide), connecting Honshu on the north with Kyushu and Shikoku on the south. The sea is known for its islets and beauty and is valued for its fishing.

Japan's Temperate Climate

Japan straddles the same latitudes as the east coast of the United States from mid-Maine to north Florida. The two areas have much in common. Because they're both temperate, they generally have similar temperatures and rainfall patterns, usually about 40 to 50 inches. Unlike the United States, though, Japan has two annual rainy seasons, with occasional coastal pockets getting as much as 100 inches of precipitation per year.

The People of Japan

Perhaps in no other developed country do the values and traditions of the past exert as much impact on the present as they do in Japan. The nation is extremely cohesive: It has one race (Mongoloid, similar to the Chinese and Korean people but smaller in stature), one language (Japanese), one culture, and one highly developed—and technology oriented—system of education that produces virtually 100 percent literacy.

Until 1947, the state religion was Shinto, in which the emperor was worshipped as a god. Now freedom of religion is guaranteed to all. The Shinto religion, indigenous to the Japanese islands, involves the worship of the divine forces of nature. Ancestor worship also factors strongly into Shinto practice. Although most Japanese are still Shinto, almost 35 percent are Buddhist. Many adhere to both religions because in practice the two largely overlap.

The U.S. occupation after World War II brought westernization and transformation to Japan. Although patterns of rural life have changed little, it's in the cities, where 79 percent of Japan's people live, that change is most evident. The uniquely Japanese mix of modern and traditional includes Western dress (but still kimonos and obis for special events), fast food (but still rice and raw fish), and a passion for Western movies and music, skiing, and baseball. Japanese workers are enjoying far more leisure time than they did in the past, and family living arrangements are changing. Traditional multi-generational families living under one roof are becoming a thing of the past.

The Japanese had a brief baby boom after World War II, but birth rates had decreased from 34 per thousand to 10 per thousand by 2002. With a drastic decrease in the death rate (life expectancy is now up to 79 years), Japan's population growth rate is only 0.3 percent, one of the lowest in the world.

In the News: Japan's Future

In the 1980s, the world, especially the United States, seemed to become obsessed by the coming Japanese threat—not military, but economic. Japanese goods were everywhere and the country was riding the crest of an unprecedented wave of growth and prosperity. The wave came crashing down in the 1990s. Although today Japan is still the second most technologically powerful economy in the world after the United States, its economy is no longer the second largest; that rank is now held by China. Even more disturbing for Japan has been its inability to shake off a more than decade-long period of recession and stagnation. Layoffs and rising unemployment eroded two of the givens of Japanese business life: the guarantee of lifelong employment and the tight cooperation and synchronization among manufacturers, suppliers, and distributors.

As it continues to try to turn itself around, Japan faces major obstacles: the lack of resources that has always plagued it; a banking system in serious need of reform; a growing scarcity of usable land for farming, industry, and people; and a population that's aging and barely growing. Japan will soon face a labor deficit, which will be difficult to remedy given Japan's traditional aversion to immigration and reluctance to fully empower women in the workplace.

On the more positive side, the famous Japanese work ethic is still more or less intact, as is their leadership in the field of high technology. Japanese breakthroughs in robotics also hold great promise for the future. Japan is one of the world's principal economic engines, so it is in everyone's interest that the Japanese find a way out of their current dilemma.

The Least You Need to Know

- Japan is an island nation with one of the highest population densities in the world.

- Japan's use of its military might to acquire new resources and markets by force ended in catastrophic defeat at the end of World War II.

- In the aftermath of World War II, democratic reforms and an infusion of U.S. aid put Japan back on its feet and on the road to becoming an economic superpower.

- Japan today is the world's second most technologically powerful economy, but is struggling to snap out of more than 10 years of economic stagnation.

Chapter 12

East Asia's Economic Tigers

In This Chapter

- ◆ A region unlike any other
- ◆ Colonial legacies
- ◆ Surveying the countries
- ◆ The Korean tinderbox

As its title indicates, this chapter is a little different in that it covers a region that is more economic than typically geographic. East Asia's four economic heavy-hitters, or Tigers, as they've been dubbed, are South Korea, Taiwan, Hong Kong, and Singapore. Although they do share certain geographic similarities—they are all located on the eastern fringe of Asia and they're all islands, peninsulas, or both—the reason for lumping them together is the strength and reach of their interconnected economies. Japan could have been included in this chapter as well, as it is closely tied with all four Tigers by trade and history. But because of the size and importance of its economy, Japan has been treated as a separate region. The "Little Tigers" of Southeast Asia—Thailand, Malaysia, and Indonesia—could also have been included in this chapter's mix, but their economies are not quite up to the level of the four Tigers. The Little Tigers are examined in Chapter 22.

Tracking the Tigers

As you can see on the following map, South Korea occupies the southern half of the Korean peninsula, just about 100 miles west of Japan. South Korea is separated from China by North Korea, with which it shares the peninsula, and by the Yellow Sea. In fact, South Korea is actually twice as far from China as it is from Japan, The island of Taiwan (formerly called Formosa) lies off the southeast coast of China. Hong Kong occupies the tip of a peninsula and several islands in southeastern China. At the southern tip of the Malay Peninsula, 1,500 miles southwest of China, is the island state of Singapore.

All four Tigers are highly industrialized and urban and enjoy high levels of economic development. As city states, both Hong Kong and Singapore are virtually 100 percent urban. South Korea and Taiwan are about three-fourths urban. Although the Tigers lost some of their roar in the Asian financial crisis of 1997–1999, the per capita *gross domestic product* (GDP) of Hong Kong and Singapore is the same as that of the most developed countries of Europe. The per capita GDP of South Korea is higher than that of European Union members Greece and Portugal; Taiwan's is just slightly below.

The four "Tigers."

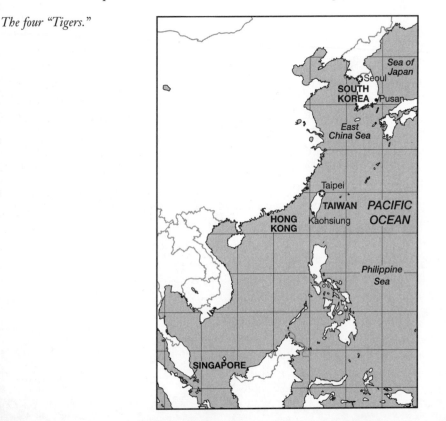

Individually, the four Tigers are the world's biggest exporters outside of North America, Japan, and the European Union. Each country annually exports between $120 billion (Singapore and Taiwan) to almost $200 billion (Hong Kong) worth of goods, mostly to the United States, China, Japan, and each other. In 2001, the four Tigers exported nearly $600 billion worth of goods, making them as a bloc the world's third largest exporter, behind only the United States and Germany but ahead of Japan, which came in fourth.

Because the islands and peninsulas of this "region" are not well endowed with natural resources, each country (or city, in the case of Hong Kong) must import raw materials, perform the necessary manufacturing and assembly operations, and then export the finished products. The four Tigers have some of the best, and best located, harbors in Asia and often serve as trade agents, shipping out other countries' goods. Throughout the "region," low-wage, labor-intensive industries are increasingly being outsourced and replaced by more technology-intensive industries.

Complex Intertwined Histories

As you've seen by now, South Korea, Taiwan, Hong Kong, and Singapore don't make up a typical geographic region. The ties that bind them are mainly economic, but also historical. All have been colonies at one time: Hong Kong and Singapore, of Britain; Taiwan and South Korea, of Japan. And all four were occupied by Japan during World War II. Both China and Japan have weighed heavily on the culture and development of South Korea, and their relationship with China is the dominant political and economic issue for both Taiwan and Hong Kong.

South Korea

Korea is an ancient land with a turbulent history. By the first century B.C.E., three separate kingdoms had emerged on the Korean peninsula. One of them, the Silla kingdom, unified the peninsula in 668 C.E. Korea's final dynastic period, the Choson, began in 1392, and despite frequent invasions by Mongols, Manchus, and the Japanese, lasted until 1910, when Japan took over Korea as a colony and ruled it brutally until the end of World War II. After the war, the peninsula was split at the 38th parallel: The north went to the Soviet Union, and the south to the United States. (If you were paying attention in Chapter 4, you might remember that *parallels* are lines of latitude that run east–west around the globe.)

Terra-Trivia

Although the Korean peninsula is an extension of the Chinese mainland, the Korean language is more closely related to Japanese.

The northern half became the communist Democratic People's Republic (usually called North Korea), and the southern half became the capitalist Republic of Korea (usually called South Korea). In 1950, North Korea set off the Korean War by invading the south in an attempt to force unification. In 1953, a truce was signed and a no-man's land was set up just north of the 38th parallel, replacing the border between the two states. More than 50 years later, the participants in the Korean War have yet to sign a formal peace treaty. (For more on North Korea, see Chapter 16.)

Since the war, South Korea has achieved extraordinary economic growth and prosperity, and fully integrated itself into the modern, hi-tech global economy. Its political development, however, has been difficult. Although democratic in name, the early post-war governments of South Korea were corrupt and repressive, often sacrificing democracy in the name of national security. Today, however, South Korea is a much more open, less repressive society, as evidenced by the election in 1997 of former dissident and political prisoner Kim Dae-Jung as president of the Republic of Korea. This marked the first transition of power from the ruling party to an opposition party in Korea's modern history.

Taiwan

The modern history of Taiwan begins in the sixteenth century when Portuguese navigators sighted the island on their way to Japan and named it *Ilha Formosa*, or "Beautiful Island." The name stuck through multiple settlements, invasions, and conquests by the Spanish, Dutch, Chinese, Manchus, and finally the Japanese, who took over the island from China at the end of the Sino-Japanese War in 1895. At the end of World War II, Formosa was returned to China, then under Nationalist Party rule.

Terra-Trivia

Since its expulsion from the United Nations in 1971, Taiwan has been in a difficult and ambiguous diplomatic situation. China regards it as a renegade province and pressures its trading partners to sever ties with Taiwan. Most have done so. The United States broke off diplomatic relations with Taiwan in 1979, but maintains "unofficial" commercial and cultural relations with Taiwan though the American Institute on Taiwan.

In 1949, Communist forces under Mao Zedong defeated Chiang Kai-shek's Nationalist government, triggering an exodus of Nationalist forces to the island of Formosa, which was renamed Taiwan. Economic and military assistance from the United States enabled the Nationalist government to survive and begin to turn the island's economy around. For two decades, both the People's Republic of China (on the mainland) and the Republic of China (on Taiwan) claimed to be the true China. In 1971, however, with the assent of the United States, Taiwan was expelled from the United Nations and its seat given to the People's Republic of China. The following year President Richard Nixon made his historic visit to mainland China.

Despite the political setback, Taiwan continued to make great strides economically. Although it is regarded by China as a renegade province and is officially recognized by few countries, Taiwan has trading relationships around the world. It is also a major investor throughout Southeast Asia and, despite ongoing tensions and conflicts, in China itself. On the internal political front, Taiwan has loosened the grip of the Nationalist Party and opened the political process to native-born Taiwanese. In 1988, Lee Teng-hui became the first Taiwanese-born president of the country. His reelection in 1996 was the first direct democratic presidential election held in Chinese territory. In 2000 Taiwan elected its first opposition party president, the pro-independence candidate Chen Shui-bian.

Hong Kong

Before the British took it over, Hong Kong was a small fishing community and a haven for opium smugglers and pirates. At the end of the first Opium War in 1842, China ceded Hong Kong to the British in perpetuity. Both the Chinese and the British soon began to realize the importance of this trading post as the best harbor between Shanghai and Indochina. In 1898, China ceded an adjacent part of the mainland, called the New Territories, to Britain for 99 years.

In the late 1940s, when the Communist leader Mao Zedong was conquering all of China, he easily could have seized Hong Kong but chose not to because it was his gateway to the world for the goods his country desperately needed for economic survival. In 1984, as the New Territories lease was nearing expiration, the British agreed to give back all of Hong Kong in 1997 in return for a guarantee from China that it would preserve the economic system and civil liberties of its former colony.

In July 1997, Hong Kong became the Hong Kong Special Administrative Region (SAR) of China. Although it was hard-hit by the Asian financial crisis of the late 1990s and by the global slowdown of the early 2000s, Hong Kong remains one of the world's great trading economies. Although it is certainly in China's interest to maintain Hong Kong's economic autonomy and vibrancy, it remains to be seen how willing it will be to preserve the former colony's political freedoms.

Singapore

Little is known about the early history of Singapore, although records of a trading post at the site of modern-day Singapore date back to the fourteenth century. By the following century, Singapore was part of the Malacca Sultanate. The island city's modern history began with the establishment of a trading post there by the British under Sir Thomas Raffles in 1819. Five years later, the sultan formally ceded the island to the British.

Singapore enjoyed two key advantages that helped turn it into one of the world's great ports: its strategic location at the mouth of the narrow passage connecting the Indian Ocean to the South China Sea and its tax-free status. With the opening of the Suez Canal in Egypt in 1869, Singapore became a major port of call for steamships taking the new shortcut from Europe to the Far East. Throughout the nineteenth and early twentieth centuries, immigrants from China, Malaysia, and India flocked to Singapore drawn by its increasing prosperity. During World War II, Singapore was occupied by the Japanese, who treated the Singaporeans brutally. In 1945, the British liberated Singapore and the next year made it a crown colony.

Terra-Trivia

Singapore is famous for its extreme (some would say oppressive) orderliness and tidiness. It is an immaculately clean, graffiti-free city, where people are encouraged to be kind, courteous, and quiet. Its crime rate is virtually zero. You're also unlikely to step on chewing gum in Singapore as importing it into the country is against the law. Other "crimes" against public order include picking flowers in a public park, letting your hair grow long (if you're a man), and eating anything on the subway.

In 1959, Singapore became a self-governing state, but four years later, in 1963, it joined with Malaya, Sabah (Northern Borneo), and Sarawak to form a new country, Malaysia. The union was short-lived. In 1965, Singapore separated from Malaysia and became an independent country and a member of the United Nations. In 1990, Chok Tong Goh became Singapore's second prime minister, replacing Lee Kuan Yew, who had served as prime minister for 31 years and remains a dominant political figure in this tightly controlled, virtually one-party state.

The Big Players in the Game

South Korea and Singapore are separated by about 2,900 miles (approximately the width of the contiguous United States). The four countries and territories in this economic region are scattered over this vast expanse, each with its own politics, economics, and character.

South Korea

South Korea is blessed with the best agricultural land on the Korean peninsula; North Korea, on the other hand, struggles to feed its people adequately and has experienced catastrophic food shortages in recent years. The bulk of the peninsula's raw materials, however, are in the north. It is one of the many tragic ironies of the continuing division of the Korean peninsula that the two halves with their complementary resources cannot find common ground to work together.

Although South Korea is known as a major producer of semiconductors and electronic products, its industrial output also includes passenger cars (Hyundai), cement, pig iron, crude steel, chemicals, and textiles. South Korea is also one of the world's largest shipbuilders. Like everyone else in this region, South Korea was shaken by the Asian financial crisis of the late 1990s, but it has weathered the global economic slowdown of 2001–2002 well, thanks to the continuing strength of its industrial and construction sectors.

South Korea's soul is Seoul! Between one-fourth and one-third of the entire population of the country lives in or around this capital city. The next-largest and next-most-important city is Pusan, South Korea's largest port city and also a major manufacturing center.

Taiwan

Taiwan's export-driven economy has generated a significant trade surplus and one of the world's largest supplies of foreign reserves. Although Taiwan's modern economic development was based on the production of steel, machinery, paper, cement, and textiles, today the emphasis is increasingly on high-tech items, such as personal computers, telecommunications equipment, radios, TVs, VCRs, and calculators. Most of these goods go to the United States, Hong Kong, Japan, and Europe. Hong Kong (and indirectly China) accounted for 21 percent of Taiwan's exports in 2001. Despite continuing tensions over Taiwan's status in relation to

Terra-Trivia

Taipei's National Palace Museum is considered to be one of the four greatest museums in the world, ranking up there with the Louvre in Paris, the British Museum in London, and the Metropolitan Museum of Art in New York. The National Palace Museum houses more than 700,000 pieces of art, some dating back 5,000 years, collected by Chinese emperors and stored in Beijing's Forbidden City.

China, in 2001 Taiwan eased long-standing restrictions on direct investment in China. It is estimated that there are some 50,000 Taiwanese businesses operating on the mainland.

Taiwan's capital, Taipei, is also its preeminent city. More than 6.5 million people live in the Taipei metropolitan area, on the northern part of the island. The second-largest city is the southwestern port of Kaohsiung.

Hong Kong

As you've seen, Hong Kong became part of China once again in 1997. Why then is it being covered separately here, rather than with China in chapter 16? Rest assured, there is a logic to this. Under the "One Country, Two Systems" formula agreed to by China and the United Kingdom before the hand-over of Hong Kong, the territory would retain its economic and political autonomy in all things except foreign affairs and defense policy. Whether or not China can refrain from imposing its political and judicial system on Hong Kong remains to be seen, but at least for now, Hong Kong is, for all intents and purposes, a separate entity from China.

Then there is the matter of economics. The countries surveyed in this part of book all have fully developed economies. Although Hong Kong, with one of the world's highest per capita GDPs, clearly falls into this category, China, despite is huge and booming economy, clearly does not, at least not yet.

Hong Kong recovered quickly from the Asian financial crisis of the late 1990s, but has been severely affected by the more recent global slowdown. Even Hong Kong's sizzling retail sector has felt the effect of the global chill. Still, Hong Kong remains one of the great tourist and shopping destinations of East Asia, with an array of shops and wares that few other places in the world can match.

GeoJargon

An **entrepôt** is a port that specializes in the *transshipment* of goods from one country (not its own) to another country (or other countries). Hong Kong plays this role, and Singapore does it in spades.

Singapore

Space, or a lack of it, is a key to understanding Singapore. Much of its food and most of its water must be imported. Most housing is high-rise, and Singapore doesn't have enough of it, especially for middle- and low-income families. Given this lack of space, plus its strategic location and superb harbor, it's no surprise that Singapore developed into a major *entrepôt*.

Crude oil from Southeast Asia is imported, refined, and then transshipped to other Asian countries.

Malaysian products (rice, spices, and timber) are also exported through Singapore, as are Malaysian and Indonesian rubber. The list of products, goods, and raw materials shipped through Singapore by less-developed neighbors with poorer port facilities is endless. Singapore also imports products for transshipment to other Asian countries (machinery and autos, for example).

Electronics and financial services top the list of Singapore's domestic industries. Others include shipbuilding and repair, oil refining, rubber processing, food processing, chemicals, pharmaceuticals, plastics, and clothing. Its main export partner is Malaysia, followed by the United States, Hong Kong, Japan, Taiwan, Thailand, China, and South Korea. Singapore has also become an important financial center with a prominent stock market.

The Lay of the Land

The four East Asian Tigers are all located on islands or peninsulas. South Korea and Taiwan are both full-size countries with varied terrain and physical features. Hong and Singapore are essentially cities with outlying areas, so the geographic discussion about them is somewhat limited. This section looks at the natural side of these economic giants.

South Korea

In area, South Korea is by far the largest country in this region—it's almost three times the size of the next-largest country, Taiwan. Even so, it's only about the size of Indiana. South Korea is a rugged, mountainous country, especially on its eastern side. As you might guess, because of the mountains, most of the population of more than 48 million tends to be distributed on the western side of the peninsula.

South Korea has cold winters and warm-to-hot summers; the waters surrounding the peninsula have a moderating influence on South Korea's climate. The country gets most of its rain on its southern coast, where agricultural land is most plentiful. Because of the temperate climate, more than one crop per year can be grown, and rice cultivation is of primary importance. South Korea is also vulnerable to an occasional typhoon (hurricane-like) deluge.

Taiwan

The island of Taiwan is only a little more than 100 miles off the southeast coast of China. Located on the edge of the Philippine tectonic plate, it's extremely vulnerable

Terra-Trivia

The Tropic of Cancer is a parallel of latitude at about 23½ degrees north of the equator (see Chapter 4). Marking the northern boundary of the tropics, this area has temperatures that remain about the same year-round. Technically, the Tropic of Cancer marks the northernmost point where the sun can appear directly overhead at noon, as it does on the summer solstice, on or about June 21.

to earthquakes. And as if that weren't enough, Taiwan also gets its share of tropical typhoons.

Although Taiwan measures only 13,900 square miles, it has almost 22 million people. Hilly and mountainous in the east, it has a long coastal plain in the west. The Tropic of Cancer crosses the island, which means that it's at the edge of the tropics. The result is minimal seasonal temperature variation—Taiwan has hot summers and warm winters. Rainfall is heavy throughout the year.

Because of the country's warm temperatures, multiple cropping is practiced on the western half of the island. The fields are intensively farmed, largely in small subsistence plots. The dominant crop in Taiwan, as in South Korea, is rice. Tropical rain forests grow on the eastern, mountainous side of the island.

Hong Kong

Hong Kong is located on approximately the same latitude as southern Taiwan, on China's southeastern coast. The former British territory is made up of the New Territories and Kowloon on the tip of a peninsula and several islands, chiefly the island of Hong Kong, scattered across one of Asia's busiest and best harbors.

Although Hong Kong is in the tropics, the southwestern monsoon makes the weather subtropical. This monsoon is a warm, moist equatorial wind that gives Hong Kong a rainy season in early summer. The temperature range is 59 to 82 degrees F. The area is naturally hilly and rugged, and the views from Victoria Peak are unbelievably spectacular, day or night.

Singapore

Singapore is a city–state, a republic, and an island off the southern tip of the Malay Peninsula. Although Singapore is one of the largest ports in the world, it's the smallest of the Pacific Rim states in both size (only about 250 square miles) and population (3 million). If Singapore were in Europe, it would vie for microstate status, as described in Chapter 10.

Unlike the other countries in the Pacific Rim, Singapore is not mountainous. Despite years of dredging, in fact, it's still somewhat swampy. Because Singapore is only one degree north of the equator, it lies in the heart of the tropics; because it's almost on the equator, it's tropical, wet (averaging 95 inches of precipitation annually), hot (81–86 degrees F), and humid. If you don't like it hot and sticky, don't go to Singapore!

The Ethnic and Religious Mix

South Korea's people are 99.9 percent Korean, and all speak Korean. More than half are atheists, one-quarter are Buddhists, and about one-fifth are Christians. More than three-fourths of South Korea's people are urbanites.

In Taiwan, most of the people live in the north and the west. Highly urban, the bulk of the population is Taiwanese (84 percent), and almost all the rest are mainland Chinese, either Nationalists who fled from China in 1949 or their descendants. Mandarin Chinese is the official language. The main religion is a mix of Confucianism, Taoism, and Buddhism; about 5 percent of the population is Christian.

The people of Hong Kong are 97 percent Chinese, and English and Cantonese Chinese are its official languages. The prevalent religion is a blend of Confucianism, Taoism, and Buddhism; about 8 percent of Hong Kong people are Christian.

The smallest of the group in area, Singapore is the most ethnically diverse. About 78 percent of the population is Chinese; 14 percent, Malay; and 7 percent, Indian. Interestingly, Singapore has four official languages: English (the language of administration), Mandarin Chinese, Malay, and Tamil. Its primary religions are Buddhism, Islam, Hinduism, Confucianism, Christianity, and mixtures of all five.

In the News: Crisis in Korea

This chapter has examined each of the four East Asian Tigers and given you a glimpse of their histories, locations, climates, and peoples. You should have a better understanding now of how they managed to overcome considerable obstacles to become global economic heavy hitters. Each country or territory, however, faces formidable challenges in the coming years, none more so than South Korea. The danger to South Korea, of course, comes from its radical neighbor to the north. Since being included in the "Axis of Evil" by President

Terra-Trivia

If an armed conflict with North Korea were to break out, South Korea's capital and its huge population cluster would be extremely vulnerable. Seoul is only about 25 miles from the demilitarized zone (or DMZ), which marks the border between North Korea and South Korea. Although this zone is one of the most heavily defended lines on earth, the threat is grave and real.

George W. Bush in 2002, and especially in the aftermath of the Iraq War of 2003, North Korea has revived its nuclear weapons program and has admitted to having at least one.

The tension on the Korean peninsula is further complicated by a growing anti-American sentiment among the people of South Korea. July 2003 marked the fiftieth anniversary of the armistice that stopped the fighting in Korea. (Officially, though, the war has not ended, because a peace treaty has not yet been signed.) Whether the Korean peninsula can go without war for another 50 years is a question with no clear answer at the present time.

The Least You Need to Know

- South Korea, Taiwan, Hong Kong, and Singapore are four small, very industrialized trading states located on the eastern fringe of Asia. Because of their economic strength and prowess, they have been called the East Asian Tigers.

- Taken together, the four Tigers are the world's third-largest exporter.

- The Republic of China on Taiwan owes its birth to the victory of the Communist Revolution in China, which forced the Nationalists to flee to the island then known as Formosa.

- Hong Kong's future depends on how well its union with China goes and whether China keeps its promise to maintain a "One Country, Two Systems" policy.

- Singapore is the smallest country in the region, a very prosperous city–state whose economy is based on export and transshipping.

Chapter **13**

Australia and New Zealand: Looking Down Under

In This Chapter

♦ Looking back at kiwi history

♦ Exploring the land of volcanoes

♦ Getting to know the kiwis

♦ Digging into Australia's past

♦ Trekking in the outback

♦ Getting to know the Aussies

Australia and New Zealand, which are neighbors despite being separated by 1,300 miles of Tasman Sea, form a region in the far South Pacific. Both countries are firmly based in English tradition, language, and rule of law and government, despite their 12,000-mile distance from Europe. Young relative to many other nations, literate and prosperous, New Zealand and Australia share many similarities.

Don't be deceived by their whimsical outlook, odd-looking animals, general informality, and macho cowboys and surfers. The people of this area are serious, hard-working, independent souls, not to be pushed around. The region is also serious about its "no-nuke" policies. The 1985 Treaty of Rarotonga declared the South Pacific to be a nuclear-free zone.

This region contains a number of islands, but only two countries. Although Australia is by far the larger of the two in both area and population, let's start with the little guy, New Zealand.

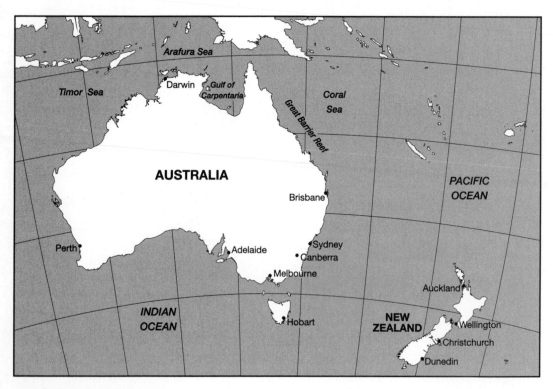

Australia and New Zealand.

Land of the Kiwi: New Zealand

If New Zealand were placed in the Northern Hemisphere, it would be at approximately the same latitude as the stretch from Los Angeles, California, to Seattle, Washington. The country is farther south than Australia and cooler—closer to the South Pole. New Zealand's Stewart Island hosts lots of penguins (to say nothing of the world's best oysters.)

First Peoples: The Maoris

New Zealand's first settlers were the ancestors of today's Maoris, a Polynesian people who began arriving in the islands as early as the ninth century C.E. A mass migration of Polynesians to New Zealand is thought to have taken place in the mid-fourteenth century. The early Maoris traveled to the land they called Aoearoa, "land of the long white cloud," in large dugout canoes filled with seeds, plants, dogs, and rats. Before the arrival of the Maoris there were no mammals on the islands of New Zealand, but plenty of flightless birds. The giant moa, a 10-foot-tall flightless bird, became an important food source for the Maoris, who eventually hunted them to extinction. The similar but much smaller kiwi is New Zealand's national bird and enjoys protected status.

Terra-Trivia

Competitive tribalism was fundamental to the Maori way of life, and their inter-tribal wars—often waged for the sake of *utu* (revenge)—were frequent and vicious. The Maoris also practiced ritual cannibalism, in which enemies killed in battle were sometimes eaten.

The Flood of Europeans

The Dutch explorer Abel Tasman was the first European to "discover" New Zealand in 1642. Faced with a decidedly unfriendly Maori welcoming party, Tasman set sail for Indonesia without actually setting foot ashore. For 127 years, New Zealand remained free of European visitors, until the great English navigator and explorer Captain James Cook sailed to New Zealand in 1769. The botanist on board Cook's ship *Endeavour* was Sir Joseph Banks, who in addition to describing the local flora and fauna in great detail, noted the "… almost certainty of being eaten as you come ashore!"

European, American, and later French whalers and traders began arriving in New Zealand in the 1790s, drawn by the islands' abundance of whales, seals, and timber. The flax trade soon grew in importance as well. The North Island coastal settlements established by these early arrivals soon turned into lawless frontier towns, where many New England whalers finished their lives far from their homes in Plymouth or New Bedford. The inter-tribal wars common among the Maori became far deadlier with the introduction of European muskets.

In 1839, the New Zealand Company sent its first three settler ships from Britain to New Zealand. Earlier that year, company agents began acquiring land illegally and unscrupulously from the Maoris, who didn't fully grasp the European concept of land ownership and sale. In an effort to bring order to a chaotic situation and to impose its sovereignty on the islands before the French could do so, the British government negotiated the Treaty of Waitangi with more than 500 Maori chieftains in 1840. In exchange for recognizing the sovereignty of Queen Victoria, the Maoris would retain their lands and fishing areas and receive the same rights and privileges as all other British subjects. Although the treaty promised that land that had been acquired illegally would be restored, the New Zealand Company settlers had other plans.

Terra-Trivia

Despite the promises made by the Treaty of Waitangi, the Maoris controlled only 5 percent of what had been their land by 1900. The issue of land rights is still on New Zealand's docket. Current claims alone will take some 50 years to straighten out. In question is half of New Zealand's land as well as much of its offshore fishing grounds.

Continuing disputes over land rights and the interpretation of the Waitangi treaty resulted in a series of settler-Maori wars, known as the New Zealand Wars, spanning a period of nearly 30 years, from 1845 to 1872. Despite putting up a valiant and fierce resistance, the outmatched Maoris were finally defeated by the superior forces of the settlers.

In the meantime, the settlers had been granted self-government by Britain in 1853. A gold rush in the 1860s swelled the population of South Island, while the advent of refrigeration in the 1880s stimulated exports of meat (especially sheep and mutton, for which New Zealand is famous) and dairy products. The foundations of New Zealand's version of the welfare state were set down around the turn of the twentieth century. In 1907, New Zealand ceased being a British colony and joined the ranks of Canada and Australia as a dominion, or a self-governing territory within the British Empire. Today, the British monarch is also New Zealand's head of state and is represented by a Governor General.

In 1951 New Zealand joined Australia and the United States in the ANZUS defense pact. Both New Zealand and Australia signed the 1985 Treaty of Rarotonga, which established the South Pacific nuclear-free zone. New Zealand's decision to ban U.S. nuclear-powered or nuclear-armed ships from its harbors in 1986 caused the United States to pull out of its ANZUS defense commitments to New Zealand.

Trembling Land

New Zealand is more than 1,000 miles long and about 1,300 miles southeast of Australia. The country is sandwiched between the Tasman Sea on the west and the South Pacific on the east. It's a mountainous nation consisting of two main islands, North Island and South Island. (A third island, Stewart, located south of South Island, is significantly smaller than the two main islands.)

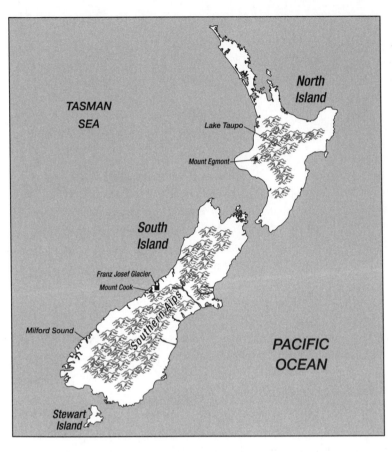

New Zealand—physical features.

Although dwarfed by Australia in size, New Zealand is still slightly larger than the United Kingdom, the European source of the most of its population. Separating the two main islands is Cook Strait, which thins to 16 miles at its narrowest point. North Island is slightly smaller than South Island but is more geologically active.

Unsteady Ground on North Island

Volcanically speaking, North Island is active indeed. There are three major volcanoes in the island's center, and one of them, Ngauruhoe, erupted in 1995. All three volcanoes are part of Tongariro National Park. To the west in Egmont National Park, extinct Mount Taranaki (Egmont) rises more than 8,000 feet above the rich surrounding plains and vineyards.

With geysers shooting 100 feet high and fumaroles (a surface hole from which volcanic gases gush out) spewing steam and gas, New Zealand derives about 7 percent of

Eye on the Environment

Like other Pacific Ring of Fire nations, New Zealand has abundant geothermal resources, most of which remain untapped. It has been estimated that using current technologies about 75 percent of the country's peak electricity demand could be met by high-temperature geothermal sources. Today only a fraction of that potential has been developed.

its power from geothermal plants. South Island's rushing streams are also used to generate hydroelectrical energy, which is transmitted by undersea cable across Cook Strait to the more heavily populated North Island.

In the heart of North Island is Lake Taupo, the island's largest lake, teeming with fish. To the far north is a long peninsular proruption (long extension of land) called the Northland. This mostly low-lying area is famous for its 90-mile beach north of Auckland. The Northland also has remnant groves of the enormous native kauri tree. Most of the ancient kauri forests were turned into huge war canoes by the Maori and into fishing fleets by European whalers.

The Alpine South

With a surface area of 58,000 square miles, South Island is larger than North Island and has a more varied terrain. Its dominant physical feature is the Southern Alps (Fjordland), a massive mountain chain that runs along the west coast of the island, providing spectacular alpine scenery and 16 miles of uninterrupted ski runs famous the world over. New Zealand's highest mountain, Mount Cook, or *Aorangi*, rises 12,349 feet near the center of the Southern Alps.

The Southern Alps are cloaked in glaciers that shrink and expand depending on weather patterns and the rate of snowmelt. The Franz Josef Glacier is unique in that it ends in a temperate rain forest. In the southwest portion of the island are Milford Sound's Mitre Peak and its matching fjord. Here 6,000-foot mountain faces rise from the water, high valleys unleash thundering falls (Sutherland Falls is fourth highest in the world at more than 1,900 feet high), and seals play on the shore.

GeoJargon

Aorangi, the Maori name for Mount Cook, means "cloud piercer."

South Island's west coast was the site of the 1860's kiwi gold rush, which at its peak in 1866 produced more than 700,000 ounces of bullion. The area is also a rich source of greenstone, or jade.

The Kiwis

North Island is home to three-quarters of New Zealand's 3.4 million people, including most of the country's indigenous Maoris. North Island also boasts the country's largest city (Auckland), its capital (Wellington), three-quarters of its industry, most of its dairy production, and more than half of its huge sheep population.

Because of the mountainous nature of New Zealand's interior, the majority of its population lives in coastal cities and towns (where most of the country's rail lines and roadways are also found). Nearly 80 percent of the population is of European descent, while Maoris account for about 15 percent. English is the most widely spoken language, but both English and Maori are official languages.

Protestant religions prevail in New Zealand, although a sizable minority of New Zealanders (about 15 percent) are Roman Catholic. New Zealand's Catholic population is largely comprised of descendants of Irish immigrants who arrived in the country in the 1800s and brought their religion with them.

The Maori

New Zealand's indigenous Maoris are unrelated to Australia's indigenous people. Maoris are a Polynesian group while Australia's Aborigines are an Austroloid group. With the arrival of the *pakeha* (white man), the Maori population shrank from an estimated (and disputed) high of 250,000 to about 40,000 in 1900, thanks in large part to diseases and firearms introduced by the European settlers. Maoris have enjoyed parliamentary representation since 1867 and their population recovered during the coarse of the twentieth century. Today, the Maoris live mostly in cities and have integrated fairly successfully in New Zealand society, while still retaining their cultural traditions and identity.

Livelihoods and Pastimes

Agriculture and food exports are major components of New Zealand's economy. The fertile Canterbury Plains on the eastern side of South Island are the country's breadbasket. It has enough good land to grow 90 percent of the country's grains. Hops, apples, and raspberries thrive. Mention New Zealand fruit and the indigenous fuzzy green kiwi fruit immediately comes to mind, thanks to its now worldwide popularity.

The economic restructuring that began in the mid-1980s has helped transform New Zealand

Terra-Trivia

Nearly 30 percent of New Zealand is forested, while more than 50 percent is given over to sheep and cattle grazing and farming (New Zealand has about 40 million sheep, roughly 10 times its human population.) About 30 percent of the nation's total area is publicly held protected land administered by the Department of Conservation.

into a more industrialized and service-oriented free-market economy. Major industries include food processing, the manufacture of wood and paper products, textiles, machinery, banking, and tourism.

The spectacular natural beauty of the islands makes backpacking a favorite pastime. Queenstown, on South Island, is the trekking center of the island, where hikers stock up, load their swags (backpacks), rest, and scramble for transport to the mountains beyond. Queenstown's lawn bowling tournaments and rose gardens provide less strenuous entertainment for the less energetic.

New Zealand's Cities

New Zealand is highly urbanized: 85 percent of its people live in the cities. The country's largest city by far, Auckland proper has about 380,000 people, while the greater Auckland area numbers more than a million people. The population of this bustling North Island city includes the largest percentage of Polynesians and Asians in New Zealand.

Wellington is New Zealand's capital, seat of government, and second-largest city. Located on the southwestern tip of North Island, Wellington straddles a fault, much like San Francisco, and is similarly earthquake-prone. The largest city on the South Island is Christchurch, known as New Zealand's "English City" as well as its "Garden City."

The Land Down Under: Australia

Australia's earliest inhabitants, the distant ancestors of today's Aborigines, reached the continent as early as 75,000 years ago by traveling over the land bridges and shallow seas that connected Ice Age Asia to Australia. Like other Stone Age peoples, Aborigines lived and hunted in isolation, using stone (not metal) tools. More than 300,000 Aborigines in 500 tribes and speaking 300 languages inhabited Australia when Captain Cook arrived in 1770 and claimed the land for Britain. Although the Dutch had sailed to Australia in the seventeenth century, and the Portuguese might have sighted it as early as 1522, it was the British who opened up Australia for colonization in the late eighteenth century.

Unlike New Zealand, whose first English settlers came voluntarily, most of Australia's pioneering settlers were convicts sent over by the British government to relieve overcrowding in its prisons. The famous First Fleet consisted of 11 ships, carrying 1,487 people, of whom 736 were convicts. After anchoring in Botany Bay in January 1788, the fleet moved on to what Cook had named "the finest harbor in the world"—Sydney.

Free settlers soon followed the convicts in 1793 to take up farming, mining, and trading. The introduction in 1797 of a hardy Spanish breed of sheep with superior wool had a huge and positive impact on Australia's economic future. Farmers pushed into the interior across Australia's eastern mountains to "squat" on land in view of eventually taking possession of it. Displaced Aborigines were forced even further into the interior as bloody conflicts escalated over land rights and access to scarce food supplies.

Gold was discovered in New South Wales in the 1850s, and the resulting gold rush helped swell Australia's sparse population. Friction with Chinese workers in the gold fields in 1861 led to immigration restrictions and a "white Australia" policy. The various Australian colonies became a federated self-governing commonwealth within the British Empire in 1901. After World War II, Australia benefited significantly, both economically and culturally, from an influx of non-British European immigrants, while in the 1980s it took in large numbers of Southeast Asian refugees, especially from Vietnam and Cambodia. Australia today has become a multicultural melting pot: Half of all Australians are foreign-born or have a foreign-born parent.

Terra-Trivia

The indigenous people of Tasmania (known as Tasmanoids) didn't survive the European onslaught of the nineteenth century. Treated with contempt by whites, the Tasmanoids died of disease, starvation, and outright butchery. A population of 6,000 in 1804 dwindled to 300 by 1830 and to none in 1876, when the last Tasmanoid died and an entire group of humans became extinct.

The Big Dry: Australia's Land

Although Australia is a huge place (the world's sixth-largest country), it's the world's smallest continent. Roughly the size of the United States minus Hawaii and Alaska, Australia is the land of superlatives. It's the lowest and the flattest continent. Except for Antarctica, Australia is also the driest continent and has the sparsest population. Covering nearly 3 million square miles, it's surrounded by one sea after another (Coral, Tasman, Timor, and Arafura) and the Pacific and Indian oceans, as shown on the following map.

Mountains and Outback

Australia's main physical regions are the Great Western Plateau, the Central Lowlands, and the Eastern Highlands. The Western Plateau, with no trees or rivers, but lots and lots of sand, encompasses about two-thirds of Australia. Sand ridges that reach 60 feet high, sand hills, sand dunes, and sand plains make up the Great Sandy Desert, the Gibson Desert, and the Great Victoria Desert, which cover the bulk of central and western Australia. Central Australia is also the site of the country's most famous natural landmark, Uluru, or Ayers Rock.

Australia—physical features.

Terra-Trivia

Ayers Rock, the largest monolith in the world, rises 1,143 feet above the Great Western Plateau. The huge, red, sandstone outcrop adds interest to "gibber" plains (huge, flat stretches covered with pebbles). This unmistakable landmark in the outback has been a sacred place to the Aborigines for countless centuries.

The Central Lowlands extend from the Gulf of Carpentaria (the big notch in the north) through the pastureland of the Great Artisan Basin. The area is laced with intermittent streams that sometimes swell with the help of runoff from mountains to the east. Much of this water is too salty for crops and is used only to irrigate grazing land. The best farmland is located in the southeast, where the Murray, Darling, and Lachlan rivers flow; sediment in this area has also yielded opals.

The Eastern Highlands, or Great Dividing Range, separate the dry interior from the fertile coast.

The area is tropically sticky and hot in the north and humid and subtropical in the south. Steep mountains (such as 7,310-foot Mount Kusciusko, Australia's highest) rise in the southeast and are the source of a great deal of hydroelectric production. The mountains in the south feed the long, gentle rivers that water the area and make possible the "fertile crescent" filled with apple and pear orchards, farms, and sheep stations.

Tasmania, Australia's island to the south, lies 150 miles across the Bass Strait and is the mountainous southern extension of the Eastern Highlands. Dairy farming, sheep grazing, oil drilling, and hydroelectric power generation are big industries in Tasmania. Native wildlife species on this wet and isolated island include wombats, of which few remain, and the Tasmanian wolf.

The Great Barrier Reef

The Coral Sea off Australia's northeast coast contains one of the world's most splendid natural wonders: the Great Barrier Reef. It consists of 600 islands and atolls that extend some 1,250 miles in shallow water 10 to 100 miles off shore. The size of England and Scotland combined, this immense living mass of coral reef, the earth's largest, covers 80,000 square miles. It's the home of 400 types of coral and 1,500 varieties of fish. Three million terns and boobies cluck and squawk above, and green sea turtles, sharks, and giant clams live below.

The Great Barrier Reef Marine Park was established in 1981 in an attempt to save the marine life of the reef from the depredations of (turtle soup) canning factories, oil drilling, bombing practice by the Australian navy, and reckless tourists. The park is the world's largest protected marine area, and it plays a crucial role in preserving the health and splendor of the reef for future generations.

Terra-Trivia

Known as undersea rainforests for the diversity of life they support, coral reefs are produced by tiny coral polyps that secrete calcium carbonate to form their own external skeletons. Reefs are fragile living organisms that can be destroyed by oil spills, strip mining residues, phosphate guano deposits, polluting runoff, and silt from farming.

Australia's Climate, Flora, and Fauna

In Australia, south is cold and north is hot. The top third of the country is tropical, and the seasons are opposite of what they are in the Northern Hemisphere. Santa sweats in his beard and red suit, and revelers eat plum pudding with brandy sauce on a hot December day, with "beer and barby" (barbecue) on the beach the next day.

Tasmania and the south coast of the mainland have a marine climate. Elsewhere, Australia's climate is fairly mild—perfect for tennis and surfing all year. Most of this vast country is arid or at least semiarid.

The "land down under" supports an amazing variety of plants that grow only in Australia—honey flower, spear lily (12 feet high with red blossoms), 600 kinds of orchids, aromatic eucalyptus trees, more than 500 types of gum trees, and acacia, the sweet-smelling plant known as the "wattle tree." (Settlers used acacia saplings for wattling, or interlacing, the frame for mud and plaster walls.)

Australia's wildlife is also quite unique. Kangaroos and other marsupials make up half the continent's native mammals. Marsupial females carry their young in a pouch until the babies are fully developed. Many types of kangaroos hop with surprising speed across the open plains. An endearing symbol of Australia, the cuddly koala bear is a marsupial and not a bear at all. It spends its time clinging to eucalyptus trees, its only food source. But the prize for oddest of all Australian animals goes to the duck-billed platypus, a web-footed mammal that lays eggs. (New Zealand, on the other hand, can claim only the bat and eel as native critters.)

Eye on the Environment

Whether deliberately or by accident, European settlers introduced a wide variety of non-native species to Australia, often with devastating consequences for the environment and for the survival of native species. Lacking natural predators, introduced species can often out-breed, out-eat, and outlast native species.

Cats and mice were brought to Australia, probably unintentionally, by ship. The fox, however, was deliberately introduced by English settlers looking to add a touch of the old country to the alien Australian landscape. Rabbits and hares were introduced for hunting purposes. The cane toad was brought over to control the cane beetle, which it then failed to do. All of these species, however, found a productive environment in Australia and began breeding, well, like rabbits. And while these newcomers were proliferating, some 19 native Australian small mammal species became extinct.

Eliminating feral (wild) animals is always a controversial matter. In 1995, Australian scientists developed a species-specific virus that has dramatically reduced the wild rabbit population in some areas of the country. Similar genetic engineering techniques might work as well with other introduced species turned environmental pests.

People of the Southern Continent

Australians are extremely urbanized: 85 percent of them live in the cities, which are located primarily along the coasts. The same population distribution pattern occurs in

New Zealand, but there it was the mountains that discouraged interior settlement; in Australia, it was the inhospitable desert.

Australia's population is almost completely literate and English-speaking. About one-quarter of the population is Anglican, one-quarter is Roman Catholic, and another quarter is of mixed Christian faiths. Aussies have a prosperous economy and enjoy a high standard of living, with a gross domestic product per capita of $27,000, slightly higher than the four dominant European economies (Germany, the United Kingdom, France, and Italy).

The Aborigines

The arrival of British settlers in Australia proved a disaster for the Aborigines. Their numbers dwindled due to the introduction of European diseases and conflict with the settlers. Those who survived were relegated to menial subsistence jobs and doomed to a life of poverty and ill-health. Between 1910 and 1970, some 100,000 Aboriginal children were taken from their parents in a misguided attempt to force assimilation. Lighter skinned children were placed with white families; black children were relegated to orphanages. Australia is still struggling with the consequences of a social policy that created what is now known as the "stolen generation."

Conditions are still difficult for Aborigines in Australia, although there have been improvements and attempts to redress past wrongs. Full citizenship was finally granted to Aborigines in 1967, segregated schools are closed, and welfare benefits have increased. Many Aborigines live and work as graziers (sheep farmers) near sheep or cattle stations in the willie woolies (outback, or back country) or in the mines. Those living in towns tend to speak their indigenous language (or a creole mix) and cling to tribal ways. Alcoholism and drug abuse are serious and widespread problems in Aboriginal communities.

The Australian States, Territories, and Cities

Australia is divided politically into six states and two territories. To the west is—you guessed it—Western Australia, in north central Australia is Northern Territory, and south of that is South Australia. In the northeast is Queensland, and south of that, in the southeast, is New South Wales. To the far south is Victoria. Off the southern coast is Tasmania, and the small governmental area carved out of New South Wales is called the Australian Capital Territory.

Australia is a country of city dwellers. Its capital, Canberra (population: 330,000), is the only major Australian city not located on the coast. With a population of 4.2 million, Sydney is Australia's largest city and a major tourist destination with more than 70 beaches. Its spectacular harbor is dominated by the world-famous Sydney Opera House. Melbourne, located at Australia's southern tip across the Bass Strait from Tasmania, is the country's second largest city (population: 3.6 million), followed by Queensland's lush, subtropical capital, Brisbane, home to some 1.5 million people. Australia's Indian Ocean gateway is the thriving, if still somewhat remote, port city of Perth with some 1.4 million residents.

The Least You Need to Know

- ◆ Both Australia and New Zealand have indigenous populations, the Aborigines and Maoris, respectively.

- ◆ Both countries share an English heritage of language, law, and government.

- ◆ Australia and New Zealand are isolated by distance from the rest of the world but are not isolationist.

- ◆ The population of both countries is highly urbanized and distributed mainly along the coasts.

- ◆ Australia has extremely varied flora and fauna that, because of the country's remote location, are found nowhere else in the world.

Part 3

A Regional Look: Nations in Transition

All of the countries you'll visit in Part 3, which is brand new to this edition, share one very important characteristic: They either had or still have a communist political system. And with varying degrees of success, most of them are making the difficult but critical transition from a government-controlled socialist economy to a more open, free-market capitalist economy.

Each of the three regions covered in this part—Eastern Europe, Russia, and China—reflects a different stage in the transitioning process. By the time the next edition of this book is published, many of these countries—especially those in the European Union—will probably have moved into the Developed World category. Others, however, have a tougher challenge ahead.

Chapter 14

The New Eastern Europe

In This Chapter

◆ Sorting through the Soviet legacy

◆ Plotting new lines on the map

◆ Joining the EU: The new Eastern European member states

◆ No longer in the USSR: Belarus, Ukraine, Moldova

◆ European flashpoints: The former Yugoslavia and Albania

◆ Understanding the physical features of Eastern Europe

Eastern Europe is a collection of "formers": former eastern bloc countries that were, until recently, satellites of the former Soviet Union; former Soviet republics that, until recently, were portions of the huge country of the Soviet Union; former pieces of Yugoslavia that, until recently, were united under a single totalitarian government. In addition to the word *former*, the phrase *until recently* pops up frequently whenever the countries of Eastern Europe are discussed.

To say that Eastern Europe has undergone drastic changes since the collapse of the Soviet Union in 1991 is the ultimate geographic understatement. Even for geography aficionados, it was sometimes difficult in the 1990s to keep track of all the changes that were being made to the map of this volatile region. And although the border situations might be calmer now, change continues as eight of the former Soviet republics and satellites join the ranks of European Union member states in 2004 and two more, Romania and Bulgaria, prepare to join in 2007.

A History of Turmoil

The latter part of the twentieth century was far from kind to Eastern Europe. During World War II, more than 1.3 million Eastern Europeans were killed on the battlefield. Many more civilian casualties were suffered as first the Nazis and then the Soviet Red Army carved swaths of destruction through the lands of Eastern Europe.

In addition to battle deaths, the Nazi occupation during World War II brought unprecedented genocide to the region. In Nazi concentration camps, many located in Eastern Europe, more than six million Jews (approximately two-thirds of the entire prewar European Jewish population) were exterminated. Jews, however, were not the only victims of the Nazis. Millions of Eastern European Gypsies and Slavs (Poles, Ukrainians, and Belarusians), physically disabled persons, homosexuals, political and religious dissidents, Jehovah's witnesses, and others considered "anti-social" were targeted and murdered by the Nazis.

Behind the Iron Curtain

Following World War II, the victorious Soviet Union imposed communist regimes on much of Eastern Europe. Economically, ideologically, and militarily, the Soviet Union dominated its so-called *satellites:* East Germany, Poland, Czechoslovakia, Hungary, Romania, and Bulgaria. Under its strong-willed leader Marshall Tito, Yugoslavia managed to resist Soviet domination and go its own way. The small and isolated country of Albania also embraced communism, but alone among the communist states of Eastern Europe, it aligned itself with Maoist Red China instead of the Soviet Union.

GeoJargon

A **satellite,** in the political sense, is a country that's subordinate to or dominated politically or economically by another country. Although the Soviet Union did not directly govern its satellites, it controlled them almost completely and even intervened militarily if one of them strayed too far from the party line.

The defeat of Nazi Germany in World War II and the redistribution of the territories it had conquered resulted in a dramatic redrawing of boundary lines. The entire country of Poland was shifted west (gaining territory from Germany and losing territory to the Soviet Union). The Baltic states of Estonia, Latvia, and Lithuania were absorbed into the Soviet Union and ceased to exist as separate nations. Romania lost the territory known as Bessarabia (now part of Moldova) to the Soviet Union, and Hungary lost the lands it had regained during the war. As a result of all this reshuffling, cartographers were kept very busy in the years following World War II.

Eastern Europe.

As the Cold War intensified in the 1950s and 1960s, several Eastern European countries attempted to gain greater autonomy from the Soviet Union. Uprisings and general strikes in Hungary (1956), Czechoslovakia (1968), and Poland (1960s to 1970s) were quelled by the Soviets, sometimes through brute force. Personal incomes lagged behind those in Western Europe as Eastern Europe foundered economically under the communist system.

Nations Reborn

With the final collapse of Soviet communism in December 1991, the old Soviet Union broke apart into 15 independent republics, six of them in Eastern Europe (seven if you count Russia itself), and once again cartographers got busy. In the northern part of Eastern Europe, the Baltic States of Estonia, Latvia, and Lithuania regained their independence. In what had been the western edge of the Soviet Union, bordering the former satellite countries, the newly independent states of Belarus (or Byelorussia or Byelarus), Ukraine, and Moldova emerged. Among the satellites, East Germany rejoined West Germany (see Chapter 8) and in 1993 the former Czechoslovakia split peacefully to form the Czech Republic and Slovakia.

Peaceful is not a word that can be used to describe the dissolution of the former Yugoslavia. Created after World War I by artificially patching together various Yugoslav ("southern slavic") ethnic groups (including Croats, Slovenes, Macedonians, Serbs, and Montenegrins), Yugoslavia had always been rife with territorial competition and ethnic tension. After World War II, communist strongman Marshall Tito held the country together by force until his death in 1980. The nation struggled for more than a decade to maintain its unity, but by the early 1990s, Yugoslavia was splitting apart along ethnic lines. Although Slovenia seceded relatively easily in 1991, the rebirths of Croatia, Bosnia and Herzegovina, Macedonia, and Serbia and Montenegro were painful and violent (see "Fractures and Feuds: The Former Yugoslavia," later in this chapter).

Terra-Trivia

East Germany was the showcase economy of the Soviet bloc. The only reason East Germany is not included in this chapter's discussion is that it no longer exists as a separate entity. A year after the Wall came down, East and West Germany became a single, unified nation for the first time since the end of World War II.

The former Yugoslavia and Albania aside, most of the nations of Eastern Europe have made remarkable strides since the late 1980s in adapting western-style democracy and free-market capitalism. The big news in the region today is the expansion of the European Union into what was once Soviet-dominated Europe. To simplify the discussion of a complex region, we've divided it into three subregions: the new and projected EU members, the remaining former Soviet Republics, and the still troubled nations of the Balkan Peninsula.

Joining the West: The New and Future EU States

The new Eastern European member states of the European Union include three former Soviet Republics—the *Baltic States* of Estonia, Latvia, and Lithuania—and five former Soviet satellite countries—Poland, the Czech Republic, Slovakia, Hungary, and Slovenia. Two other former satellite states, Romania and Bulgaria, are expected to join the EU in 2007 and will be covered in this section as well.

Estonia

The smallest of the Baltic States, Estonia enjoys the highest per capita gross domestic product of the three. It is a flat and scenic country, dotted with lakes. Its capital, Tallinn, is home to more than half a million residents, or nearly one-third of Estonia's population. Because of the large number of Russians living in Estonia (about 28 percent), Russia still has a stake in the country and has yet to sign a 1996 border agreement with its small neighbor.

Estonia's rich oil-shale reserves make it self-sufficient in energy, but exploiting those reserves took its toll on the environment, especially during the years of Soviet rule. Estonia's environmental situation has been improving steadily, however, with air pollution emissions down about 80 percent since 1980.

GeoJargon

The **Baltic States** take their name from their position on the Baltic Sea, a long arm of the Atlantic Ocean by way of the North Sea. Poland, Germany, Denmark, Sweden, and Finland also share a coastline on the Baltic Sea.

Only the narrow Gulf of Finland separates Estonia from Finland, and not surprisingly, the two nations have a great deal in common. The Estonian language is related to Finnish (technically speaking, it is a Finnic language of the Uralic family), and Finnish is widely understood and spoken in Estonia (so are Russian, German, and English). As in Finland, Lutheranism is the dominant religion in Estonia, followed by Orthodoxy.

Latvia

Like Estonia, Latvia has a large Russian minority (about 30 percent of the population) and unresolved border issues with Russia. It also has substantial Ukrainian and Belarusian minorities. With nearly 750,000 residents, the Latvian capital, Riga, sometimes called "the Paris of the Baltic" for its striking urban landscape, is the largest city in the Baltic States.

Although Latvia is heavily dependent on imported energy and raw materials, about 70 percent of its electricity is generated by domestic hydroelectric and thermal plants. The growing importance of the service sector in Latvia's economy has had the added benefit of easing many of its Soviet-era environmental problems.

The country's official language is Latvian, a Baltic language related to Lithuanian. English, German, and Russian are the most widely spoken foreign languages. Lutheranism, Catholicism, and Russian Orthodoxy are the main religions.

Lithuania

Lithuania is the southernmost Baltic State and, with a population of about 3.6 million, the most populous. Its capital, Vilnius, has some 600,000 residents. Through the centuries, Lithuania and its neighbor to the south, Poland, forged close cultural ties with each other. Like Poland, but unlike its fellow Baltic States, Lithuania is heavily Roman Catholic.

This gently rolling, forested land has a strong agricultural base. Lithuania's trade is increasingly oriented to the West rather than to Russia, and it has largely completed the process of privatizing its enterprises in preparation for joining the European Union. Lithuania is famous for its amber, the fossil resin that resembles a semi-precious stone.

Lithuania is more ethnically homogeneous than either Latvia or Estonia. The Lithuanian language is a Baltic language closely related to Latvian. Polish and Russian are also widely spoken.

Poland

Poland is the largest of the six former Soviet satellites in land area (more than 120,000 square miles) and population (38.6 million). Much of Poland is a vast plain with the only highlands in the south where the Carpathian Mountains run through the country's southern border with Slovakia.

Poland is an ancient country with a proud if tormented history that has seen it swallowed up by neighboring powers on several occasions. The rise of the Polish independent trade union Solidarity in the 1980s played a key role in the fall of communism in Eastern Europe and Russia. Reforms in the early 1990s revitalized Poland's economy and helped turn it, for a time, into one of the strongest in the region. Today, however, it faces considerable hurdles as it prepares to join the European Union in 2004. Heavy industry remains a major, if weakening pillar of the economy, and a hefty 27 percent of the workforce is involved in agriculture.

The Poles speak Polish, a Slavic tongue, and are overwhelmingly Roman Catholic in their faith. Warsaw, with a population of 1.6 million people, is Poland's capital, largest city, and cultural heart.

The Czech Republic

After the disintegration of the Austro-Hungarian Empire at the end of World War I, the closely related Czech and Slovak peoples joined to form the country of Czechoslovakia. After World War II, a territorially reduced Czechoslovakia became a Soviet satellite. In 1968, a Warsaw Pact invasion put an end to the famous Prague Spring, Czechoslovakia's brief, but brave, attempt to liberalize communism. The country finally attained its freedom in 1989, and in 1993 peacefully split along ethnic lines to form the two new countries of the Czech Republic and Slovakia. The Czech Republic consists of the old Czech lands of Bohemia and Moravia.

Geographically Speaking

The Cold War pitted two enormous military alliances against each other. The Warsaw Pact was a mutual defense alliance created by the Soviet Union in 1955 in response to the perceived threat from the North Atlantic Treaty Organization (NATO), the mutual defense alliance of the Western democracies created in 1949. The Warsaw Pact nations included the Soviet Union and all of the Eastern European communist countries except for Yugoslavia. After the collapse of the Soviet Union, NATO has expanded to include several former Warsaw Pact nations.

The Czech Republic is the more urbanized, industrialized, and westernized part of the former Czechoslovakia. It is also one of the most economically stable and prosperous countries of Eastern Europe, second only to Slovenia in per capital gross domestic product. The heart of the republic is the splendid capital of Prague (population 1.2 million), one of Europe's big tourist magnets. The people speak Czech, a Slavic language close to Slovak and Polish. Roman Catholicism is the predominant religion in the republic, although acknowledged atheists slightly outnumber Catholics.

Slovakia

The other half of the former Czechoslovakia is not to be confused with Slovenia, the small country to the south that was once a part of Yugoslavia. Slovakia is smaller in both size and population than the Czech Republic, and somewhat less developed, although it, too, is successfully making the transition from a centrally planned economy to a free-market economy. Slovakia is a land of rough mountains, forests, and high pastures.

Slovakia has a significant ethnic Hungarian minority, whose status has caused some tension with Hungary. Although Slovak is the official language, Hungarian is also spoken in many areas. The capital of Slovakia is Bratislava, with about 450,000 inhabitants. Roman Catholicism is the principal religion, although significant minority religions include Protestantism, Eastern Orthodoxy, and Judaism.

Hungary

Independent Hungary emerged from the ruins of the Austro-Hungarian Empire at the end of World War I. Having sided with Germany in World War II, Hungary suffered a particularly harsh occupation at the hands of the victorious Soviets. The Hungarian uprising of 1956 was brutally put down by Moscow. In 1968, Hungary began a process of liberalizing its economy, in what came to be known as "goulash communism," after the country's trademark spicy stew.

Hungary is a land of vast fertile plains, and prior to World War II its economy was primarily agricultural. Post-war industrialization was badly managed by the government and large portions of the country's heavy industries collapsed after the end of communist rule in the early 1990s. Although the initial period of transition between a controlled and a free-market economy was difficult, Hungary has weathered the crisis and today enjoys the third highest per capita gross domestic product in the region, after Slovenia and the Czech Republic. Hungary's capital, Budapest (population 1.8 million), was the second city of the Austro-Hungarian Empire and rivals Vienna in its grandeur.

Hungarians are descendants of an ancient Asian tribe called the Magyars, and their language is neither Slavic nor Germanic. Hungarian comes from the Ugrian branch of the Finno-Uralic family of languages, and is most closely related to Lap and Finnish. Almost 70 percent of Hungary's people are Roman Catholic, while Protestants account for another 25 percent.

Slovenia

A major Eastern European success story, Slovenia is the first former Yugoslav country to become a member of the European Union. Its economy is the strongest in the region, with the same per capita gross domestic product as Portugal. Slovenia's historic ties to the West explain, in part, its successful transformation into a Western-style capitalist democracy. After the collapse of the Austro-Hungarian Empire in 1918, the Slovenes opted to join with the Serbs and Croats to form what would become known as Yugoslavia. After a brief 10-day war in 1991, Slovenia became the first part of the unraveling Yugoslavia to become independent.

About the size of Massachusetts, Slovenia is a largely mountainous country, about half of which is forested. Its population is predominately Slovene, with small Serb and Croat minorities. The overwhelming majority of Slovenes speak Slovenian, a South Slavic language closely related to Serbian and Croatian, which most Slovenes also speak. About 70 percent of the population is Roman Catholic. The political and cultural heart of the Slovene nation is the capital, Ljubljana, a charming and highly cultured city of some 258,000 people.

Romania

Both Romania and Bulgaria are expected to join the European Union in 2007, but both also have formidable economic and political hurdles to overcome before achieving that goal. The communist regime imposed on Romania by the Soviet Union after World War II gave rise to Nicolae Ceausescu, who ruled brutally from 1965 until 1989, when he and his wife were executed in a bloody coup. It wasn't until 1996 that former communists were voted out of power.

Romania inherited from the communist era a largely outmoded industrial base and a huge agricultural sector that accounts for a whopping 40 percent of its workforce. Although Romania has enjoyed strong growth in recent years and has considerable mineral resources and oil and gas reserves, it continues to be plagued by poverty, bureaucratic red-tape, and corruption.

Romania's center is dominated by the rugged Carpathian Mountains that run north–south and the Transylvanian Alps that run east–west. West and north of these ranges lies Transylvania, the area made famous, or infamous, by Count Dracula (who did exist and was no humanitarian, although the vampire stuff may be somewhat exaggerated).

The capital of Bucharest is by far Romania's largest city, with about 2.2 million people. The population is almost 90 percent Romanian, but includes a sizable Hungarian minority and a smaller Gypsy minority. Unique in Eastern Europe, Romanian is a Romance language of the Italic group, closely related to Italian, Spanish, and Portuguese. The main religion in Romania is Romanian Orthodox.

Bulgaria

Bulgaria is an ancient land that for much of its modern history (from the late fourteenth century to 1878) was under Turkish rule. In the twentieth century, it found itself on the losing side of both World Wars and under Soviet domination from 1945 to 1990. Although Bulgaria has experienced stability and economic growth since 1996 and aspires to join the EU in 2007, it still has to cope with high rates of poverty and unemployment.

Bulgaria is a mountainous country bisected by the Balkan Mountains. It is one of three Eastern European countries to have a Black Sea coastline (the other two are Romania and Ukraine). Bulgaria has a sizable Turkish minority (9.5 percent) and a smaller Gypsy minority. Bulgarian, a Slavic tongue, is the language of the land, with Turkish spoken in Turkish enclaves. As is true of most Eastern European capital cities, Sofia (population 1.1 million) is several times larger than Bulgaria's next largest city.

Out on Their Own: Belarus, Ukraine, and Moldova

As you've seen, six of the Eastern European nations were once part of the Soviet Union. Three of them—the Baltic States of Estonia, Latvia, and Lithuania—have made a successful transition to Western-style capitalist democracy. The remaining three—Belarus, Ukraine, and Moldova—have lagged behind the "EU-ready" nations in the region, for a variety of historical, economic, and political reasons, which will be briefly explored in this section.

Belarus

Terra-Trivia

Before its demise in 1991, the Soviet Union was a huge nation, consisting of 15 Soviet Socialist Republics (SSRs). The six European former SSRs are covered in this chapter. Russia, the biggest and most important of them all, gets its own chapter (Chapter 15). The other eight former SSRs—Georgia, Armenia, Azerbaijan, Kazakstan, Uzbekistan, Turkmenistan, Kyrgyzstan, and Tajikistan—are described in Chapter 21.

Sandwiched between the Baltic States to the north and Ukraine to the south is the former Soviet Socialist Republic of Belarus. Although it declared its independence in 1991, Belarus maintains closer cultural, economic, and even political ties with Russia than any of the former SSRs or satellite nations. In fact, Belarus and Russia signed a treaty in 1999 that set the framework for increased political and economic integration. Implementation of the treaty, however, has been slow in coming.

The economy of Belarus remains closely linked to Russia's and isolated from the West. Unlike its Baltic neighbors, Belarus has resisted free-market reforms in favor of "market socialism." The result has been increased government intervention in the private sector and high rates of inflation and poverty.

Landlocked Belarus is low-lying, with many rivers and lakes, and lots of swamps (its peat marshes are an important energy resource). About one-third of the land is forested. The country's main population and industrial center is its capital, Mensk (or Minsk), with nearly 1.8 million residents. The population of Belarus is predominantly Belarusian, but includes a large Russian minority (more than 11 percent). The people speak Belarusian (White Russian), a Slavic language closely related to Russian. Eastern Orthodoxy is the primary religion.

Ukraine

Although Ukraine is technically one of Europe's newest countries, it has an ancient and proud history. What is now Ukraine was the birthplace of the very first Slavic state, Kievan Rus, whose glory days were back in the tenth and eleventh centuries. By the end of the eighteenth century most of Ukraine had been taken over by the Russian Empire. After the fall of the Romanoff dynasty in 1917, Ukraine enjoyed a brief period of independence before it was swallowed up again by the Soviet Union in 1921. During the Soviet era, Ukraine endured some of the blackest periods in its history, with millions of its citizens dying of starvation in the 1920s and 1930s and at the hands of Russian or German armies in World War II.

With abundant natural resources and vast expanses of fertile soil, Ukraine was one of the chief industrial engines of the old Soviet Union as well as its primary breadbasket. Ukraine today remains overly dependent on Russia as a trading partner, and its major industries are still mostly state-controlled. Ukraine's inability to put the Soviet past behind it and embrace free-market reforms, coupled with corruption and technical instability, are serious obstacles to its economic development.

> **CAUTION**
>
> ### Eye on the Environment
>
> An especially troubling legacy of the Soviet era is the nuclear power plants on which Ukraine depends for about 43 percent of its electricity. Radiation from the Chernobyl nuclear power plant, which exploded with such devastating consequences in 1986, is still contaminating northeastern Ukraine and parts of Belarus. (For more on the Chernobyl disaster and its lingering effects, see Chapter 27.)

With an area of 233,000 square miles, Ukraine is the largest country entirely in Europe. (Russia is larger, but it straddles Europe and Asia.) Nearly 78 percent of Ukraine's population is ethnic Ukrainian, and about 17 percent is Russian. The language of the country is Ukrainian, a Slavic tongue similar to Russian and Belarusian. The vast majority of Ukrainians are Eastern Orthodox; about 13.5 percent practice Ukrainian Catholicism. The capital and largest city is Kyiv (or Kiev), with more than 3.3 million people.

Moldova

What had been known in the fourteenth century as the principality of Moldavia was taken over by the Ottoman Turks in the sixteenth century and has been ruled at various times since then, in whole or in part, by the Turks, the Russians, and the Romanians. Even after Moldova became independent in 1991, Russia maintained a presence on the eastern fringe of the country, east of the Dniester River, to support the majority Russian and Ukrainian population there.

Moldova today is one of the poorest countries in Europe and in 2001 had the dubious distinction of being the first former Soviet Republic to elect a communist president. Its dependence on the former Soviet republics for its raw materials and trading relations is a great economic vulnerability. Despite its many problems, Moldova has managed to privatize its farmland (with U.S. help) and has been showing steady growth in the early 2000s.

Moldova is a small, but densely populated country. The capital and largest city is Chisinau, with 750,000 inhabitants. Moldovan/Romanians are the majority population, although Slavs, mostly Russians and Ukranians, are the majority east of the Dniester River (in the disputed Transnistria region). Moldovan (almost the same as Romanian) and Russian are official languages. Moldova's people are largely Romanian Orthodox in their faith.

Ethnic Powder Kegs: The Former Yugoslavia and Albania

The third subregion of Eastern Europe encompasses five countries on the western side of the Balkan Peninsula, as shown on the map of Eastern Europe. (Although Slovenia is also a former Yugoslav country, it has been grouped in this chapter with the more stable and prosperous Eastern European nations that are the newest members of the European Union.) The fractured, mountainous landscape reflects the ethnic splintering that has long dominated the history of this subregion. Although a description of the many ancient land claims and seething rivalries in the former Yugoslav countries and Albania is beyond the scope of this book, a brief summary follows.

Fractures and Feuds: The Former Yugoslavia

As you've seen, Yugoslavia was patched together after World War I out of an assortment of disparate and often conflicting states, territories, and ethnic and religious groups. The Nazi occupation of the country during World War II was particularly harsh and bitterly fought by opposing partisan guerilla armies—one monarchist, the other communist under the ironfisted leadership of Marshall Tito. After the war, Tito imposed a unique form of communism on Yugoslavia that combined rigid political control with a certain amount of economic and cultural flexibility. In 1948, Tito broke with the Soviet Union and went his own way. Yugoslavia never became a Soviet satellite or a member of the Warsaw Pact.

After Tito's death in 1980, a rotating presidency attempted to hold the country together and prevent ethnic rivalries from boiling over. The system worked for a while, but by the early 1990s, with Soviet communism about to collapse, Yugoslavia began to fall apart. Slovenia, Croatia, and Macedonia were the first to secede in 1991. Bosnia and Herzegovina followed suit in 1992. A rump Yugoslavia, consisting of Serbia and Montenegro, reconstituted itself as the Federal Republic of Yugoslavia (FRY), under the presidency of Serbian nationalist Slobodan Milosevic. In the conflicts that followed, atrocities would be committed by all sides, but it was the FRY that controlled the bulk of the former Yugoslav army, and Milosevic used it ruthlessly to achieve his goal of creating a "Greater Serbia."

Croatia and Bosnia and Herzegovina were the first major battlegrounds. In Bosnia, ethnic Serbs, supported by the FYR, launched a civil war with the goal of dividing the country along ethnic lines. The Dayton Peace Accords, signed in 1995 and guaranteed by the presence of NATO peacekeepers, put an end to three years of bitter fighting. But Milosevic wasn't quite done. In 1998, he turned his sights on the separatist movement in the mostly Muslim Serbian province of Kosovo. The "ethnic cleansing" campaign unleashed in Kosovo was so vicious that it finally prompted NATO to intervene militarily in March 1999. After 11 weeks of air strikes, Serbia signed a UN-brokered peace agreement, and NATO peacekeepers entered Kosovo. Today, Slobodan Milosevic, who was finally forced out of power in September 2000, is being tried by the International Criminal Tribunal for the former Yugoslavia in The Hague, the Netherlands, for crimes against humanity.

Terra-Trivia

The 1984 Winter Olympics were held in Sarajevo, Yugoslavia (now Bosnia and Herzogovina). The world watched while smiling faces and elaborate ceremonies proclaimed the oneness of the human spirit and the cooperation of all nations. Who was to know that within 10 years Sarajevo would be in ashes, the victim of Serbian artillery and ethnic hatred?

Essentially nothing remains of the former Yugoslavia today. In 2003, the rump Yugoslavia became a loose federation of its two constituent republics, and is now called Serbia and Montenegro. Slovenia, Croatia, Bosnia–Herzegovina, and Macedonia are all now fully independent countries. Although the fighting has stopped, border disputes linger and the region remains an extremely complex patchwork of ethnic pockets.

The Former Yugoslav Nations: An Overview

Geography, history, language, and religion have all played a role in the *balkanization* of the Balkans. The entire region is dominated by the Dinaric Alps, whose rugged mountains and isolated valleys provide the framework for the development of individual cultures in isolation. Because the mountainous divides hindered cultural exchange, groups tended to develop separately. The result is today's mix of neighboring, but often very distant, ethnic groups—Serbs, Croats, Slovenes, Montenegrins, Macedonians, and Bosnians.

GeoJargon

The Balkan Peninsula includes the land area below an approximate line drawn from the northern Adriatic Sea to the northern Black Sea. It derives its name from the Balkan Mountains in Bulgaria. The term **balkanization** has come to mean the breaking up of a country or region into smaller, hostile units, which certainly describes the history of the Balkan Peninsula.

Terra-Trivia

Despite regional and ethnic difference, the predominant language throughout the former Yugoslavia used to be called Serbo–Croatian. No longer. It is now called either Serbian or Croatian or Bosnian, depending on the speaker's politics and ethnicity. There's not even a shared alphabet. Croatians and Slovenes use the Latin alphabet, while Serbs, Bosnians, Montenegrins, and Macedonians use the Cyrillic alphabet.

The most common languages in this subregion are Serbian, Croatian, Bosnian, Slovenian, and Macedonian. Religion has also been an extremely divisive issue in the former Yugoslavia, with Eastern Orthodox Serbs pitted against Roman Catholic Croatians or Muslim Bosnians. Here's a rundown of the new countries that emerged from the dissolution of Yugoslavia:

- **Bosnia and Herzogovina:** Capital: Sarajevo (Mostar is the traditional Herzogovinian capital). Ethnic groups: Bosniak (48 percent), Serb (37 percent), Croat (14 percent). Languages: Bosnian, Serbian, Croatian. Religions: Islam (40 percent), Orthodox (31 percent), Roman Catholic (15 percent).

- **Croatia:** Capital: Zagreb (population: 685,000). Ethnic groups: Croat (78 percent), Serb (12 percent), with small Bosniak, Slovenian, and Hungarian minorities. Languages: Croatian, Serbian. Religions: Roman Catholic, Orthodox, Islam.

- **Macedonia:** Alexander the Great's old stomping ground. Capital: Skopje (population: 587,000). Ethnic groups: Mostly Macedonians with a sizable Albanian minority and some Turks, Gypsies, and Serbs. Languages: Macedonian and Albanian are official languages; Turkish is also spoken. Religions: Eastern Orthodox and Islam.

- **Serbia and Montenegro:** Capital: Belgrade (with 1.7 million residents, this is the largest city in the entire area). Ethnic groups: Mostly Serb (more than 62 percent), with a large Albanian minority (about 16 percent). Languages: mostly Serbian, some Albanian. Religions: Serb Orthodox, Islam.

◆ **Slovenia:** For a full description of Slovenia, the success story of the former Yugoslavia, see the section on this country earlier in this chapter.

Albania: Chinese Ties and Isolationism

The last of the Eastern European countries described in this chapter is also one of the poorest and most backward in Europe. Albania emerged from more than 400 years of Turkish rule in 1912 only to be taken over by Italy in 1939. Near the end of World War II, the communist guerilla leader Enver Hoxha imposed a repressive Stalinist regime on the country. In 1961, he broke with Moscow and aligned himself with Maoist China. After Mao's death in 1978, Albania completely isolated itself from the rest of the world and sank into economic ruin. The end of Albania's Communist regime in 1992 initiated a difficult decade of transition and adjustment, marked by crime, corruption, high unemployment, and involvement in its neighbors' conflicts, especially in Kosovo. A reformist government elected in 2002 is attempting to control crime and stimulate the economy.

Albania is a mostly mountainous country, with most of its farmlands and population concentrated in the western coastal lowlands. The country is heavily dependent on agriculture, which accounts for about 50 percent of its gross domestic product, and on money sent home by Albanians working abroad, mostly in Italy and Greece. Both Albanian and Greek are widely spoken, and Islam (70 percent) and Albanian Orthodoxy (20 percent) are the main religions. The capital, Tirana, has a population of about 350,000.

Plains, Mountains, and Rivers

This chapter has taken you on a tour of the countries of Eastern Europe and noted some of their problems and achievements. To conclude the overview, here's a capsule description of the region's physical features.

The Landforms

Large expanses of plains dominate Eastern Europe's northern and eastern sections. Although the area's flat plains have historically invited hostile armies to march across them with depressing ease and frequency, they have also been a blessing. With no intervening mountain range, the warmer air from the Atlantic Ocean and Baltic Sea helps to moderate the climate. Although still quite cold, temperatures in the plains are warmer than might be expected for their latitude. The fertility of the plains is also responsible for the region's agricultural abundance.

Moving south and west through Eastern Europe, the plains give way to mountains. The Carpathian Mountains and Transylvanian Alps are located in the center of the region, the Balkan Mountains are in the southeast, and the long chain of the Dinaric Alps are in the southwest. These mountains and their many valleys and basins have helped nurture and separate the region's diverse cultures.

Eastern Europe—physical features.

The Great Waters

Eastern Europe is sandwiched between three large bodies of water: To the north, it's bounded by the Baltic Sea; to the southeast, the Black Sea; and to the southwest, the Adriatic Sea. All except six of the region's countries have a coastline on one of these three seas. (Belarus, the Czech Republic, Slovakia, Hungary, Moldova, and Macedonia are all landlocked.)

The region is also laced by several important rivers. The northern plains are drained by Poland's Vistula River, which empties into the Baltic Sea. In the east, waters from the vast plains of Belarus and Ukraine drain south through the Dnieper River to the Black Sea. The region's most important river, the Danube, winds its way through most of the countries in the south to exit in the Black Sea. It runs through or touches more countries than any other river on Earth.

In the News: The Soviet Environmental Legacy

The defunct Soviet Union left its former republics and satellites with a dark double legacy: shattered economies and a blighted environment. During the Soviet decades, the region's state-run economies were focused on increasing industrial output no matter what the cost to the environment or to the health of its citizens. Aging, substandard, and largely unregulated factories, mines, and power plants spewed toxic fumes into the air and dumped pollutants into the water supply, with dire consequences for the health of humans, wildlife, and forests. The deadliest symbol of the Soviet Union's failed industrial and environmental policies is the Chernobyl nuclear power plant, whose shuttered ruins are still a threat to the people of Ukraine and Belarus.

Environmental cleanup doesn't come cheaply, and the nations of Eastern Europe have been struggling to undo the damage inflicted on their lands. Membership in the European Union, however, is a powerful incentive to get the job done. It has been estimated that the eight Eastern European nations that joined the EU in 2004 will have to spend some $100 billion to comply with existing European Union environmental regulations.

The Least You Need to Know

- This region's 18 countries include former Soviet republics, former satellites, and new offshoots from the former Yugoslavia.
- All of the countries in this region are undergoing a transition from communist economies to Western-style free-market economies, some more successfully than others.

- The eight most successful economies of Eastern Europe—Estonia, Latvia, Lithuania, Poland, the Czech Republic, Slovakia, Hungary, and Slovenia—are the newest members of the European Union.

- The post-communist breakup of Yugoslavia was accompanied by ethnic violence, civil war, and foreign intervention.

- One of Europe's great rivers, the Danube, flows through the region's southern tier before emptying into the Black Sea.

Russia in Transition

In This Chapter

- ◆ Tracking the tsars: A look at Russian history
- ◆ Assessing the new Russia—without the republics
- ◆ Touring the tundra and taiga
- ◆ Meeting the Russian people
- ◆ Looking toward the future

It's been more than a decade since the disintegration of the Soviet Union in 1991 and the reemergence of Russia as a separate country. Yet there is still good reason to call this region the "new" Russia, for Russia is still very much in the process of defining itself and its new place in the world. Much about the country still seems new and not 100 percent permanent yet: personal freedoms, a market economy, nuclear disarmament, the rebirth of religion, and of course the presence of fast-food restaurants all over the place.

This chapter looks at Russia's origins, the rise of the Soviet state, the emergence of a new Russia, and its prospects for the future. Even without its former republics, Russia is a huge country. In any place as large, you would expect variety, and Russia has it. Varying landscapes, varying resources, and varying cultures—each of these topics is explored in this chapter on the world's largest country.

Tsars, Communists, and Free Markets

The first Russian dynasty was started in the ninth century by Vikings who navigated Russia's rivers inland from the Baltic Sea to Novgorod—the oldest city in Russia, even though its name means "new town." By the eleventh century, eastern Russia was under the domination of Kiev, the most powerful city in the region at the time and now the capital of Ukraine. Kiev's hold was not to last, however, as the Mongols swept across the *steppes*, or plains, of Russia and destroyed the city in 1240. The Mongols were not expelled until the late fifteenth century.

GeoJargon

Steppes are Russia's semiarid plains, where short grasses are the predominant vegetation.

Russia.

The Tsarist Centuries

In the sixteenth century, Ivan the Terrible became the first tsar (the Russian term for emperor; tsars ruled Russia until the overthrow of the Romanoff dynasty in 1917). Another noted tsar, Peter the Great, founded Saint Petersburg in 1704 and began the process of westernization in a Russia that was still locked in medieval ways. Later in the eighteenth century, the tsarina, Catherine the Great, continued Peter's modernizations and further expanded the empire.

In the early nineteenth century, Napoleon led his armies onto the steppes in a foolhardy effort to conquer Russia. He was horribly defeated, as much by the Russian winter as by the Russian army (a history lesson Adolf Hitler failed to heed in the twentieth century). The last tsar was Nicholas II, who reigned over an increasingly corrupt regime until repeated defeats in World War I led to urban rioting and eventually the Bolshevik Revolution of 1917. The tsar and his family were murdered by the Bolsheviks in 1918, bringing to an end 300 years of Romanoff rule. The Communists, led by Vladimir Lenin and Leon Trotsky, emerged victorious from a violent civil war and by 1920 had complete control over Russia. The Union of Soviet Socialist Republics, or USSR, was born in 1922.

Back in the USSR: The Age of Communism

After Lenin's death in 1924, Communist Party General Secretary Joseph Stalin expanded his power in a brutal manner, eliminating his opponents either by killing them in political purges or sending them to forced labor camps. Millions died of starvation that resulted from the forced collectivization of farmlands. It has been estimated that Stalin was responsible for some 20 million deaths.

As Germany began gobbling up its neighbors in the run up to World War II, Stalin decided to get in on the action and signed a nonaggression pact with his sworn enemy, Nazi Germany, in 1939. With no fear of retaliation, at least for the moment, from Germany, Stalin felt free to invade Poland and Finland and to annex large chunks of Romania and all of the Baltic States of Estonia, Latvia, and Lithuania (see Chapter 14). In 1941, Hitler, true to form, double-crossed Stalin by launching an invasion of the Soviet Union.

Although the Nazis made rapid initial advances across Russia's western plains, like Napoleon, they were slowed by Soviet resistance and the Russian winter. In 1942 and 1943, the tide of the "Great Patriotic War" turned against the Nazis in Stalingrad. Stalin pushed the Red

Terra-Trivia

The Nazis exacted a horrible toll on the Soviet people. Soviet battlefield deaths were estimated at more than 13.6 million. Civilian deaths, including those in labor and concentration camps, totaled another 7.7 million. Almost every family in the Soviet Union lost a loved one in the "Great Patriotic War," as World War II is known in Russia.

Army west with a vengeance, grabbing territory, establishing satellites (politically subordinate countries), and exacting retribution along the way.

After World War II, Stalin made increasingly aggressive moves toward the West, beginning with an attempt to isolate the Allies (Western powers) in west Berlin in 1948. The Allies responded with the Berlin airlift, which delivered massive amounts food and supplies from Western Europe and America to the beleaguered city. In 1949, the Allies established the North Atlantic Treaty Organization (NATO), a mutual defense pact designed to safeguard members from Soviet aggression. The Soviets countered in 1955 by creating the Warsaw Pact (see Chapter 14). Europe was now divided by an "iron curtain," and the Cold War was in full bloom.

During the Cold War years, the Soviet Union engaged in an economically devastating arms race with the West (specifically, the United States). Both sides stockpiled nuclear weapons, and the entire world figuratively held its breath during the Cuban missile crisis of 1962. In 1979, the Soviet invasion of Afghanistan caused U.S.–Soviet relations to hit another low.

The rise to power of Mikhail Gorbachev, who became communist party general secretary in 1985, marked the beginning of a process that within six short years would lead to the demise of the Soviet Union and a new beginning for Russia. In 1987 Gorbachev launched a campaign to modernize Soviet communism through political and economic restructuring (perestroika) and increased openness and personal freedom (glasnost). But instead of modernizing communism, Gorbachev unleashed the forces that would bring it down.

The election of Boris Yeltsin as president of the Russian Republic in 1991 and the failure of a communist-orchestrated coup against him in August of that year led to the disbanding of the communist party after more than 70 years in power and the dissolution of the USSR. On December 21, 1991, the USSR was replaced by the new Commonwealth of Independent States, an economic and security union currently consisting of Russia, now officially known as the Russian Federation, and eleven of the newly independent former Soviet republics.

The Loss of the Former Republics

Just how geographically different is Russia from the Soviet Union of yesterday? This section will help you compare the old and the new. The Union of Soviet Socialist Republics consisted of 14 republics and one Soviet Federated Socialist Republic (Russia), as shown on the following map:

Former Soviet republics.

Armenia

Azerbaijan

Belarus

Estonia

Georgia

Kazakstan

Kyrgyzstan

Latvia

Lithuania

Moldova

Russia

Tajikistan

Turkmenistan

Ukraine

Uzbekistan

Today, all 15 former Soviet republics are independent nations and members of the United Nations. Twelve of them are also members of the Commonwealth of Independent States (CIS). (The former Soviet republics of Estonia, Latvia, and Lithuania refused to join the CIS and are on their way to becoming members of the European Union.)

Before the breakup, the USSR's land area of 8,649,489 square miles made it the largest country in the world. On its own, Russia is still the world's largest country, with a land area of 6,592,800 square miles—more than 11 percent of the earth's total land surface. Nonetheless, Russia lost more than 2 million square miles, or almost a quarter of its former land area.

The losses in terms of population were even more dramatic. The population of the former Soviet Union was about 289,000,000. Russia's population is now 145,000,000, almost a 50 percent drop. It's interesting that Russia's own population has declined in recent years. With a birthrate around 10 per 1,000 and a death rate around 14 per 1,000, Russia is experiencing negative population growth, as are many of the former Soviet republics and satellites. Ethnic Russians represented only about 52 percent of the population of the former USSR. Today, they constitute nearly 82 percent of the population of post-Soviet Russia.

> **Terra-Trivia**
>
> Until 1867, the area that is now Alaska was Russian territory. In a deal driven by then Secretary of State William Seward, the United State purchased Alaska for just $7.2 million. Although critics of the purchase dubbed the new territory "Seward's Folly," Alaska's gold and oil alone have earned back its purchase price countless times over.

Despite the loss of its East European republics, Russia still spans the same 11 time zones that the Soviet Union did. The reason for this is that after the breakup of the Soviet Union, Russia retained a small, fragmented area, called the Kaliningrad Oblast, squeezed between Poland and Lithuania on the Baltic Sea, in what was once part of East Prussia. Kalingrad adds an extra time zone to Russia's 10.

The result of the disintegration of the USSR is that Russia remains the largest country in the world in terms of land area. In terms of population, the USSR was the world's third-largest country (behind China and India). Russia today ranks sixth in the world in population (behind China, India, the United States, Indonesia, and Brazil).

An Unforgiving Climate

All it takes is a quick look at a globe to see just how far north Russia lies: Almost all of it is located above 50 degrees north latitude, roughly equivalent to Canada's latitude. Saint Petersburg is located at about 60 degrees north latitude, placing it less than 7 degrees south of the Arctic Circle and about 2 degrees north of Alaska's capital, Juneau.

Although Russia's latitude parallels Canada's, Russia is colder. Northern Canada encompasses the huge Hudson Bay, which has a moderating effect on climate. The fact that Arctic Canada consists mainly of islands and waterways also helps to temper the climate somewhat. Russia, on the other hand, occupies the extreme north of the world's largest uninterrupted continental landmass, Eurasia, with no large, interspersed body of water to moderate the harsh temperatures.

During the northern Siberian winter, temperatures plummet and the sun never rises above the horizon for months on end. High pressure is indicative of cold, dry air. The world's highest sea-level air pressure was recorded at Agata, Russia, in central Siberia.

With no major body of water to moderate temperatures, Siberia is also subject to tremendous ranges in annual temperature. In parts of central Siberia, the normal January temperature of less than –50 degrees F rises to an average of 50 degrees F in July—an astonishing 100–degree range.

Because the temperatures are so low, the air can hold little moisture. The result is precipitation totals of less than five inches between November 1 and April 30. Although the summer months bring slightly more precipitation, especially in the western parts of Russia, the bulk of Russia is quite dry. As though cold and dry weren't bad enough, much of Russia's northern lands are covered with permafrost.

Permafrost, or permanently frozen subsoil, poses all sorts of problems to permanent structures and transportation lines. If it remains frozen, it provides a hard surface. Beware, however, if it thaws. Road-building equipment, houses, and chunks of road surface have simply disappeared into the quicksand-like ooze of thawed permafrost. Rail lines laid across permafrost and warmed from the friction of train traffic or solar radiation look like they have been strung across a giant washboard.

So there you have it, cold, dry, permafrost—what more could a farmer dream of? Russia is not, of course, agricultural nirvana. Feeding its people has always been a problem. Even the USSR, with the more fertile lands of Belarus, Ukraine, and Moldova, had trouble producing enough food. The scenario in Russia is worse now that these lands are independent and no longer part of the country.

By Land or by Sea

As you might expect in a country as large as Russia, the physical realm, as shown on the following map, includes just about everything (except anything having to do with the tropics). Vast mountains, vast plains, vast forests—Russia is sort of a vast-fest. It also has some of the world's largest lakes and rivers. Russians are a little like Texans: They think that anything having to do with their homeland is the world's largest. And quite often, they are right.

The Land—and Lots of It!

In the West, Russia's European territory is dominated by the Russian Plain, an extension of the Northern European Plain. The northern part of the Russian Plain is primarily woodland, while the southern reaches contain Russia's primary agricultural lands. To the south, the plains terminate between the Black and Caspian seas in the Caucasus Mountains. The eastern edge of the plain (and the end of continental Europe) is marked by the long north–south chain of the Ural Mountains. The Ural chain is a primary source of the fuels and minerals that supply Russia's industry.

Russia—physical features.

Terra-Trivia

Russia possesses large gold reserves and more than 40 percent of the world's platinum. It's also stocked with copper, nickel, iron, and tungsten and is extremely rich in fossil fuels. All this underground wealth, however, comes with a catch—it's located in some of the most inaccessible and inhospitable places on Earth. Extracting, refining, and transporting Russia's wealth is a daunting and expensive task.

Beyond the Urals lies the often frozen expanse of Siberia. The first thousand miles or so consist of the Siberian Lowlands. They're dominated by needleleaf evergreens and tens of thousands of square miles of swamps. East of the lowlands, the terrain begins to rise. First comes the Central Siberian Plateau, and, on Russia's Pacific coast, the mountainous Eastern Highlands. Much of the plateau and highlands are covered with needleleaf deciduous trees called *taiga*.

As inhospitable as this land might seem, it contains great wealth. Beneath the frozen surface lie tremendous energy reserves. Russia has more than 16 percent of the world's known coal reserves (second to the United States, with 23 percent). It's a world leader in uranium reserves, and ranks sixth (or seventh, depending on who's counting) in the world in proven petroleum reserves. Perhaps the most staggering statistic is that Russia has more than one-third the world's total known natural-gas reserves.

Frozen Waters

Russia is bounded by several large bodies of water. Starting in the north and going clockwise from there are the Barents Sea and the frigid Arctic Ocean. To the northeast, Russia is separated from the United States (Alaska) by the Bering Strait. In the east are the Bering Sea, the Sea of Okhotsk, the Sea of Japan, and the Pacific Ocean. To the south there is mostly land, but Russia also has a southern coastline on the Caspian Sea and the Black Sea. In the west, Russia has access to the Baltic Sea via the fragmented bit of territory called Kaliningrad Oblast, as well as a coastline on the Gulf of Finland.

Russia has only a couple of lakes that you need to remember. It borders on the Caspian Sea, which is the world's largest lake in terms of surface area. (Because the Caspian is landlocked, it's considered a lake, even though it's called a sea and contains saltwater.) Lake Baykal, known for its cold, crystal-clear water, is the earth's sixth-largest natural fresh-water lake (in surface area). It is also by far the earth's deepest lake, with a maximum depth of 5,316 feet (more than a mile deep!).

Russia also has its share of major rivers, but the most important of them by far is the Volga. Europe's principal river in terms of length, volume of water, and size of drainage basin, the mighty Volga and its tributaries flow through the heart of the populous Russian Plain before finally draining into the Caspian Sea. But although the Volga might be a big deal in Europe, it is outranked in terms of length, volume of water, and size of drainage basin by four of Russia's Asian rivers. The Volga is 2,194 miles long. Going from west to east, the four longer rivers in Russia are the Ob'–Irtysh (3,362 miles), the Yenisey–Angara (3,100 miles), the Lena (2,734 miles), and the Amur–Shilka (2,744 miles).

People of the Steppes

To understand the distribution of Russia's population, think west and think south. The overwhelming mass of the Russian population lives west of the Ural Mountains on the great Russian Plain, which is also the country's agricultural heartland. The population in this area is concentrated mostly in the southeast and gradually thins as you move to the north. East of the Urals, the huge expanse of Siberia is largely uninhabited. The hardy souls who do subsist in this frozen realm are clustered primarily along its southern reaches.

Geographically Speaking

Though Russia has 11 cities with populations of more than one million, you have to remember only two of them: the capital, Moscow, and the former capital, Saint Petersburg. In addition to being Russia's largest cities, they are also the twin centers of its politics, history, and culture.

Russia's population is more than 80 percent ethnic Russian; the remaining 20 percent includes a variety of other ethnic groups including Tatars and other Turkic peoples such as Chuvash and Bashkri, Ukrainians, Mongolians, Caucasians, Moldovians, and Finno–Ugrians. Although Russian is the official and almost universal language of the land, many citizens who are not ethnic Russians also speak their native language.

Terra-Trivia

Religion in Russia was suppressed during the Communist years. The official "religion" of the state was atheism, and many Russians still claim no religious affiliation. Although the predominantly Muslim Central Asian republics of the old USSR are now independent, Russia still has a large Muslim minority, about 19 percent of the population.

Since the breakup of the Soviet Union, Russia has experienced a tremendous religious resurgence. Most of the growth has occurred in the country's ancient and predominant religion, the Russian Orthodox Church. Russian Orthodoxy has deep roots here and a history that dates back more than 1,000 years. The onion-domed churches that the Soviets converted to museums are now houses of worship once again, and the medieval icons and sacred rituals have new importance in people's lives.

Russia in the Twenty-First Century

Russia today is a country of enormous potential and equally enormous problems. It is a major industrial power, but its factories and plants are outmoded, inefficient, and often decrepit. Russia is immensely rich in minerals and fossil fuels, but extracting them from remote, inhospitable areas can be difficult and costly. In addition, Russia's economy is now overly dependent on exported raw materials, making it highly vulnerable to fluctuations in market prices. Even though it controlled the rich agricultural lands of the Ukraine, the old Soviet Union often had trouble feeding its people. In today's truncated Russia, less than 8 percent of the land is arable and the agricultural sector remains in crisis mode, plagued by a crumbling rural infrastructure and farm privatization that has been slow, difficult, and controversial. Despite the many problems, however, Russian grain production is rising and the terrible Soviet-era food shortages seem to be a thing of the past, at least for now.

Unlike its former satellites in Eastern Europe, Russia was slow to implement free-market reforms and as a result saw its economy contract for five years after independence. After a brief period of growth in 1997, Russia was hard hit by the world financial crisis of 1998, which forced the government to devalue the ruble (the Russian currency) and default on its debts. But Russians have proven themselves to be nothing if not resilient, and today, in the early years of the twenty-first century, the Russian economy is growing once again. Still, the new Russia has many serious hurdles to overcome before it completes its transition to a free-market economy. The long list includes pervasive corruption, organized crime, high levels of poverty, shaky banks, a less than independent judicial system, and an environment scarred by decades of unregulated industrialization and neglect.

CAUTION Eye on the Environment

The Soviet era was an environmental disaster not only for Eastern Europe (see Chapter 14), but perhaps even more so for the Soviet heartland itself. The legacy of the relentless communist drive to industrialize agrarian Russia at all costs is an ugly one: lakes of spilled oil on the fragile tundra, widespread deforestation and soil erosion, air choked by coal-burning factories with scrubberless stacks, and waters fouled by radioactive and chemical spills. Cleaning up its environmental nightmare is a daunting challenge for a cash-strapped Russia, but it must find the way to get the job done if it is to take its rightful place among the fully developed nations of the world.

Perhaps one of the greatest challenges facing Russia today is defining its role in the new post-Cold War world order. Russia is no longer the superpower it was just over a decade ago. Even its space program, the centerpiece of Soviet science and technology, is teetering on the brink of collapse for lack of funds. This loss of power and prestige does not sit well with the Russian people. How they cope with their transition and go about fulfilling their country's vast potential will be a matter of concern over the coming decades not only for Russia and its neighbors, but for the entire world.

Geographically Speaking

On paper at least, economic cooperation between Russia and Japan makes a lot of sense. Japan is a major, if stagnating, economic power with the business savvy and the technical resources Russia lacks, and Russia has the raw materials that Japan lacks. But there's a hitch, of course. At the end of World War II, Stalin grabbed Japan's Kuril Islands. Japan wants the islands back; Russia doesn't want to give them up. This lingering political impasse has kept the two nations from, believe it or not, signing the peace treaty ending World War II and forging an economic alliance that would be so beneficial to both parties.

In the News: Chechnya

For much of the 1990s and into this century, the name Chechnya has been a fixture on newspaper front pages around the world. During the Soviet era, Chechnya was part of the mostly Muslim Chechen-Ingush Autonomous Soviet Socialist Republic located in the Caucasus Mountains. The Chechens suffered bitterly at the hands of Stalin, who deported thousands of Chechens to Siberia at the end of World War II for siding with Germany against the Soviets. When the Soviet Union collapsed in 1991, Chechnya declared itself independent, a move that then Russian president Boris Yeltsin refused to recognize. Tensions escalated into violence, Russian troops were sent to Chechnya, and a bloody war broke out in 1994. Two years, and an estimated 70,000 casualties later, Yeltsin agreed to a ceasefire and pulled out his troops from what was now an essentially autonomous Chechnya. Although this was perceived as a victory for Chechnya and a humiliation for the Russian army, the war resolved nothing.

In 1999, with Chechnya falling into anarchy, the Russian-Chechen conflict took on a new and more dangerous turn. Taking their cue from other radical Islamic groups, Chechens began moving the battle into the Russian heartland through a series of terrorist bombings in Moscow and other cities. Chechen forces invaded neighboring Dagestan in support of Islamic fundamentalists there, prompting Russian Prime Minister, later President, Vladimir Putin, to launch a massive Russian invasion, which he labeled an "anti-terrorist operation." Despite claims that the military phase of the Chechen conflict is over, the situation in the separatist republic remains highly volatile. Presidential elections held in Chechnya in September 2003 were widely criticized for irregularities and fraud.

The Least You Need to Know

- Russia's emperors were called tsars. The last tsar was overthrown in the communist revolution of 1917.

- The USSR's 15 former republics have all spun off and become separate countries (including Russia).

- Russia is the world's largest country in area and sixth largest in population; Moscow and Saint Petersburg are its largest and most important cities.

- Its high latitudes give Russia an extremely cold climate.

- Most of the region's people speak Russian and are Russian Orthodox.

- Russia is in the process of making a slow and difficult transition from a rigid state-controlled economy to a free-market economy.

Chapter 16

China: The Awakened Giant

In This Chapter

- ◆ Grasping the vastness of China
- ◆ A walk through China's history
- ◆ China today: Calculating the pluses and minuses
- ◆ A return to forgotten Mongolia
- ◆ North Korea: East Asian flashpoint

One of the most difficult decisions involved in organizing the regional sections of this book by level of economic development was where to place China. In many ways, this huge and hugely important country is still a developing nation. In terms of poverty and unemployment rates, living standards, environmental conditions, and percentage of the work force involved in agriculture, China has a long way to go to catch up with the developed nations of the world.

On the other hand, China's gross domestic product has now surpassed that of Japan, making it the world's second largest economy after the United States. Like the other transition countries covered in this part of the book, China is undergoing a major transformation from a rigidly state-controlled economy to a more open, Western-style free-market economy. The difference between China and the former Soviet republics and satellites is that

China politically, if not economically, is still a rigidly controlled Communist state. Whether or not China can maintain its phenomenal economic growth in a politically repressive environment is one of the great unanswered questions for the coming decade.

By dint of its sheer size and dominance, China is the primary focus of this chapter. Also covered, however, are Mongolia, China's remote and physically isolated neighbor to the north, and North Korea, China's politically isolated and increasingly dangerous communist neighbor to the northeast.

Asia's Megastate: The People's Republic of China

However you look at it, China is huge. The world's third-largest country in area (behind only Russia and Canada), China occupies 3.7 million square miles of Eastern Asia. It has more than 3,400 offshore islands, Hainan being the largest of them. With almost 1.3 billion people, China is the world's most populous country. Fifty of its cities have more than a million people apiece; nearly 1 out of every 5 people on Earth is Chinese.

Although everyone knows where China is, few could name its neighboring countries and surrounding bodies of water, at least not without looking at a map. To the north are Mongolia and Russia; to the northeast is North Korea; to the east are the Yellow and East China seas; to the south is the South China Sea, Vietnam, Laos, Myanmar, India, Bhutan, and Nepal; to the west, China touches Pakistan, Afghanistan, and Tajikistan; to the northwest are Kyrgyzstan and Kazakhstan. Whew!

China is so extensive that it includes almost every type of climate, every degree of rainfall, and every topographical feature imaginable. Its surface is 43 percent mountains, 26 percent plateaus, 19 percent basins, and 12 percent plains. The only way to get a handle on such diversity is to break the country up into smaller areas and describe each one in detail. To simplify matters, this chapter divides China into north-central, southern, northeastern, northwestern, and southwestern sections. (The southwestern section is composed primarily of Tibet.) We'll start with the north-central section because that's where China began.

In the Beginning: North-Central China

The ancient cradle of Chinese civilization is the 2,900-mile Huang He (Yellow River) basin. It is here that what is thought to have been China's first dynasty, the Xia dynasty, developed by the year 2000 B.C.E. In its lower reaches, the Huang River flows through the fertile and densely populated North China Plain, one of the country's most important agricultural regions. Located in the northeastern corner of the plain are China's capital, Beijing, home to some 6.6 million people, and the important industrial city and commercial port of Tianjin, with another 4.3 million people. Surrounding these two megacities is a major industrial and oil-drilling area.

The last Xia ruler was overthrown by the founder of the Shang dynasty in about 1700 B.C.E. During the period of Shang rule, which lasted until 1027 B.C.E., China developed a writing system and the technique of bronze metalworking. The Shang were followed by China's longest-lasting dynasty, the Zhou dynasty, which ruled for more than 800 years.

China and its neighbors.

The short-lived Qin (pronounced Chin) dynasty (221–207 B.C.E.) is famous for several achievements: It gave China the name by which it is known to the rest of the world; it created the model for imperial rule that would last in China until 1912; and it standardized China's writing system and legal codes. But perhaps most famously, the Qin dynasty consolidated and extended existing defensive walls into a 3000-mile-long complex now called the Great Wall of China. The wall provided central China with at least psychological protection from the Mongolian hordes to the north. Although the wall ultimately failed to prevent future Mongol invasions, the vast resources and manpower required to build it led to popular revolts that helped topple the Qin dynasty in 207 B.C.E.

Terra-Trivia

Around 500 B.C.E., China's greatest philosopher, Confucius, began to work as an itinerant teacher and soon became known for his learning and character. He deplored the breakdown of political institutions in his day and urged a return to ancient concepts of morality. Practical and ethical rather than religious in nature, the teachings of Confucius still exert a powerful influence on Chinese society.

Under the militaristic Han dynasty. China expanded westward, almost to its current northwestern border, and annexed parts of northern Korean and northern Vietnam. China's westward expansion made it possible to secure the "Silk Route," the series of caravan trails across central Asia that enabled China to export silk to the Roman Empire via Middle Eastern ports. Technological advances during the Han period included the invention of paper and porcelain. So great was the impact of the Han dynasty on Chinese life and history that China's majority ethnic group refers to itself as the "people of Han." Mounting political corruption and the sheer burden of managing so vast and complex an empire caused the Han dynasty to collapse by 220 C.E.

China's Furnace: The Southern Section

South of the Northern Plain is a varied landscape of mountains, plateaus, hills, and river basins, dominated by the 3,900-mile-long Chiang Jiang (Yangtze River), China's (and Asia's) longest river and the third longest in the world. The Chian Jiang basin is China's major rice- and tea-growing region and home to about a third of its population. The Chiang Jiang drains into the East China Sea near Shanghai, China's most populous city, home to about 9 million people in the city proper and as many as 12 million in its metropolitan area. Shanghai is also China's largest port and a commercial and financial center now rivaling Hong Kong in importance.

Upriver from Shanghai is the city of Nanjing, which became China's capital in 317 C.E., during the period of civil war and political upheaval that followed the end of the Han dynasty. This so-called Era of Disunity lasted nearly 400 years until the Sui dynasty reunified China in 589 C.E. Like the earlier Qin dynasty, the Sui dynasty came to a quick end, after only 36 years, thanks in part to another monumental, and costly, engineering project: the completion of the Grand Canal linking Beijing to the south. The busy artificial waterway is the oldest and longest canal in the world.

The following Tang dynasty, which lasted nearly 300 years, from 618 to 907, was a cultural highpoint in the history of China. The arts flourished, Buddhism became fully Chinese, and block printing was invented. Military defeat at the hands of Arabs in Central Asia in 751, however, initiated a long period of fragmentation and decline. The Song dynasty (960–1279) reunified much of China and presided over a revival of Confucianism, an expansion of trade and commerce, and the development of cities as commercial centers. The Song emperors also set the pattern for a centralized civil service run by scholar-officials. In the latter part of the period (1127–1279), however, the Song emperors were unable to resist Mongol invaders from the north and were forced to abandon northern China.

Geographically Speaking

On a tributary of the Zhu Jiang (Pearl River) in China's southern Guangxi Province is the favorite tourist destination of Guilin, famous for the limestone pinnacles that jut up dramatically from an otherwise flat terrain. Part of a geologic phenomenon known as *kharst* topography, the pinnacles are created by flowing subsurface water that cuts through existing limestone forming underground caverns and sinkholes. As caverns collapse and more sinkholes are formed, the remaining exposed limestone rock takes the shape of sharp pinnacles. Long regarded as the most scenic spot in China, Guilin has been inspiring poets and artists for centuries.

China's deep south is known as the "furnace" for its extreme heat and humidity. The Zhu Jiang (Pearl River) basin area is tropical in all respects: heavy rainfall, high temperatures, and lush vegetation. Two, sometimes even three, rice crops are harvested in a year here. The Zhu Jiang flows through Guangdong Province, birthplace of China's modern-day economic miracle, before draining into the South China Sea. Located in the mighty Zhu Jiang delta are Hong Kong, the economic powerhouse and former British colony now restored to China (see Chapter 12); Macao, the former Portuguese colony now also part of China once again; and Shenzhen's Special Economic Zone, one of China's first experiments in modernizing its economy. At the north end of the Zhu Jiang delta is the great trading metropolis of Guangzhou (Canton). Capital of Guangdong Province and China's southern gateway, Guangzhou has a population of about 6.6 million people. China's largest island and smallest province, Hainan, lies just off Guangdong Province's southwestern Leizhou Peninsula.

China—physical features.

The Top Half: Northeastern and Northwestern China

Northeastern China is half of historic Manchuria (which used to include Mongolia). This extensive area, consisting of a large central plain bordered on the east and west by mountain chains, is China's "Siberia": It's cold and has a short growing season. The great Manchu dynasty originated in this area.

Northwestern China comprises three large desert basins, each surrounded by great mountain chains. The entire area is sparsely settled. Aside from the Silk Route, not much else has passed through this area.

As you've seen, the Great Wall turned out to be not so great after all—it certainly didn't prevent Mongol invaders from conquering China in the mid-thirteenth century. Kublai Khan, grandson of Genghis Khan, defeated the Southern Song and established the first foreign dynasty to rule all of China. The Mongol Yuan Dynasty lasted less than a century (1280–1368), but during this time, the city of Beijing was rebuilt and contacts and cultural exchanges with Europe multiplied. It was also during the Yuan Dynasty that many Chinese in the western part of the country converted to Islam. Yuan rulers discriminated against their Han subjects, and a combination of popular discontent, natural disasters, and infighting within the imperial family led to the fall of the dynasty and the restoration of Chinese rule in 1368 under the Ming dynasty.

The Ming ruled for three centuries, during which time they contained the Mongol threat to the north, reconquered northern Vietnam, and sent maritime expeditions as far away as the east coast of Africa and, at least according to some theories, possibly all the way to the Americas (see Chapter 6). In the mid-1600s, the weakened Ming dynasty gave way to another group of foreign invaders, this time coming out of Manchuria. Under the militaristic Manchu, or Qing, dynasty, China reached its greatest extent. Outer Mongolia was added to the empire in the late seventeenth century. A protectorate was established over Tibet in the eighteenth century. The island of Taiwan became part of the empire for the first time under the Qing.

Terra-Trivia

It was during the Yuan Dynasty that the Venetian adventurer Marco Polo visited China (or Cathay, as the Europeans called it). He traveled to China twice and crisscrossed the country six times (including once across the Gobi Desert). During most of his 26 years in Asia, Marco Polo was employed by the great emperor Kublai Kahn. He even served as governor of a Chinese city for three years!

Unlike the Mongols, the Manchus had been greatly influenced by Chinese culture even before the conquest, and as rulers, retained many Chinese governing institutions and court customs. Qing rule, however, was also despotic and discriminatory against the majority Han population. Popular resentment was never too far below the surface. But it was the Qing dynasty's own corruption and its inability to deal with the challenges posed by encroaching Western powers throughout the nineteenth century that led to its overthrow in 1911 and the end of China's millennia-old empire.

Ancient Tibet: China's Reluctant Southwest

Tibet (or Xizang, its Chinese name) is the largest, least populated (2.2 million people), and most isolated of China's provinces (it's actually an autonomous region rather than a province). The Tibetan capital is Lhasa, renowned for the Potala Palace, the traditional home of the Dalai Lama. Tibet is located in southwestern China, just northeast of India. It is called the "roof of the world" and with good reason: With an average elevation of more than 16,000 feet, it is the highest region on Earth. Tibet's central plateau (also the world's highest with an average elevation of 15,000 feet) is surrounded by mountains to the north and west and by the mighty Himalayan Mountains to the south. Tibet's share of the Himalayas includes Mount Everest, the world's highest peak at 29,028 feet, which straddles the Tibet-India border.

Tibet's plateau is Asia's principal watershed. It's the source of the Huang (Yellow), Chang Jiang (Yangtze), Mekong, Ganges, Indus, and Brahmaputra rivers. Tibet is semiarid (it receives only 15 inches of precipitation annually) and is generally cold and windy. Its people are mostly Tibetan (although a sizable Chinese minority lives there) and speak Tibetan. They are strongly Lamaist (Tibetan Buddhist), although China encourages atheism. Most Tibetans survive through herding and subsistence farming.

Terra-Trivia

The Dalai Lama, the spiritual and temporal leader of Tibet and Tibetan Buddhists across the world, fled Tibet in 1959 when he was in his mid-20s. He now lives in exile in India but lectures all over the world in nonviolent opposition to Chinese rule in Tibet (for which he received the 1989 Nobel Prize for peace).

Tibet's history is obscure until the seventh century, when Buddhism was introduced from India and Tibet emerged as a regional power. After a period of expansion that included a successful invasion of China, the Tibetan kingdom disintegrated in the mid-ninth century. In the confusion and strife that followed, the influence of Buddhism began to recede, only to reassert itself in the tenth and eleventh centuries. New schools of Tibetan Buddhism sprang up, and the country became a primary focal point of the faith.

The Mongols invaded Tibet in the mid-thirteenth century and made Tibetan Buddhism the state religion of their empire. After the fall of the Mongol empire a century later, Tibet was ruled by its religious leaders, the lamas. China made Tibet a protectorate in the early eighteenth century and exerted varying degrees of influence or domination for the next 200 years. Britain and Russia also began setting their sights on Tibet in the nineteenth century. Britain invaded Tibet in 1904, but signed an accord with China two years later recognizing Chinese sovereignty over Tibet.

After the fall of the Manchus, Tibet enjoyed a period of independence that lasted until 1950, when China's communist leader Mao Zedong ordered an invasion of the country. The Chinese communists set about transforming Tibet along Marxist lines, often with disastrous consequences for the Tibetan people and their culture. A failed uprising in 1959 forced the Dalai Lama and thousands of his followers to flee to India. Since then, negotiations for his return to Tibet have led nowhere. The Tibetan people's aspirations for independence continue to be suppressed by China, despite the sympathy and support of a prominent and vocal international "Free Tibet" movement.

The Birth of Modern China

The isolationist policies of the late Qing dynasty meant that China missed the industrial revolution that transformed the West in the late eighteenth and nineteenth centuries. An economically backward China was not prepared to fend to off ambitious Western powers that were eager not only to trade with China but to encroach on its sovereignty and territory. Britain, France, the United States, and other nations established "spheres of influence" on Chinese soil. In the course of the nineteenth century, a series of wars and uprisings, including two Opium Wars with Britain and the anti-foreigner Boxer Rebellion of 1900, whittled away the power and autonomy of the Qing dynasty. China also lost possessions to foreign powers: Japan took the Ryukyu Islands and, later, Taiwan; France took Vietnam; Britain took Burma (now Myanmar); Russia took Manchuria and Northern Mongolia; and North Korea gained its independence. By 1911, China's last dynasty was ripe for overthrow by the republican revolution of Sun Yat-sen.

Terra-Trivia

Imperial China left behind a rich legacy of poetry, music, drama, dance, opera, and art. Many of the emperors' treasures are on display at the National Palace Museum in Taipei, Taiwan. No tourist lucky enough to have visited China, however, can ever forget the Great Wall or the imperial palace compound in Beijing called the Forbidden City (forbidden to commoners, that is).

The Republic of China

Sun Yat-sen became the first president of the Provisional Chinese Republic in 1912, but it wasn't until 1928, after years of civil war between rival warlords and various republican factions, that nationalist leader General Chiang Kai-shek was able to unite most of China under a right-wing nationalist government. Civil strife between nationalists and communists continued through the 1930s. While Chiang consolidated his regime, communist organizer and theoretician Mao Zedong increased his control over the Communist Party. In 1934, after losing several battles to the Nationalists, Mao led communist forces and party officials on the famous 6,000-mile Long March back to north China to regroup. At the end of the ordeal, Mao was the unquestioned leader of the Chinese communists.

In 1931, Japan had conquered Manchuria and set up a puppet regime there. Six years later, Japanese forces attacked China from Manchuria, and within a year had most of northeastern China under its cruel control. Facing a common enemy, the nationalists and communists joined forces to fight the Japanese. With the defeat of Japan at the end of World War II, however, the civil war resumed, but this time it was Mao's turn to take control of China. In 1949, he proclaimed the new People's Republic of China. Chiang was on the run, and the nationalists withdrew to Taiwan.

The People's Republic of China: The First Half-Century

Under Mao, China's communist government strictly controlled all aspects of its citizens' lives. Religious expression was suppressed and a cult of personality developed around Mao. Millions died in misguided attempts at reengineering society, such as the "Great Leap Forward" of 1958 and especially the Cultural Revolution of 1966 to 1969. During this three-year reign of terror, brigades of Red Guards were unleashed on the country to wipe out all traces of bourgeois ways and enforce "Mao thought" and ideological purity.

U.S. President Richard Nixon's visit to China in 1972 was the beginning of an often-difficult reconciliation between the two countries that culminated in the establishment of full diplomatic relations in 1979. After Mao's death in 1976, Deng Xiaoping eventually assumed the leadership of the country and introduced sweeping economic reforms in 1978 that helped turn China into the economic giant it is today. China continues to relax economic controls, but political controls are another matter. The violent suppression of the pro-democracy Tiananmen Square demonstrations by armored People's Liberation Army troops in 1989 drew the indignation of the world on China and continues to cast a shadow over China's achievements in the economic sphere.

China Today

China has the potential to outdo anything it has done in its remarkable past. Remember that because China is so big, whatever it does, it does in a big way!

The Economic Miracle

Since introducing market reforms in 1978, China's gross domestic product (GDP) has quadrupled, and is now second only to that of the United States. China's enormous population, however, means that its per capita GDP is considerably less robust. In 2002, it was on the same level as Paraguay and El Salvador and just above Albania. Still, China's potential as a producer of goods and services and as a market for the rest of the world's goods and services has barely been tapped.

Over the past several years China's economic growth rates have been in the extraordinary 7 to 8 percent range. Even factoring in a tendency to exaggerate such figures,

China has maintained an enviable growth rate, despite regional financial crises and global slowdowns. Perhaps even more impressive is that China's growth has been in both the industrial and the agricultural sectors, the latter being of particular importance to a country historically prone to devastating famines. China's admission to the World Trade Organization in 2001 was not only an acknowledgement of its economic reforms, but also a boost to its growth and an incentive to continue liberalizing not only its economy, but its government and legal system as well.

On the energy front, China has vast coal deposits and is well endowed with petroleum, natural gas, and oil shale. It also has the potential to be the world's biggest producer of hydroelectric power. China is known to have reserves of 153 different minerals and many of them in abundant quantities. Its supplies of iron, copper, aluminum, tin, zinc, lead, and others are among the biggest in the world.

Eye on the Environment

China's fabled Chiang Jang, better known in the West as the Yangtze River, is both a national lifeline and a flood-prone killer. More than a million people are estimated to have died in Yangtze floods over the past 100 years. Now, in what has been called the nation's largest public works project since the Great Wall, China proposes to tame the mighty Yangtze with what will be, when it's completed in 2009, the world's largest hydroelectic dam, the controversial Three Gorges Dam.

Rising amid the Three Gorges canyons, one of China's scenic wonders, the mile and a half wide dam will create a reservoir hundreds of feet deep and more than 400 miles long that will enable ocean-going freighters to penetrate into central Sichuan province, some 1,500 miles inland from the coast. The dam's 26 turbines are expected to generate one-ninth of China's electrical power. (In contrast, the Itaipu Dam on the Paraná River between Brazil and Paraguay, currently the world's largest hydroelectric dam, has only 18 turbines.)

The monumental project, however, is not without a serious environmental and social downside. The reservoir is expected to flood nearly 400 square miles of territory, including some of China's most fertile farmland, precious archeological sites, and highly polluted industrial sites. By the time it's all over, the Three Gorges Dam project will have caused between one and two million people to relocate to higher ground.

The Challenges

Where to begin? Perhaps one should start with the people or, more to the point, their numbers. As has been noted many times before, China's population is huge. By stressing (and sometimes unacceptably enforcing) one-child families, late marriages, and all kinds of birth control, China has reduced its population growth rate to a more manageable 0.6 percent. But even that figure, multiplied by the country's population of 1.3 million, translates into an increase of nearly 8 million people per year—plenty of hands to do the work, but also plenty of mouths to feed, bodies to house, and children to educate.

Which brings us to the second major challenge—food. Some of the worst famines in human history have occurred in China. And although those days are mercifully in the past, feeding such a huge population is still a major undertaking. China has more acres of farmlands, forests, and grasslands than almost any other country on Earth, yet its per capita acreage is about a third of the world's average—a worrisome statistic in a country whose major rivers are particularly prone to devastating floods and that is losing significant amounts of arable land to erosion and development. According to some estimates, China has lost as much as one-fifth of its farmland since 1949.

Although China continues to modernize and privatize its economy at a remarkable pace, it has not been a smooth, continuous process. China's leadership has a tendency to backtrack periodically, reimposing centralized controls. In many ways, China suffers the worst of both its systems: the inefficiency and bureaucracy of the state controlled sector and the sudden windfalls and income disparities of unregulated capitalism. Unemployment and underemployment are huge, if somewhat concealed, problems in China, especially in rural areas. In 2002, the urban unemployment rate was 10 percent, but this does not include the millions who have already been sent home (at reduced wages) from closed or downsized state factories. The official figure also doesn't include members of the "floating" migrant population who have left the private rural sector in search of either work or a better life.

And finally, like Russia, China still needs to fully define its position in the world today. Modern China lives and dies by trade and it needs a cooperative global environment to stoke its massive economic engine. Retreating into hostile isolation is not an option any longer, neither is baiting its major trading partners. China has a host of international issues to resolve. The Chinese government is still under international scrutiny for the Tiananmen Square attacks. And the world is watching closely to see how China manages its new relationship with Hong Kong, whether—and how—China follows up on its promise to take back Taiwan, and what steps China will take to diffuse the ticking time bomb across its border in North Korea.

A Quick Look at China's People

About 91 percent of China's population is ethnic Han Chinese. China's 55 other minority groups make up the remaining 9 percent. The most numerous of these groups are the Zhuang, Hui, Uygur, Yi, Miao, Manchu, Tibetans, Mongolians, Buyi, and Koreans. Mandarin and Cantonese are the most commonly spoken of the dozen or so Chinese dialects. For 3,000 years, China has had only one written language, which not only eases communication but also contributes immeasurably to a common culture. Both internally and through the United Nations, China is pushing the use of Pinyin, a system of phonetic spelling, for personal and place-names (Beijing for Peking and Mao Zedong for Mao Tse-tung, for example).

The communist government officially discourages religion and has periodically suppressed various beliefs and denominations. The most recent example was the campaign against Falun Gong, an extremely popular, and apolitical, religious movement that combines elements of traditional Chinese religions with martial arts. After the traditional Chinese religions, the most common religions in China are Christianity and Islam.

> **Terra-Trivia**
>
> Most people in China who are religious practice the "traditional Chinese religions" or the Three Teachings, a mix of Confucianism, Taoism, and Buddhism. Taoism, unique to China, centers around the teachings of Lao-tzu, a sixth-century B.C.E. mystic who espoused the *Tao,* or "the Way." In many ways, the Chinese belief system is more philosophical than religious.

The Forgotten Republic of Mongolia

Mongolia occupies an isolated and desolate spot in central Asia. (The last time Mongolia made front-page news was in the thirteenth century.) This lonely place is bounded on the north by Russia and on the east, south, and west by China. Though Mongolia is a large country, it has a population of only 2.7 million people. With a density of only 4 people per square mile, Mongolia is the world's most sparsely populated country. Its capital is Ulan Bator, with a population of about 800,000.

The country's most productive region is its northern plateau. To the west are the Altai Mountains. In the south is Mongolia's most prominent feature, the Gobi Desert. The Gobi—the coldest, northernmost desert in the world—covers more than 500,000 square miles. This dry, rocky, foreboding place does not conform to the classic desert image: Only 5 percent is covered by sand dunes.

Mongolia's climate is extreme—extremely cold in winter and extremely hot in summer. (Because it's in central Asia, no nearby large water body moderates climate extremes.) Most Mongolian people are rural nomadic herders living in yurts (felt tents that protect against the intense cold and heat). Mongolia is poor: Its annual gross domestic product per capita is only $1,770, which puts in the same bracket as the Gambia and just above Bangladesh and Haiti.

The country's history, however, is fascinating. In the thirteenth century, Genghis Khan and his sons and grandsons (most notably Kublai) forged the Mongol empire, history's largest land empire, consisting of China, India, Eastern Europe, Russia, most of the Middle East, and all of central Asia. The Mongols, fierce and devastating in war, were known as superb horsemen and excellent archers. Strong discipline and leaders with brilliant military strategies led to unrivaled military successes.

GeoJargon

The Mongol invaders are also known by other names that have lived on in infamy. At various times and in different places, the Mongols were referred to as the Golden Horde and the Horde from hell.

At the height of their rule, they lost only once on land—to Burmese elephants that sent the Mongol horses fleeing in fear. They always lost at sea—twice while trying to invade Japan and once while trying to conquer Java. In the fourteenth century, the empire fragmented, and its weakened armies were driven back to their Mongolian plateau. The great Mongol empire was history.

Mongolia's subsequent history has been somewhat less dramatic. The Mongol homeland became part of Manchu China in the late seventeenth century. After the fall of the Manchu Qing dynasty in 1912, what was then known as Outer Mongolia became an independent Buddhist theocracy. With the help of Soviet forces, Mongolian revolutionaries took Ulaan Bator in 1921, and in 1924 declared a Mongolian People's Republic, the world's second communist nation. As the Soviet Union was falling apart in 1990, pro-democracy voices were also making themselves heard in Mongolia. In 1996, the Mongols elected their first noncommunist government since 1921. But former communist party members maintained their grip on the nation's affairs and slowed the pace of reform. In the 2001 elections, the former communist party took control of both the presidency and Parliament, although promising to implement much needed reforms.

The Democratic People's Republic of North Korea

North Korea occupies the upper half of the Asian peninsula just west of Japan. It is home to about 22 million people, all of whom speak Korean. North Korea's climate is temperate and its terrain mountainous. Its capital is Pyongyang, a city of more than 3 million people. North Korea is a Marxist, government-controlled state that makes China look liberal! For a brief summary of the history of the Korean peninsula before its division, see the section on South Korea in Chapter 12.

As noted in Chapter 12, at the end of World War II, the Korean peninsula was divided at the 38th parallel, with the northern part going to the Soviet Union and the southern part to the United States. In 1948, the northern half became the communist Democratic People's Republic (North Korea), and the southern half became the capitalist Republic of Korea (South Korea). Two years later, North Korea invaded South Korea and attempted a military unification. With help from the United States, North Korea was pushed back almost to the Chinese border; with Chinese help, South Korea was pushed back to the 38th parallel, where a no-man's land was set up in the 1953 truce.

While South Korea was developing into one of East Asia's economic Tigers, North Korea was racing downhill economically. The nation's founder, the despotic Kim Il-song, and his son and successor in 1994, Kim Chong-il, seemed determined to drive North Korea into poverty and starvation, while maintaining one of the largest standing armies in the world. After decades of economic mismanagement, North Korea's industrial base is a shambles and the country is heavily dependent on foreign food aid. A four-year famine in the late 1990s is thought to have claimed as many as 3 million lives. Although the food situation has eased in recent years, the North Korean population is still subject to malnutrition and deteriorating living standards.

In the News: The North Korean Nuclear Threat

Tension has existed between the two Koreas since the end of the Korean War, but has escalated dramatically since U.S. President George W. Bush, in his 2002 State of the Union message, lumped North Korea with Iraq and Iran in the "axis of evil." With the United States threatening war against Iraq to dismantle a "weapons of mass destruction" (WMD) program that might or might not have existed, North Korea admitted in October 2002 that it had violated a 1994 agreement to freeze its nuclear-weapons program and had in fact been developing WMD. Later that year, North Korea expelled United Nations weapons inspectors, and in January of 2003 withdrew from the Nuclear Non-Proliferation Treaty. Much delayed multi-lateral talks among the two Koreas, Japan, China, and the United States were held in Beijing in August 2003 but failed to produce a breakthrough. It's possible that North Korea's nuclear-arms waving is just a crude ploy to acquire much needed international assistance. But with Japan in range of North Korean missiles and with U.S. troops stationed just across the border in South Korea, North Korea's nuclear weapons are more than just a Korean problem.

The Least You Need to Know

◆ With 1.3 billion people (one-fifth the world's population), China is the world's most populous country and the third-largest country in area.

◆ China has the world's second largest economy. Since the introduction of economic reforms in 1978, China has experienced tremendous economic growth.

◆ Unlike Russia and the former Soviet satellites in Eastern Europe, China is still a communist country politically. It has eased controls on its economy, but has not liberalized its government.

◆ Although Tibet has a distinct people, language, and religion and wants to regain independence, China won't give up its control.

◆ Mongolia is a remote and isolated country of nomads and herders.

◆ Staunchly communist North Korea lags significantly behind South Korea economically, and its nuclear weapons program is a source of global tension.

Part 4

A Regional Look: The Developing World

By definition, the developing countries are less wealthy and less technologically advanced than the developed countries (Part 2) and most of the transitioning countries (Part 3). Yet such distinctions can be deceiving. Many of the countries in this group are indeed desperately poor, verging on hopeless. But some are actually on the brink of joining the economic A-list, while others have enormous potential, thanks to their as yet untapped natural and human resources. It's also true that some very poor countries are extremely rich culturally and/or spiritually. And of course, some of the most beautiful places on earth are to be found in the developing world.

Chapter 17

Middle America: Bridging Two Continents

In This Chapter

- ◆ The Middle American bridge to the past

- ◆ The mainland connection

- ◆ Mexico: Middle America's heavy hitter

- ◆ Central America's turbulent past

- ◆ A cruise through the enchanted Caribbean islands

As you saw in Chapter 6, the continent of North America includes two regions: developed Northern America (Canada and the United States) and developing Middle America (Mexico, Central America, and the islands of the Caribbean). Geographically, Middle America serves as a bridge between the rest of North America and South America, two giants that are both connected and separated by Middle America. Actually, as you can see in the following map, Middle America is a dual bridge: the funnel-shaped mainland in the west (this is what physical geographers refer to as a *land bridge*) and the arc of islands in the east. The twin spans of the Middle American bridge are separated by the Gulf of Mexico and the Caribbean Sea.

Middle America has been the stage for mighty empires, bold conquests, and many political upheavals. This chapter describes the geographic and historical importance of a region that is not just a bridge between continents but a significant destination in and of itself.

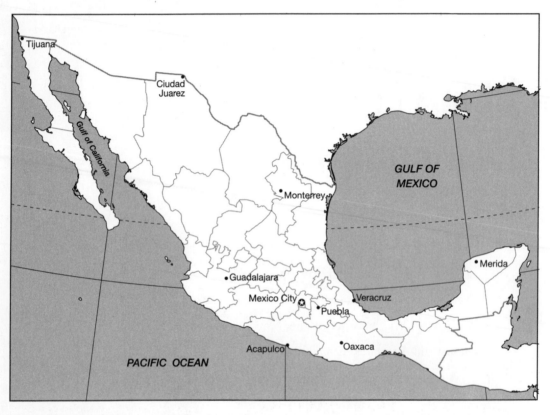

Middle America.

Empires of the Past

Mexico in Middle America and Peru in South America were the main hubs of Spain's vast New World empire. But long before Christopher Columbus arrived in the Caribbean and set in motion the Spanish conquest of the region, Middle America had given rise to advanced and powerful Indian civilizations, whose ruined monuments today are among the world's most spectacular archaeological sites.

The Amazing Maya

Of the great Middle American civilizations, the Mayans stand out for the magnificence of their architecture and the sophistication of their science and art. The Mayan civilization developed in the lowlands of what are now Honduras, Guatemala, Belize, and the Yucatán Peninsula in southeastern Mexico. At its peak, the Mayan population was between two and three million people. Although no one can be certain about exactly when Mayan civilization began, it dates back some 3,000 years and reached its greatest heights in the classic period between 250 and 900 C.E.

Unlike the later Aztec Empire, Mayan civilization was based on autonomous city-states with their own kings and ruling classes. Agriculture was the basis of the Mayan economy. Maize (corn) was the principal crop, but they also grew many other crops, including beans, squash, chilis, tomatoes, avocados, and cotton. The Maya were also known for their fine cloth, pottery, jewelry, and, perhaps their most remarkable accomplishment, the Mayan calendar. It was the most accurate calendar in the world until the advent of the Gregorian calendar that is now the global standard.

The ruins of the great Mayan cities and ceremonial centers, some of them amazingly well preserved, are the most visible and spectacular legacy of the Maya. These centers, like others in ancient Mexico, consisted of vast open plazas surrounded by palaces, ritual ball courts, astronomical observatories, and pyramidal mounds surmounted by temples.

Throughout their ancient history, the Maya would suddenly abandon a city for reasons that remain mysterious. This process began to accelerate at the end of the classic period, around 900 C.E., when the southern Mayan cities began to collapse. The northern Mayan cities of the Yucatán continued to prosper until they, too, began to be abandoned in the thirteenth century.

Population pressures, drought, disease, uprisings, and invasion have all been proposed as likely causes for the decline of Mayan cities. Despite their decline, however, the Maya put up fierce resistance to the Spanish invaders when they arrived in Mayan lands in 1517. Today, the Yucatán, Guatemala, and Belize are still the Mayan heartland, where descendants of the ancient Maya preserve their cultural and tribal traditions, with large numbers of them still speaking Mayan dialects rather than Spanish.

> **Terra-Trivia**
>
> Maize was of vital importance to the Maya and to all the peoples of the Middle American mainland. In addition to being their chief food crop, corn was believed to be the material out of which the gods created humankind.

The Mighty Aztecs

About the time that the Mayan civilization had reached its peak, the civilization of the Toltecs was beginning to form around present-day Mexico City. The warlike Toltecs extended their domination south to the Yucatán Peninsula, where they conquered the declining Maya. The Toltec empire, however, was relatively short-lived; by 1200 C.E., it was in decay.

The Toltecs were eventually supplanted as the overlords of central Mexico by an equally warlike Amerindian group from the north, called the Aztecs. In 1325, the Aztecs founded a settlement named Tenochtitlán on an island in one of the many lakes that dotted central Mexico. Tenochtitlán, today's Mexico City, would become the capital of the mighty Aztec empire and not only the greatest city in prehispanic America but also one of the largest in the world at the time of the Spanish conquest in 1521.

Terra-Trivia

The Aztecs were also known as the Mexica (pronounced "me-shee-ka"), a name that was later applied to the capital of the Spanish vice-royalty of New Spain and eventually to the entire country after it gained its independence from Spain. "Mexico" is also the name of the small state that almost completely surrounds the Federal District that contains the Mexican capital.

The Aztec religion called for human sacrifice and even cannibalism. As their empire expanded, the Aztecs acquired slaves to work their fields and satisfy the seemingly insatiable needs of their gods for human sacrifice. The peoples conquered by the Aztecs naturally grew to fear and hate their cruel masters.

The Spanish Empire in North America

The Spanish conquistador Hernán Cortés capitalized on the hatred felt for the Aztecs by their subject peoples when he invaded Aztec territory in 1519. Anti-Aztec Indian tribes joined Cortés' army as he set out to overthrow the Aztec king, Montezuma II. Although the Aztecs were ferocious warriors, they were no match for Cortés' horses, cannon, and armor, or for the diseases that the Spanish armies brought with them and against which the Indians had no defenses. Tenochtitlán fell in 1521, and on its ruins Cortés began the construction of the great capital of New Spain, Mexico City.

The Spanish imposed Catholicism and a harsh quasi-feudal economy on the defeated Indians of Middle America. They also extracted massive amounts of wealth from the silver mines of Mexico and from the forced labor of the indigenous population throughout the region. With the passage of time, a native-born creole elite began to feel greater allegiance to its American homeland than to the mother country of Spain, while the mixture of Spanish-born settlers and Indians resulted in the creation of a new *mestizo*, or mixed-race, population. By the beginning of the nineteenth century, resentment against Spanish rule was reaching a boiling point, and after a long struggle, Mexico and the Central American provinces declared their independence from Spain in 1821.

The Geography of Middle America's Mainland

The Middle American mainland bridge extends from the United States-Mexico border down to the narrow isthmus of Panama that connects to South America. Only 40 miles wide at its narrowest point, the *isthmus* gradually widens as it progresses in a northwesterly direction. The Middle American mainland has two major bulges on its eastern side: the eastern Nicaraguan and Honduran hump and the Yucatán Peninsula. The Baja California peninsula on the west juts south for more than 750 miles from the point where the Middle American bridge (now about 1,400 miles wide) joins with the rest of North America.

The mainland funnel has as many geographic contrasts as it does similarities. The great Rocky Mountain chain of Northern America continues south through Middle America under many local names until it reaches the massive South American Andes Mountains. Most mountains in Middle America are volcanic in origin, and many are still active.

GeoJargon

An **isthmus** (try to say it three times fast) is a narrow strip of land with water on both sides that connects to two larger land bodies—which pretty much sums up Panama.

The combination of volcanic eruptions and earthquakes, which are also frequent in this region, make the Middle American mainland one of the more dangerous areas in the world. In most of the countries of this subregion, mountains run through the length of the country with highland plateaus in the middle and lowland coastal forested belts or plains to the west and east. The climate is affected by the altitude: The lowlands and coastal belts are tropically hot and extremely rainy (as much as 150 inches a year), while the highlands are temperate, much cooler, and less rainy.

The Middle American mainland funnel, more than 3,800 miles long, is anchored by giant Mexico in the north. At 762,000 square miles, it is by far the largest country in the region—almost twice the size of all the others combined—and the 14th largest in the world. Because of its size and world importance, Mexico by itself is treated as one of the four subregions in Middle America, along with Central America, the Greater Antilles, and the Lesser Antilles.

Mexico: Middle American Dynamo

Mexico's turbulent post-independence history has played out under the shadow of its powerful, and often interventionist, neighbor to the north, the United States. The Mexican-American War of 1846 to 1848 was triggered by the decision of Texas, which had broken away from Mexico in 1836, to join the United States. Mexico lost the war and was forced to cede what are now the states of California, Nevada, and Utah, most of Arizona and New Mexico, and parts of Wyoming and Colorado. The United States again intervened militarily in Mexico in 1916 during a period of revolutionary strife and civil war.

Mexico's twentieth-century history was dominated by a single political party, the PRI (from the Spanish for Institutional Revolutionary Party). In July 2000, 71 years of one-party rule came to an end with the election of President Vicente Fox of the opposition National Action Party. Six years earlier, Mexico had taken an equally giant step in the economic sphere by joining the United States and Canada in the North American Free Trade Agreement (NAFTA). This landmark and still controversial treaty will phase out all tariffs among the three nations over a fifteen-year period and help integrate Mexico's economy more fully with those of the two North American powerhouses.

> **CAUTION**
>
> ## Eye on the Environment
>
> Mexico City is one of the largest and most vibrant cities in the world, and a key center of Latin American economic and intellectual life. But it is also a city plagued by severe environmental issues—a massive population, terrible air and water pollution, the gradual sinking of heavy buildings due to the lowering of the water table, and an extremely high risk of earthquake.
>
> But more than anything else, Mexico City has been notorious for the noxious quality of its air, caused mainly by the constant outflow of noxious pollutants from millions of vehicles and tens of thousands of area factories. The situation is worsened by the city's location 7,350 feet up, in the middle of a basin surrounded by mountains that not only trap pollution but also spew it out in the form of volcanic ash and gasses.
>
> Fortunately, Mexican authorities have made significant efforts in recent years to combat the capital's pollution problems. Drivers of private cars are now required to install catalytic converters or stay off the road one day a week. Incentives have been created to encourage the purchase of newer and cleaner vehicles and the conversion of industrial plants to low-sulfur fuels. Although Mexico has a long way to go to clean up its environment, Mexico City is no longer at the very top of the list of the world's most polluted cities.

Mexico, officially The United Mexican States, is composed of 31 states and a federal district. It has a population of nearly 105 million people, and its largest cities are the enormous capital of Mexico City and the industrial centers of Guadalajara, and Monterrey. The majority of the population is *mestizo* (Amerindian-Spanish) or either pure or predominantly Amerindian. Nearly 90 percent of Mexicans are at least nominally Roman Catholic. The primary language of the country is Spanish, although many indigenous languages are also spoken, sometimes exclusively. Mexico today is still plagued by high levels of poverty, underemployment, and inequitable income distribution. Yet it is also a major oil and gas exporter with a modernizing economy that has made extraordinary strides since the currency devaluation and recession of the mid-1990s.

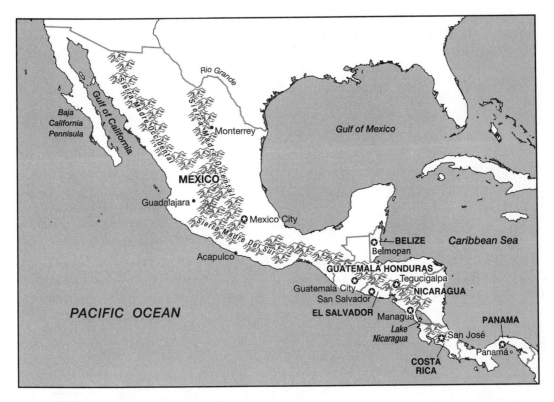

Mexico and Central America—physical features.

The Nations of Central America

South of Mexico on the Middle American land bridge is the subregion of Central America, consisting of seven countries: Guatemala, Belize, Honduras, El Salvador, Nicaragua, Costa Rica, and Panama.

Spanish is either the official or the main language of all of the Central American countries, except Belize, where it's English. The countries' populations are generally of mixed race, either mestizo (European and Amerindian) or mulatto (European and black African). Most Central Americans are Roman Catholic.

Central America declared its independence from Spain on September 15, 1821, only to be annexed by the short-lived Mexican Empire of Agustin Iturbide in January 1822. When a republic was declared in Mexico the following year, the Central American provinces opted to separate from Mexico and form the United States of Central America. The union began to dissolve in 1838. Following is a brief overview of the seven Central American nations, going from north to south.

Guatemala

With more than 13 million people, Guatemala is the most populous country in Central America. With more than 2.6 million people, the capital, Guatemala City, is Guatemala's and Central America's largest city. A poor, mostly agricultural country, Guatemala has had a tormented history marked by civil strife and military dictatorships. A peace agreement signed in 1996 ended a 36-year guerilla war—the longest in Latin American history—that left some 100,000 people dead (200,000 according to some estimates) and created about a million refugees.

Belize

Belize is the only Central American country with a British background (it was called British Honduras until 1981 when it gained its independence from Britain) and the only one with no Pacific coastline. The second-smallest country in the subregion, it's also one of the least populated, with only 263,000 people. The economy is heavily dependent on tourism, followed by agriculture (sugar cane, citrus fruits, and bananas).

Honduras

Honduras, the second-largest country in Central America, is near the bottom in terms of wealth. Only Nicaragua is poorer. Since its independence from Spain in 1821, Honduras has also had a history of dictators, revolutions, and coups. During the 1980s, the country served as a base for anti-Sandinista guerrillas fighting Nicaragua's Marxist government and assisted the Salvadoran government in its fight against leftist guerrillas. The economy of Honduras is based on agriculture; coffee and bananas are the main exports.

El Salvador

El Salvador, the smallest country in Central America, is the only one with no Caribbean coastline. Following independence in 1821, El Salvador was involved in several wars with its neighboring countries during the rest of the century. The first few decades of this century saw some stability and growth in the economy, only to be followed by a series of strong but oppressive dictatorships. A costly 12-year civil war finally came to an end in 1992. Honduras, El Salvador, and Nicaragua were all badly affected by Hurricane Mitch in 1998, and powerful earthquakes and drought struck El Salvador in 2001. A mainstay of the economy is money sent home by Salvadorans working in the United States. The U.S. dollar is the country's legal tender.

Nicaragua

With a gross domestic product per capita (the value of domestic goods and services divided by the number of residents) of only $2,500, Nicaragua is not only the largest country in Central America but also the poorest. Corruption and political upheaval are largely to blame for the country's economic difficulties. The Marxist Sandinista regime came to power in 1979, but was opposed throughout much of the 1980s by the U.S.-sponsored guerrilla fighters known as the *contras* ("those against" in Spanish). Beginning with the free elections of 1990, the Sandinistas have been repeatedly defeated at the polls.

Costa Rica

Costa Rica's gross domestic product per capita of $8,500 is the highest in Central America. Its indigenous people resisted colonization, and, with no obvious wealth, Spanish subjugation came late and then only indirectly through Nicaragua. Costa Rica achieved independence in 1821, and its transition to democratic rule was relatively smooth. With no army and a thriving middle class, the entire country enjoys a standard of living that's quite high by Middle American standards.

Panama

In many ways, Panama is a creation of the United States. With U.S. backing, it seceded from its South American neighbor, Colombia, and promptly signed a treaty with the United States allowing it to build the Panama Canal and take possession of the Canal Zone. As in most of the other Central American nations, Panama's history has been plagued by dictators and coups. U.S. forces invaded Panama in 1989 to oust the most recent dictator, General Manuel Noriega, who was subsequently tried and convicted on drug and money-laundering charges. Panama's stable, service-based economy is the second-strongest in Central America after Costa Rica.

Geographically Speaking

The dream of linking the Atlantic and Pacific Oceans by building a canal through Central America (Nicaragua was a candidate at one point) was a very old one. After a French attempt to cut a canal through Panama in the 1880s failed, the United States engineered Panama's independence and took over the project. Generally recognized as one of the greatest engineering feats of all time, the Panama Canal was built between 1905 and 1914 by the U.S. Army Corps of Engineers. The canal is 40 miles long and has lock chambers that lift and lower ships 85 feet during the passage between oceans. By going through the canal instead of around South America, a ship sailing from the East Coast of the United States to Japan saves about 3,000 miles. Under the terms of a 1977 treaty, the United States ceded full control of the Canal to Panama on December 31, 1999.

America's Caribbean Island Paradises

The names of the various island chains in or around the Caribbean can be confusing. Here's a list to help clarify the situation for you as you look at the map of the Caribbean:

- **West Indies:** The catch-all term for all the island groups in the area: the Bahamas, Turks and Caicos Islands, Greater Antilles, Cayman Islands, Lesser Antilles, and Netherlands Antilles

- **Greater Antilles:** The big islands of Cuba, Hispaniola (Haiti and the Dominican Republic), Jamaica, and Puerto Rico

- **Lesser Antilles:** The arc of small islands in the eastern and southern Caribbean: the Leeward Islands, Windward Islands, and Netherlands Antilles

- **Leeward Islands:** The northern half of the island arc that is sheltered from the trade winds: the Virgin Islands, Saint Eustatius, Montserrat, Saba, Antigua–Barbuda, Saint Kitts–Nevis, and Guadeloupe

- **Windward Islands:** The southern half of the island arc: Dominica, Martinique, Saint Lucia, Barbados, Saint Vincent and the Grenadines, Grenada, and Trinidad and Tobago

- **The ABC Islands:** Just off the northern coast of South America, the Dutch islands of Aruba, Bonaire, and Curaçao

Geographically Speaking

A problem shared by all the countries surrounding the Caribbean Sea and the Gulf of Mexico (as well as the Atlantic-coast states of the United States) is the annual threat of deadly hurricanes. These immense storms, spawned in the tropical waters of the southern Atlantic, the Caribbean Sea, and the Gulf of Mexico, are rated from 1 to 5: 1 means minimum winds of 74 m.p.h.; 5 means winds exceeding 155 m.p.h.

From a center eye of about 15 miles in diameter, hurricane winds circle out to about 150 miles in diameter. Gale winds (32 m.p.h. and higher) at the outer edge of the storm enlarge the circle to about 300 miles in diameter. A hurricane's combination of intense winds, torrential rains, and high seas can cause severe loss of life and untold property damage. In 1998, Hurricane Mitch claimed the lives of an estimated 10,000 people in Nicaragua and Honduras. That same year, Hurricane Georges killed more than 600 people in the Caribbean region and caused nearly $6 billion worth of property damage in the United States and Puerto Rico. The peak Atlantic hurricane season lasts from June through the end of October.

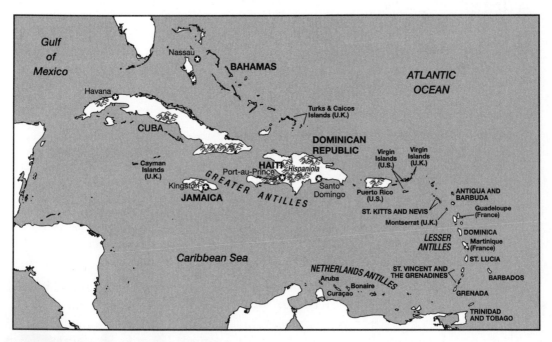

Islands of the Caribbean—physical features.

The Greater Antilles: The Big Islands

This subregion of Middle America consists of four main islands: Cuba, Hispaniola (Haiti and the Dominican Republic), Puerto Rico, and Jamaica. Cuba is by far the largest island in the Greater Antilles group. On the east, Cuba is separated from Hispaniola by the Windward Passage. Hispaniola, the second-largest island, contains two nations: Haiti, which occupies the western third of the island, and the Dominican Republic, which occupies all the rest. South of Cuba is the third-largest island, Jamaica. The smallest island in the group, Puerto Rico, lies east of Hispaniola.

These four islands have in common their large size (compared to the Lesser Antilles) and their relative adjacency (they're all close together in the north Caribbean). They also share an important historical fact: All were discovered by Christopher Columbus and therefore started their colonial existence under the Spanish flag and the Roman Catholic cross.

The people Columbus found in the Greater Antilles were a gentle Arawak people called the Tainos. Although far less advanced than the great Maya and Aztec civilizations of mainland Middle America, the Tainos at the time of the conquest lived in settled communities and had developed a political organization and social hierarchies. The Tainos, unfortunately, were the first Native Americans to bear the brunt of European colonization. Within about 50 years of Columbus's arrival, the Tainos had been almost completely wiped out due to overwork in Spanish mines and plantations, murder, suicide, and Spanish diseases against which they had no defenses.

Although all the islands of the Greater Antilles were originally colonized by Spain, other European powers were eager to get in on the action as well. Britain conquered Jamaica in 1655. In 1697, Spain officially ceded the western third of Hispaniola, today's Haiti, to France. What is now the Dominican Republic remained in Spanish hands until 1795, when it, too, was ceded to France. Despite periods of French and Haitian rule in the first half of the nineteenth century, the Dominican Republic is still Spanish speaking, as are Cuba and Puerto Rico, which remained Spanish colonies until 1898. In addition to sharing a common language, Cuba, the Dominican Republic, and Puerto Rico are closely bound by history and culture.

Although similar geographically in many ways, the Greater Antilles are also full of contrasts. Cuba, for example, is relatively flat and only about one-quarter mountainous, while the other islands are very mountainous. All have tropical or semitropical climates, but there are climatic contrasts as well. In Jamaica's rain forest, the amount of rainfall can exceed 200 inches per year, while parts of Haiti get only 20 inches of rain annually. All the islands in this group are blessed with miles of spectacular beaches that have made them meccas for tourists from around the world.

The Greater Antilles have witnessed bloody battles for independence and more than their share of political upheaval, dictatorships, and violence. Following is a brief overview of the four countries and one commonwealth in this subregion.

Cuba

Cuba has a population of 11.2 million people, all Spanish-speaking and most with a mixed Spanish and black African or Amerindian heritage. Its capital and largest city is Havana, once the Paris of the Caribbean, now in a serious state of disrepair. Before the victory of Fidel Castro's communist revolution in 1959, Cubans were almost all Roman Catholic, but now only one-third are, and more than half declare themselves nonreligious.

Cuba's economy has been struggling to regain its footing after the collapse of the Soviet Union and the loss of Soviet aid (about $5 billion a year) in 1991. Sugar, nickel, and tobacco (those famous Cuban cigars) are the country's main exports. Although the U.S. embargo against Cuba, which Castro blames for most of his nation's ills, prevents most American citizens from traveling to Cuba, the island is a popular destination for European tourists, and the tourism sector had been growing steadily until the post-September 11, 2001, slump.

Geographically Speaking

A particular bone of contention between the Castro government and the United States is the U.S. Naval Base at Guantanamo Bay in the southeastern tip of Cuba. The United States established the base at the end of the Spanish-American War in 1898 and leased it permanently in 1903, in a deal that Castro considers to have been illegitimate.

Guantanamo has been used at various times to house Cuban and Haitian refugees. In 2002, the United States began housing suspected Taliban and al-Qaeda prisoners taken in Afghanistan in a Guantanamo base prison called Camp X-ray. Because the base is in Cuban territory, U.S. prisoners held there do not enjoy the same constitutional rights they would if they were being held in U.S. territory. In addition, the Bush administration has labeled these prisoners illegal combatants rather than prisoners of war, thus depriving them of the protections of the Geneva Convention. The situation of the Guantanamo Afghanistan prisoners continues to be monitored and criticized by international human rights organizations.

After more than 40 years in power, Castro still dominates the Cuban story. One of the key questions for the region over the coming years is how Cubans, both on the island and in exile in the United States, will respond to the inevitable departure of Fidel Castro from the world stage.

Haiti

Haiti's history is rich and fascinating. In the eighteenth century it was the source of incalculable wealth for its French overlords. But the Haitian sugar economy was built on the backs of hideously oppressed and ill-treated slaves. In the only successful slave uprising in history, Haitians overthrew the plantation owners and went on to defeat the armies of Napoleon to become the world's first independent black republic.

Terra-Trivia

Voodoo is a religion (practiced primarily in Haiti) that combines elements of Roman Catholicism with tribal religions of western Africa. Voodoo practitioners worship the high god Bon Dieu, ancestors, the dead, twins, and spirits called Loa (frequently identified with Roman Catholic saints). The priest is called a *houngan;* the priestess, a *mambo.*

Haiti's post-independence story is far less glorious. Wars, uprisings, cruel dictatorships (most notoriously that of the Duvaliers, "Papa Doc" and his son, "Baby Doc"), foreign (U.S.) interventions, and strife between the light-skinned mulatto elite and the black majority have turned Haiti into the poorest country in the Western Hemisphere and one of the poorest in the world.

Haiti has about 7 million people, about 95 percent of whom are black. The capital and main city is Port-au-Prince. French and Creole are both official languages. Although most Haitians are Roman Catholic, many practice *voodoo,* a blend of Catholicism and African religions.

Nearly 80 percent of Haiti's population lives in extreme poverty, barely eking out a living from subsistence farming. To make matters worse, foreign aid was essentially cut off in 2000 after a legislative election that was marred by irregularities. The political crisis that triggered the aid cut-off remains unresolved in 2003.

Of the many crises facing Haiti, poverty, disease, and ecological destruction top the list. Infant mortality rates are high, life expectancy is short (under 50 years), and the country has the highest HIV and tuberculosis infection rates in the Western Hemisphere. On the environmental front, 99 percent of Haiti is now deforested and tropical rains have washed away much of the soil. The result is a land that's denuded—in many areas, down to bedrock level. What little soil remains is overworked and is used for growing export crops rather than life-sustenance crops.

Dominican Republic

The Dominican Republic shares the island of Hispaniola with Haiti, and the histories of the two countries have been closely intertwined. Haiti ruled the future Dominican Republic from 1822 to 1844, and Dominicans celebrate their independence not from Spain in 1821, but from Haiti, in 1844. The period of Haitian rule left a bitter legacy of mutual antagonisms and prejudices that persist to this day.

The Dominican Republic has also suffered from political violence, foreign (U.S.) intervention, and cruel dictatorships. Its twentieth century history was dominated by Rafael Trujillo, who ruled despotically from 1930 until his assassination in 1961. After decades of turmoil and upheavals, the Dominican Republic attained relative political stability and a sustained period of economic growth in the 1990s.

The Dominican Republic has a population of 8.7 million people who speak Spanish, have a mixed black and Spanish heritage, and are predominantly Roman Catholic. Its capital and largest city is Santo Domingo, the oldest European settlement in the New World. Although the Dominican Republic is known primarily as an exporter of sugar, coffee, and tobacco, tourism has overtaken agriculture as its main income earner.

Jamaica

After more than 100 years of Spanish rule followed by more than 300 years of English rule, Jamaica finally attained its independence 1962. Since independence, the island has experienced periods of political unrest and crime waves that have threatened the vital tourism industry.

Jamaica has 2.7 million inhabitants, consisting mostly of blacks (91 percent) and people of mixed heritage (7.3 percent), plus small English, South Asian, and Chinese minorities. The Jamaican capital is Kingston.

Most Jamaicans speak the official language, English, or a local English dialect that incorporates African, Spanish, and French elements. The majority of Jamaicans are Protestants, although the island is also the birthplace of Rastafarianism, a religious sect whose members believe in a prophetic return of blacks to Africa.

The two pillars of the Jamaican economy are tourism and bauxite. After a long recession in the second half of the 1990s, Jamaica has been experiencing negligible growth in the first years of this century as tourism continues to be affected by the aftermath of 9/11 and a slow-to-recover global economy.

Terra-Trivia

Jamaica, the homeland of Bob Marley, is perhaps best known for its favorite export, reggae music.

Puerto Rico

Puerto Rico is by far the wealthiest nation in the Greater Antilles, and the only one that is not independent. Its capital and largest city is San Juan. Its population of 3.8 million people is Spanish speaking and mostly Roman Catholic. Puerto Ricans are of predominantly white Spanish descent; about 8 percent are black and 11 percent are of mixed heritage. Industry, especially pharmaceuticals, and tourism have overtaken agriculture as the prime movers of Puerto Rico's economy.

Puerto Rico was a Spanish colony until the Spanish–American War in 1898, when it was ceded to the United States, under whose control it has remained. Puerto Ricans were granted U.S. citizenship in 1917. They have been electing their own governors since 1948, and handling their internal affairs since 1952. As a commonwealth of the United States (or as it's referred to in Spanish, a "free associated state"), Puerto Rico does not have voting representation in the United States Congress nor can island residents vote in U.S. presidential elections. (Puerto Ricans who move to the mainland can vote for president.) On the other hand, Puerto Ricans do not pay federal taxes. The future status of Puerto Rico—statehood, independence, or commonwealth— has been a matter of much debate for decades. In 1967, 1993, and again in 1998, the people of Puerto Rico voted to maintain their commonwealth status.

The Lesser Antilles: The Small Island Arc

The fourth subregion of Middle America is an *archipelago* (a large group of islands) that extends in a grand arc all the way down to South America. Consisting of the Lesser Antilles and the Netherlands Antilles, the subregion includes nine independent countries, many possessions and territories, and almost countless islands, islets, and cays (tiny islands of rock, sand, or coral).

The Lesser Antilles, all small in both population and size, were inhabited by warlike Carib Indians when Columbus first landed there. Although many of the islands were "discovered" and named by the Spanish, they were eventually taken over by other European powers, mainly France, England, and Holland. Most of the islands are lush and some have rain forests. Almost all of them have golden white (or in some cases, black) sand beaches and coral reefs. The Lesser Antilles are actually the tops of underwater mountains; some have active or recently active volcanoes.

In most of the islands of the Lesser Antilles, the population is predominantly black (descendents of Africans brought over as slaves to replace the decimated Amerindians) or of mixed European and African ancestry. An exception is Trinidad and Tobago, where immigrants from northern India and their descendents make up more than 40 percent of the population. Trinidad and Tobago is also unique in that it has a major oil and natural gas industry and tourism is less important than in most of the other Caribbean islands. Petroleum refining and transshipment are also key components of the Netherlands Antilles economy. Like the Greater Antilles, the islands of the Lesser Antilles are a tourist paradise for Americans, Europeans, and everyone else seeking tropical island warmth and beauty.

During the seventeenth and eighteenth centuries, the West Indies became the battle-ground for many conflicts originating in Europe. During that time, the islands of the Lesser Antilles changed hands frequently. One way to sort these islands is by the European power that either colonized them originally or subsequently acquired them:

♦ **British:** The British Virgins, Antigua–Barbuda, Saint Kitts–Nevis, Montserrat, Barbados, Saint Lucia, Saint Vincent and the Grenadines, Grenada, and Trinidad and Tobago. Although the Virgins and Montserrat are still dependencies of the United Kingdom, the others are independent countries that elect their own heads of government but recognize Queen Elizabeth II as their sovereign.

♦ **Danish:** The U.S. Virgin Islands. In the seventeenth century, the Virgin Islands were divided between Britain and Denmark. The Danes sold their part to the United States in 1917.

♦ **Dutch:** Part of the Kingdom of the Netherlands, the Netherlands Antilles used to consist of the ABC islands—Aruba, Bonaire, and Curaçao—and three of the Leeward Islands (actually $2^1/_2$ islands): Saba, Saint Eustatius, and St. Maarten (half of an island shared with France). In 1986, Aruba pulled out of the Netherlands Antilles to become a separate part of the Dutch kingdom.

♦ **French:** Martinique and Guadeloupe are full overseas departments of France, not colonies or dependencies. Guadeloupe is actually an archipelago of nine inhabited islands, including St. Martin, the other half of the island shared with the Dutch. Dominica has been a true political football: It started out Spanish, then became French, then English, and finally independent.

In the News: Middle America Looks North

In 1993, Mexico took a giant step toward embracing globalization by signing along with Canada and the United States the North American Free Trade Agreement (NAFTA), which went into effect on January 1, 1994. By phasing out most barriers to trade and investment among the three countries over a 15-year period, NAFTA was expected to stimulate cross-border investment, generate new jobs in all three countries, and in the process pull Mexico up to an economic level closer to those of its northern neighbors. Things haven't quite worked out that way.

From the beginning NAFTA has been strongly opposed by U.S. labor organizations and environmental groups. Labor feared that U.S. jobs would migrate to Mexico, with its lower labor costs; environmentalists were concerned that U.S. environmental standards would be lowered in order to allow U.S. companies to stay competitive with Mexico, where environmental laws are much weaker and enforcement is lax. As it turns out, both fears have been justified to an extent. Even in Mexico, NAFTA has had mixed results. Although many U.S. companies have indeed set up shop in Mexico and its trade with the United States and Canada has tripled since 1994, Mexico continues to struggle with high rates of underemployment. Mexico's small farmers have been particularly hard hit by the importation of cheaper grain produced by U.S. agribusinesses.

Despite the controversies surrounding NAFTA, the free-trade bandwagon rolls on. Mexico has signed free-trade agreements with Guatemala, Honduras, and El Salvador. The United States signed a NAFTA-like free-trade accord with Chile in December 2002, and a Central American version of NAFTA, CAFTA, was finalized among El Salvador, Guatemala, Honduras, Nicaragua, and the United States at the end of 2003.

The Least You Need to Know

- ◆ The Middle American region has four primary subregions: Mexico, Central America, the Greater Antilles, and the Lesser Antilles.

- ◆ The Middle American mainland was home to the great empires of the Maya and the Aztecs.

- ◆ Mexico is the largest and most politically and economically important country in the region.

- ◆ The Panama Canal connects the world's two largest oceans via the thin Isthmus of Panama in Central America.

- ◆ The Greater Antilles include the four largest Caribbean islands: Cuba, Hispaniola, Puerto Rico, and Jamaica.

- ◆ The Lesser Antilles consist of many small islands, where tourism has mostly supplanted sugar production as the main industry.

Chapter 18

South America: The Iberian Legacy

In This Chapter

- ◆ Surveying South America's troubled history
- ◆ Exploring the continent's countries and cities
- ◆ Traveling down the Amazon and up the Andes
- ◆ Meeting South America's mix of peoples
- ◆ A continent at a crossroads

South America is the fourth-largest continent, with 12 percent of the earth's landmass. It's south of North America and much farther east than most people realize. Longitudinally, virtually the entire continent lies east of Florida. Its eastern bulge is less than 1,700 miles from the western bulge of Africa, whereas the Atlantic Ocean measures more than 3,500 miles from New York to Lisbon.

This more easterly location of South America extends the Pacific Ocean. The distance from western South America to Australia is almost twice as much as from California to Japan. The equator passes through the northern part of South America. From east coast to west coast, the continent measures 3,000 miles; from north to south, it's 4,600 miles. South America is 900 miles south of Florida but only 600 miles north of Antarctica.

Peopling the Continent

How and when did humans get to South America? The most commonly held theory is that East Asian peoples crossed the Bering land bridge (when it connected Asia with Alaska) and then fanned out east and south, eventually passing down through the Middle American land funnel into South America. The archeological evidence indicates that humans had reached the southern tip of South America by 9,000 to 8,500 B.C.E. The discovery of coastal settlements in Peru dating back 12,000 years opens up the possibility that South America was first inhabited by people traveling by foot and, in stretches, by boat along the west coast of the Americas at a time when ice-age glaciers prevented terrestrial migration through North America. These are still just theories, of course—plenty of scientists have their own ideas about South America's first inhabitants and when they arrived.

Before the Europeans "discovered" the Americas, hundreds of native tribes lived, hunted, herded, and farmed in South America. The most famous, the Andean peoples, lived in the highlands along South America's west coast. From about 8,000 to 2,000 B.C.E. the Andean Indians developed from nomadic hunter gatherers into sedentary farmers, growing potatoes, maize, squash, peppers, some fruits, and even cotton in the fertile Andean valleys protected on both sides by high mountains.

These people eventually produced crop surpluses, which allowed for the development of permanent settlements and cities in which advanced cultures gradually developed. Stone buildings, roads, altars, and gold artifacts all began to appear. Over the centuries, a variety of civilizations succeeded each other in the Andean region until the rise of the last and greatest of them all—the Incas.

Terra-Trivia

Inca architects built dry stone walls (which still stand) to terrace hillsides. Archaeologists still look with amazement at Inca masonry: Despite the use of only massive stone blocks and no mortar, the joints are so fine that an index card cannot fit between the stones.

The Inca Empire

The Incas first arrived in the Andean highlands of Peru in about 1150 C.E., and over the next 300 years established their domination over the peoples living in the Valley of Cuzco. The fifteenth century was a period of rapid expansion for the Inca Empire, which despite extremely difficult terrain—coastal desert, mountain rain forests, and high mountain plateaus—grew into the largest empire ever seen in the Americas. At its greatest extent, the Inca Empire included all of present-day Ecuador, most of Peru and Chile, and parts of Bolivia and Argentina. Its territories measured 2,500 miles long by 500 miles wide and contained almost 20 million people.

South America.

The Inca Empire was a centrally controlled monarchy. Politically, it was arranged like a pyramid, with a godlike Inca (king) on top; a court and advisors, priests, and administrators at the next lower level in the hierarchy; and farmers, artisans, and other producers on the bottom. The Incas built fortresses, palaces, temples, aqueducts, canals for their balsa boats, thousands of miles of paved roads for people and llamas to tread (the Incas lacked the wheel and horses), amazing suspension bridges (more than 300 feet long), and grain storage towers (in case of famine). It was this vast and sophisticated empire that the Spanish adventurer Francisco Pizarro had his sights on when he began the conquest of Peru in 1531.

The Coming of the Conquistadors

After establishing a base on the coast, Pizarro sent emissaries to ask for a meeting with Atahualpa, the Inca god–king. Assuming that the fair-skinned Spaniards were returning Inca gods, Atahualpa agreed to the meeting with hordes of unarmed Incas in attendance. Pizarro asked Atahualpa to become a Christian and swear allegiance to the Spanish king. Atahualpa refused.

Pizarro arrested Atahualpa, massacred 2,000 Incas, marched into the capital city of Cuzco, and imprisoned the king in his own palace. To ransom his freedom and throne, Atahualpa offered a 17-by-22-foot room filled with gold as high as a man could reach (and then the same room filled twice again with silver). Although Pizarro accepted the ransom, a few weeks later he had Atahualpa strangled to death.

The Spanish burned Cuzco in 1535, and in 1572, they captured and executed the last Inca ruler, Tupac Amaru. With his death, the Inca Empire was history, although evidence of its culture still remains—along with two to three million Inca descendants living in the Andean highlands in much the same way as they did 500 years ago.

Exploring and Exploiting the New Continent

In 1493, Pope Alexander VI wisely realized that Spain and Portugal would be competing for the wealth of the new lands just "discovered" by Columbus, so he drew a north–south line in the mid-Atlantic that divided all newly discovered lands between Spain and Portugal. Everything west of the line was Spanish territory; everything to the east was Portuguese. Because the papal line gave Spain all of the Americas except for a tiny bit at the eastern end of Brazil, Portugal refused to go along with the deal. The following year, Spain and Portugal signed the Treaty of Tordesillas that moved the line west to about 46 degrees west longitude, giving Portugal a much larger chunk of Brazil. Over time, Portugal greatly expanded its share of South America by encroaching on Spanish territory west of the line.

Spain began to exercise control over its South American colonies by establishing viceroyalties. The only part of the continent that wasn't settled by the Spanish or the

Portuguese was the Guianas to the north of Brazil. This hot and humid area, forested down to its beaches, had little mineral wealth to loot. The Dutch were the first to establish settlements in the Guianas in 1621, followed by the English and then the French. The various Guianese settlements changed hands many times over the centuries, but what eventually became Dutch Guiana is now independent Suriname. British Guiana is now independent Guyana, while French Guiana is still part of France.

In the Guianas and the vast reaches of Brazil to the south, the Europeans didn't find any wealthy empires to loot or huge populations to conscript for "their" farms and mines, so they turned to a practice that was proving lucrative up in the Caribbean. The Spanish and Portuguese imported millions of black Africans to labor on their plantations.

> **Terra-Trivia**
>
> In the seventeenth century, the Dutch and British both occupied what is now Suriname. In 1667, the British ceded their share to the Netherlands in return for the Dutch colony of New Amsterdam, today's New York City—nice trade!

A Tidal Wave of Independence

It took almost three centuries of mineral and economic exploitation, human abuse, and political injustice for the South American colonies of Spain and Portugal to reach a boiling point. They finally did in the early 1800s. The occupation of Spain by Napoleonic France emboldened South American colonials to rise up against the mother country. South America's independence struggle took nearly 15 years and gave rise to many great leaders and heroes, the most famous of whom were José de San Martín, in Argentina and Chile, and Simón Bolívar, in the north.

In Portuguese-dominated Brazil, the same yearning for independence had been developing, but it came to a head in a different way. In 1822, Dom Pedro, son of the King of Portugal and regent in Brazil, was recalled to Portugal. Instead he declared Brazil an independent state with himself as emperor. After losing a brief war to retain its giant colony, Portugal recognized Brazil's independence on August 29, 1825.

The Guianas story is also different. Guyana (the former British Guiana) gained self-rule from Great Britain in 1961 and became a republic in 1970. Suriname (the former Dutch Guiana) didn't become independent from the Netherlands until 1975, and French Guiana, as you've seen, never gained its independence and is today an overseas department of France. Who would have guessed that France, which had the smallest South America colony, would hang on to it the longest?

The South American Countries and their Major Cities

Following is a brief overview of South America's 12 nations (and one French department).

Brazil

Brazil is South America's largest (and the world's fifth largest) country, in terms of both population and area. With about 176 million inhabitants, Brazil is home to more than half of all South Americans. The country has 13 cities with a population of more than 1 million. The two largest are São Paulo (South America's most populous city) with 10 million people (nearly 19 million in its metropolitan area), and Rio de Janeiro, with more than 6 million (more than 11 million in its metropolitan area). Brazil's capital, Brasília, is relatively small in comparison with a population of nearly 2.1 million people. The country's twentieth-century history was marked by repeated military interventions and government corruption. Brazil today has South America's biggest economy, but grossly unequal income distribution remains a problem. In 2003, Brazilians elected their first president from a working-class background.

Colombia

With 41 million inhabitants, Colombia is South America's second most-populous country. Unique among South American countries, Colombia has coastlines on both the Pacific Ocean and the Caribbean Sea. Nearly 7 million people live in the Colombian capital of Bogotá. Its second largest city, Cali, has more than 2 million inhabitants; Medellín and Barranquilla each have populations of more than 1 million people. A focal point of the international drug trade, Colombia has also suffered for decades from an increasingly violent anti-government insurgency.

> **Terra-Trivia**
>
> The dream of South America's great liberator, Simón Bolívar, was to unite the entire continent as one nation, but that was never to be. Even the Republic of Gran Colombia, which he founded in 1819, lasted barely a decade. By 1830, the republic had broken up into the three modern-day nations of Colombia, Venezuela, and Ecuador.

Argentina

Argentina is South America's third most populous country (with nearly 38 million inhabitants) and its second-largest in land area. Its capital, Buenos Aires, birthplace of the tango, has about 3 million people, but its greater metropolitan area is home to nearly a third of the country's total population. Cordoba and

Rosario are the other Argentine cities with more than a million inhabitants. Argentina's vast natural and human resources make it a natural candidate for the economic leadership of South America, but a post-World War II history of authoritarian rule, military dictatorship, and political corruption have kept this rich nation from fulfilling its potential. Although Argentina has consolidated its democratic institutions following the military dictatorship and "dirty war" of the late 1970s and early 1980s, its economy struggled through the 1990s and nearly collapsed in 2000 and 2001. Since mid-2002, Argentina's economy seems to be on the mend, but remains fragile. Despite recent setbacks, Argentina still has South America's highest per capita gross domestic product.

Peru

Peru is home to about 28 million people, about half of whom are Incas and other Amerindians living in the Andean highlands. The remaining population consists mostly of mestizos, or mixed Amerindians and whites (37 percent), whites (15 percent), and blacks and Asians (about 3 percent) living on the coastal plain. Peru's capital and largest city is Lima, with more than 6.5 million residents. Like its neighbors, post-World War II Peru has experienced its share of military dictatorships, anti-government insurgencies, and economic problems. In the 1990s, Peruvian president Alfredo Fujimori achieved notable success in suppressing the violent Marxist rebel group known as Shining Path, but his rule grew increasingly authoritarian. In 2000, Fujimori was ousted and forced into exile in Japan. Political instability had a negative impact on Peru's economy, although as of 2002 the country is experiencing strong growth. Mining and the extraction of petroleum and gas are important sectors of the Peruvian economy. Peru is also the world's largest grower of coca, from which cocaine is refined.

Terra-Trivia

One of South America's most spectacular archeological sites is Machu Picchu, an abandoned ancient Inca city on a high ridge in the Andes about 50 miles north of Cuzco. Believed to have been built in the mid-1400s as a royal estate and abandoned shortly after the beginning of the Spanish conquest, Machu Picchu was rediscovered in 1911 by the American historian Hiram Bingham.

Venezuela

Venezuela is a country of 24 million people, 3.5 million of whom live in the modern capital of Caracas. The cities of Maracaibo and Valencia both have populations of more than a million people. Venezuela has the largest oil reserves in the Western

Hemisphere and the sixth largest in the world. Its economy, however, is overly dependent on the petroleum sector and therefore highly vulnerable to oil-price fluctuations. In recent years, Venezuela's economy and society have been rocked by political instability under the divisive regime of President Hugo Chavez.

Chile

Chile is 2,650 miles long from north to south, but averages only 110 miles wide from east to west. Its population of 15.5 million people is almost entirely mestizo (mixed white and Amerindian) or white. Full Amerindians make up about 3 percent of the population. Chile's capital, Santiago, has more than 5 million people in its metropolitan area, or about one-third of Chile's total population. Chile's history over the last 30 years has been particularly traumatic. The overthrow and violent death of leftist President Salvador Allende in 1973 ushered in a repressive 17-year military dictatorship under General Augusto Pinochet. Economic reforms begun under Pinochet and continued under the democratic governments that succeeded him have established Chile as one of South America's strongest and most stable economies. In December 2002, Chile signed a NAFTA-like free-trade accord with the United States. (For more on NAFTA, see Chapter 17.)

Ecuador

Ecuador (with a *c*) straddles the equator (with a *q*). The country's population of 13.4 million is mostly mestizo or pure Amerindian. The capital, Quito, located high in the Andes, has 1.8 million people in its metropolitan area. Coastal Guayaquil is even larger, with 2.6 million people in its metropolitan area. The famous Galapagos Islands, where the naturalist Charles Darwin first began to formulate his theories on natural selection and evolution, belong to Ecuador. Ecuador was once the northern part of the Inca empire, but its modern relationship with its southern neighbor, Peru, has been far from smooth. During the twentieth century, the two countries fought several wars over territory, with Ecuador usually on the losing side. A treaty signed in 1999 finally ended the longstanding border dispute. The pillars of Ecuador's unstable and now dollarized economy (the U.S. dollar became the nation's currency in 1999) are petroleum and agriculture. Despite recent reforms, the Ecuadorian economy remains weighed down by budget deficits and a huge foreign debt.

Bolivia

Landlocked Bolivia (it lost its only corridor to the sea in a war with Chile) has a population of 8.4 million people, most of whom are either Amerindian (Quechua or

Aymara) or mestizo (mixed white and Amerindian). It has two capitals: Sucre, the legal capital, and La Paz, the administrative capital and seat of government. Each city has a population of less than one million. Bolivia's history is a long, sad tale of coups, political unrest, and poverty. Although Bolivia has major natural resources and has made moves to reform and privatize its economy, it remains South America's poorest and historically most unstable country.

> **GeoRecord**
>
> Located high in the Andes at an altitude of 11,913 feet above sea level, La Paz, Bolivia, is now the world's highest capital city. That distinction used to belong to the Tibetan capital of Lhasa (12,087 feet above sea level) before Tibet was taken over by China.

Paraguay

South America's other landlocked country, Paraguay has a population of 5.8 million, with more than one million living in and around the capital of Asunción. Almost all of Paraguay's people are mestizo. Paraguay has consolidated its democratic institutions since the overthrow of the cruel 35-year military dictatorship of Alfredo Stroessner, but its economy is held back by corruption, political uncertainty, and debt. Nearly half of Paraguay's work force is engaged in agriculture, often on a subsistence level.

Uruguay

Uruguay has a population of just more than three million, most of whom are white and half of whom live in or around the capital of Montevideo. Like so many other South American nations, Uruguay was ruled by military dictators from 1973 to 1985, when democratic government was restored. Like neighboring Argentina, Uruguay has a strong agricultural sector and a well-educated workforce, neither of which prevented its economy from going into near free fall between 1999 and 2002, mostly as a result of Argentina's economic woes. The Uruguayan situation stabilized somewhat in 2003, but full recovery still seems distant.

The Guianas

The Guianas were all plantation colonies, more closely resembling the Caribbean colonies than the South American ones. The largest of the three is Guyana (the former British Guiana), with Dutch and English roots and a population of about 698,000 people, most of whom are East Indians (whose ancestors came from northern

India) or blacks. Guyana's capital is Georgetown. Its economy is based primarily on agriculture and bauxite mining.

Terra-Trivia

French Guiana is perhaps most famous, or infamous, as the location of Devil's Island, the notorious French penal colony, whose facilities were divided between an offshore islet and part of the mainland at Kourou. France ran its Guianese prison from 1852 through 1946 (although it stopped sending prisoners there in 1938). Among Devil Island's most prominent inmates were Alfred Dreyfus and Henri "Papillon" Charriere.

The next-largest of the Guianas is Suriname (the former Dutch Guiana), with Dutch and English roots and about 436,000 people, most of whom are East Indian, Creole (mixed white and black), Javanese, and black. Suriname's capital is Paramaribo. Since gaining its independence from the Netherlands, Suriname has alternated between military and civilian governments. Its economy is heavily dependent on bauxite mining.

The smallest of the Guianas (and South America's smallest "state") is French Guiana, with only about 186,000 people, most of whom are either black or mixed black and white (whites make up about 12 percent of the population; East Indians, Chinese, and Amerindians, another 12 percent). French Guiana is an overseas department of France; its capital is Cayenne. French Guiana's economy is based on fishing, forestry, and subsidies from France.

Physical Facts

A glimpse at the following map of South America's physical features shows that the continent is dominated by two geographic wonders: the mighty Amazon River and the Andes Mountains. With its main trunk just south of the equator flowing from west to east, the Amazon dominates the northeastern portion of the continent. Running along the entire western edge of South America is the spectacular Andes Mountain chain, which acts as a huge wall dividing the continent's low interior areas from the vast Pacific Ocean.

The World's Largest River

Although the Amazon is probably second to the Nile in length (the Amazon is about 4,100 miles long while the Nile is 4,145 long), it's by far the world's largest river. The drainage basin of the Amazon encompasses more than 2.7 million square miles, or about 40 percent of South America's total surface area. Its discharge into the Atlantic Ocean measures more than 50 billion gallons per second. In one day, that's enough to meet the needs of all U.S. households for five months.

South America—physical features.

The Amazon has been estimated to be about 4,100 miles long, but its exact length can't be measured because it changes course every year. With its more than 2,000 tributaries, a massive drainage system, and an abundant rainy season, the river rises annually an average of 65 feet and overflows the jungle on each side as much as 30 miles. This phenomenon makes the Amazon almost 60 miles wide at spots in the rainy season. When the river subsides during the dry season, its location is altered in many areas along its 4,100-mile course.

The High Andes

The Andes chain is one of the world's greatest mountain systems. It is 4,500 miles long and ranges from 50 to more than 400 miles wide. It reaches heights of as much as 22,831 feet (Mount Aconcagua, in Argentina, is the highest mountain in the Western Hemisphere). The *snow line* of the Andes ranges from 4,000 feet in the south to 17,000 feet at the equator. The *tree line* ranges from a southerly 3,000 feet to an equatorial 12,000 feet.

GeoJargon

The **snow line** is the point on the landscape above which snow remains year-round. The warmer the average temperature at a mountain's base, the higher the snow line's elevation. The **tree line** is the point on the landscape above which it becomes too cold and the growing season is too short to allow trees to grow.

The Andes contain many lofty peaks, including 50 higher than 20,000 feet. Located at the juncture of the Nazca and South American plates, the Andes Mountains are subject to frequent, and often violent, earthquakes and volcanic eruptions. Of the 13 South American countries, only Paraguay, Uruguay, and the Guianas are considered non-Andean.

South America's Environments

When most people think about South America's environment, they think of tropical rain forests. It's not an illogical thought because the tropical rain forests that dominate the Amazon basin are the largest in the world. But South America has other notable environments, a few of which are discussed in this section.

In northern South America is an area of grassland and open dry forests called the *llanos* of Colombia and Venezuela. In the continent's southern cone are the famous pampas, home of the Argentine cowboys known as *gauchos*. This landscape of tall grass extends from central Argentina into Uruguay. In addition to being prime cattle country, the pampas are also a major grain-producing region.

South of the pampas is another major Argentine livestock area called Patagonia. With its higher elevations, cooler temperatures, and drier climate, Patagonia is ideally suited to sheep ranching. And finally, in a narrow strip west of the Andes in northern Chile is the Atacama Desert, one of the driest places on earth. Because the desert is on a 2,000-foot-high plateau and is favored by cool winds blowing off the Humboldt Current, its average year-round temperature is only 65 degrees F.

The Waters

With a surface area of 3,220 square miles, Lake Titicaca, between Bolivia and Peru is South America's largest lake. (And at 13,000 feet above sea level, it is also the highest navigable lake in the world.) At 5,000 square miles, Venezuela's Lake Maracaibo is larger than Titicaca, but it's not a true lake. It's actually a bay connected to the Gulf of Venezuela by a narrow channel. Other large lakes dot the continent, many of them a result of human-made river dams.

In addition to the Amazon, South America has several long and locally important rivers. The Orinoco River in the north, more than 1,700 miles long, ends in a vast delta on the Atlantic Ocean. In the south, the 2,450-mile Parana River flows from south–central Brazil. It has a confluence with the Paraguay River and then unites with the Uruguay River to form the Río de la Plata *estuary* on the Atlantic Ocean.

GeoJargon

An **estuary** is an inland arm of the sea that meets the mouth of a river. It also refers to the area where the river's current meets the sea's tides.

A few more physical features are important. In one of Venezuela's remotest spots, almost impossible to get to by land, are the Angel Falls. At 3,212 feet, they're the world's highest falls (compare them with Africa's Victoria Falls at 400 feet and Niagara Falls on the United States–Canada border, at 167 feet). On the Brazil–Argentina border are the Iguaçu Falls, one of South America's great natural wonders. In the dry season, the river drops in two 200-foot-high crescent-shaped falls. In the wet season, however, it becomes one gigantic fall about two miles wide (compare this with Victoria's one-mile width and Niagara's combined width of $^2/_3$ mile).

Geographically Speaking

Feared by generations of sailors for its treacherous conditions, Cape Horn on Horn Island in the Tierra del Fuego archipelago is South America's southernmost point. Chile owns Horn Island and the famous cape that juts out from it, making it the southernmost country in South America. (A *cape*, by the way, is defined as a strip of land jutting out into a large body of water.)

The largest island in the archipelago is the Isla Grande de Tierra del Fuego. The passage between the large island and the mainland forms much of the Strait of Magellan. The western half of the large island is owned by Chile, and the eastern half is owned by Argentina. On the Argentine half is the world's southernmost city, Ushuaia.

To South America's extreme south lies the Strait of Magellan, "discovered" by Ferdinand Magellan in 1520. The strait connects the Atlantic and Pacific oceans by winding through South America just north of the southerly archipelago, Tierra del Fuego. It measures 331 miles long and only 2 to 15 miles wide. Although the strait is difficult to navigate, it enables ships to avoid the hazardous winds, rains, currents, and waves off Cape Horn at the southern tip of South America.

GeoJargon

The **Tropic of Capricorn** is a latitude parallel 23 degrees 27 minutes south of the equator, marking the southern boundary of the tropics, where the temperature remains about the same year-round. Technically, it's the southernmost point where the sun appears directly overhead at noon on or about December 22.

A Look at South America's Weather

South America's northern area is all tropical; southern Brazil, half of Paraguay, almost all of Argentina, much of Chile, and all of Uruguay lie south of the *Tropic of Capricorn* (about 23½ degrees south latitude) and are temperate. For most of South America's tropical zone, the yearly temperature is 75 to 85 degrees F; it's cooler in the Brazilian Highlands, however, and even cooler in the Andes. In the temperate zone, winters range from the 60s to the 30s as you head south. Again, it's always cooler in the Andes. Down near Cape Horn, the average temperature is less than 32 degrees F year-round.

South America's People

South America has 12 percent of the earth's landmass but only 6 percent of its people. The large majority of these people live within 500 miles from a coast, particularly the east, central east, and northwest coasts. The vast interior is sparsely inhabited. Today's ethnic breakdown is largely a result of South America's history.

Pure Amerindians live primarily in former Inca lands—Ecuador, Peru, and Bolivia. Mestizos (people of mixed Amerindian and European heritage) are found in the Inca lands, surrounding Amerindian populations, as well as in Chile, Paraguay, Colombia, Venezuela, and in smaller numbers throughout the continent. East Indians are the largest ethnic group in both Guyana and Suriname.

African blacks, brought to South America as slaves, live primarily in the Guianas, Brazil, and Venezuela, with smaller populations in Peru, Ecuador, and Colombia. Mulatto peoples (those of black African and European stock) live mainly in Venezuela, Brazil, the Guianas, and Colombia. Whites dominate only in Argentina and Uruguay, where they represent more than 85 percent of the total population. They also make up a significant part of the populations of Brazil, Chile, and Venezuela, and less than 20 percent of the populations of Peru and Colombia.

Language also parallels the region's colonial history. Portuguese is spoken in Brazil, English in Guyana, Dutch in Suriname, French in French Guiana, and Spanish everywhere else. Not surprisingly, many indigenous languages are also spoken. In Paraguay, 90 percent of the population speaks Guaraní. In Bolivia, Quechua and Aymara are both official languages along with Spanish, which about 40 percent of the people don't speak. Quechua is also an official language of Peru and is common in Ecuador. In Guyana, you hear Hindi and Urdu, and in Suriname, a Creole language called taki-taki.

Roman Catholicism is dominant everywhere in South America, except in Guyana, where it's third to Hindu and Anglicanism. Because of South America's great diversity of ethnic groups and languages, Roman Catholicism is the continent's most universal unifying cultural force.

In the News: South America in the Global Economy

The experience of globalization has not been an altogether successful one for South America. After embracing free trade and market reforms in the 1990s, many South American countries did enjoy periods of rapid growth, only to see those gains dramatically reversed in the early years of the twenty-first century. Economic mismanagement, corruption, and the political instability that is always just under the surface, even in the apparently stable democracies, continue to hinder South America's economic progress. The promises of globalization have certainly not yet been fulfilled here, especially for the masses of South America's poor.

Nonetheless, South American countries continue to move toward greater integration and freer trade among themselves and with the rest of the world. Mercosur, the free-trade association comprising the four "southern cone" nations of Brazil, Uruguay, Paraguay, and Argentina (with Bolivia and Chile as associate members) is seeking to strengthen its international negotiating position, especially with regard to the expanding NAFTA group up north. In August 2003, Mercosur signed a free-trade agreement with Peru, which is viewed as a potential bridge across the historic divide between the Mercosur countries and the Andean countries. How successfully South America integrates itself in the global economy and whether it can finally lay to rest its ancient political demons are key questions for the region over the coming years.

The Least You Need to Know

◆ South America is home to the world's largest river, the Amazon, and one of the world's greatest mountain chains, the Andes.

◆ South America's Andean region gave rise to the great Inca Empire.

◆ Brazil, with Portuguese, black African, and Amerindian roots, accounts for half of South America's land mass and population. The rest of the region has mainly Spanish and Amerindian roots.

◆ South America has had a troubled political and economic past, and despite recent stabilization and progress, it isn't out of the woods yet.

Saharan Africa: The Sands of Time

In This Chapter

- Taking stock of ancient African rulers
- Navigating the Nile
- Discovering the Maghreb
- Traversing the transition zone of the Sahel
- Heading out on the Horn of Africa

The vast region of Saharan Africa encompasses the whole of northern Africa. It is bounded on the west by the Atlantic Ocean; on the north by the Mediterranean Sea; on the east by the Suez Canal, the Red Sea, and the Indian Ocean; and on the south by the northern tier of sub-Saharan African countries, from Senegal to Kenya (see Chapter 20).

Although the region is made up of 14 countries, it is named for and dominated by the gigantic Sahara Desert, by far the world's largest. Saharan Africa's other major geographic features include the Atlas Mountains in the northwest; the great Nile River, Egypt's lifeline, flowing through Egypt and Sudan; and the Niger River in the southwest.

Saharan Africa is united by the Islamic faith. The majority of the region's people, whether Arab, Berber, or black African, are Muslim. The northern tier nations bordering the Mediterranean are 95 to 99 percent Muslim. In the Sahel, indigenous religions and Christianity are practiced by minorities of varying sizes.

> **GeoRecord**
>
> The Nile River is the longest river in the world. Its remotest headstream (source of a river) is the Luvironza River in Burundi, 4,145 miles from the Mediterranean Sea. Its drainage basin (the area that drains to a river) is approximately one million square miles!

The Sahara is even larger than the region named for it. The desert extends beyond the Red Sea, across Arabia and the Persian Gulf, through Iraq, and into Iran. It measures more than 3.5 million square miles and includes about 80,000 square miles of *oases*, or fertile patches. Its daytime temperature in the shade can reach 130 degrees F, with nights falling into the 30s. Average yearly rainfall is less than five inches, and some areas get no rain at all for several years.

From Pharaohs to Sultans

Ancient Egypt was one of the birthplaces of human civilization. About 6,500 years ago, a complex culture began to form along the fertile banks of the Nile River and its delta. By 3100 B.C.E. a unified Egyptian kingdom had emerged, with its capital at Memphis, 12 miles south of present-day Cairo. Over the next 3,000 years, Egypt under its pyramid-building Pharaohs developed one of the most brilliant civilizations the world has ever seen. Its achievements in calendrics (the study and development of calendars), agriculture, astronomy, architecture, engineering, and metallurgy were revolutionary; its paintings, statutes, and jewelry still delight modern viewers. The ancient Egyptians also left behind enough ruins of temples and tombs to keep archaeologists busy for umpteen centuries.

> **GeoJargon**
>
> **Nomads** were not aimless wanderers. They were hunters and herders who did indeed wander, but in an established pattern known by their leaders for centuries. The pattern was dictated by water wells, seasonal winds, available animals, and personal whim. Sometimes they stayed tented for months in one camp and then, almost overnight, packed their camels and pounded the sand.

The Berbers: People of the Sand

The story of ancient North Africa is by no means limited to Egypt. About the time the unified Egyptian kingdom was forming, a people called the Berbers were settling along the western part of coastal North Africa, especially in the Atlas Mountain region. Most Berbers were traditionally small farmers and herders, although some groups, like the Tuaregs of the Sahara, were *nomadic*. The Berbers were also great traders and their trans-Saharan caravans played a major role over the centuries in linking the Mediterranean world with West Africa and its supplies of gold and slaves.

Saharan Africa.

In the ninth century B.C.E., Phoenicians, from what is now Lebanon, founded the city of Carthage in Berber lands near present-day Tunis. By the fifth century B.C.E. Carthage controlled the entire coast of North Africa to the Egyptian border and had become the leading commercial power in the western Mediterranean. After the fall of Carthage to Rome in 146 B.C.E., the Berber kingdoms of Mauretania and Numidia (roughly today's northern Morocco and Algeria) eventually became Roman provinces. In 31 B.C.E., Egpyt, too, became a province of Rome after the defeat and suicide of the last of the Pharaohs, the legendary Cleopatra.

> **Geographically Speaking** _____
>
> The Mediterranean Sea has always been one of the world's most important commercial and cultural crossroads. Its name in Latin means "middle of the earth," and that's exactly what is was in the ancient world—the place where the three known continents came together. Control of Mediterranean shipping lanes was a strategic goal of all the great powers that developed along its shores. And whoever controlled North Africa was well on the way to dominating the Mediterranean.
>
> After the discovery of the Americas, Europe's commercial focus shifted to the Atlantic. But with the opening of Suez Canal in Egypt in 1856, the Mediterranean regained much of its importance as a trade link between the Atlantic and Indian Oceans.

The Arrival of Islam

Within a century of the death of Mohammed in 632 C.E., Muslim armies from Arabia had conquered Egypt and the rest of coastal North Africa. Although the Berbers of the Maghreb (discussed later in this chapter) resisted at first, by 709 the Arab conquest of North Africa was complete. Two years later, the Berbers even joined the Arabs in the conquest of much of Spain, where the invaders became known as Moors. The conquered populations of North Africa were mainly Christians, but over time most embraced Islam.

Egyptian culture flourished again in the centuries after the conquest. Under the homegrown Mamluk dynasty (1250–1517), Cairo became the leading cultural, political, and spiritual center of the Muslim world. At the western end of the region, two Moroccan Berber dynasties, the Almoravids and the Almohads, controlled the Maghreb and Moorish Spain from 1050 to about 1250. Muslim traders brought Islam to the Sahel, where great trading empires rose and fell in what are now southern Mali, Mauritania, and Senegal.

The Colonial Era

Beginning in the early sixteenth century, the Ottoman Turks added Egypt and most of coastal North Africa to their growing Mediterranean Empire. After defeating the last of the Moorish kingdoms on its soil in 1492, Spain established military outposts on the north coasts of Algeria and Morocco, two of which, Ceuta and Melilla, remain Spanish to this day. In the late nineteenth and early twentieth centuries, as the Ottoman Empire weakened and European colonial ambitions exploded, France conquered a huge African empire that included, as colonies or protectorates, Morocco, Algeria, Tunisia, Djibouti, and all the countries of the Sahel west of Sudan. Spain began its takeover of what would become Spanish (later Western) Sahara in 1884. Italy took Libya, Eritrea, part of Somalia (the other part was British), and later, in the run-up to World War II, Ethiopia, which had never before been colonized. In 1882, Egypt became a de facto protectorate of Great Britain. Sudan, conquered by Egypt in 1874, was ruled jointly by Egypt and Britain from 1898 to 1955.

As elsewhere in the world, the end of World War II signaled the beginning of the end of the colonial system in Saharan Africa. With the exception of Western Sahara, whose status remains unresolved, most of the countries in the region became independent during the 1950s and 1960s. Throughout the region, the post-independence era has been marked by instability and turmoil, and in the case of the Sahel and the Horn of Africa, extreme poverty as well.

The Maghreb

Maghreb is the Arabic word for Northwest Africa. Technically, it refers only to the countries that share the Atlas Mountains—Morocco, Algeria, and Tunisia—but Libya is usually included in the group. (To be really technical about it, *Maghreb* refers only to the area between the Mediterranean and the high ranges of the Atlas Mountains.) In general, the terrain of the three Atlas Mountains countries consists of coastal plains, plateaus, and mountains in the north, and desert in the south. Libya is mostly flat arid plains and desert. Moving south from the Mediterranean Sea (which all four countries border) or east from the Atlantic (which only Morocco borders), the amount of annual rainfall plummets from 30 to 40 inches to less than 5 inches. Not surprisingly, 95 percent of the population lives in the coastal plains.

The Maghreb.

The original inhabitants of the Maghreb were the Berbers. Today, the populations of these countries are mostly either Arab or mixed Arab-Berber (sometimes referred to as Moors). Pure Berbers live in mountain villages throughout the area. All the people of the Maghreb speak Arabic and nearly all are Sunni Muslims (see Chapter 20).

Morocco

Morocco is a relatively stable and open constitutional monarchy that has maintained good relations with the West and a moderate stance in the Arab-Israeli conflict. Morocco has come under international fire, however, for its de facto annexation of Western Sahara (formerly Spanish Sahara) and its opposition to holding a referendum that would allow the people of the territory to determine their status. Morocco received an enormous shock in May 2003 when a terrorist bomb attack, believed to be the work of al-Qaeda, killed 24 people in Casablanca.

Agriculture is still a key sector of Morocco's economy. Other industries include mining (phosphates), textiles, food processing, and tourism, which has taken a hit recently due to the global slowdown and terrorist threats.

Morocco has a population of 31 million. Its capital is Rabat, with 1.6 million people; its largest city is Casablanca, with 3.4 million. Its tourist treasures include the trio of magnificent medieval royal cities: Fez, Meknes, and Marrakech.

Algeria

Algeria finally attained its independence from France in 1962 after eight years of bitter and violent war. If anything, the country's post-independence history has been even more violent. Since 1965, Algeria has been run essentially as a military dictatorship, despite the facade of more or less regular elections. When an Islamic fundamentalist party won the first stage of parliamentary elections in 1991, the military intervened to overturn the result, triggering a bloody civil war in which, according to some estimates, 100,000 people have been killed. Although an amnesty program in 2000 has ended the worst of the violence, Algeria is still plagued by regular terrorist attacks, rising unrest among the Berber minority, and massive unemployment.

Terra-Trivia

Algeria is famous for the *sirocco*, a hot, dry, dust-filled wind that blows north from the Sahara Desert across the Mediterranean to Italy and its vicinity. Siroccos are most common in the spring.

The one bright spot in all of this is that at least Algeria is not destitute. It has plenty of oil and even more natural gas (it's the world's second largest gas exporter). Of the country's 32 million people, 1.7 million live in the capital, Algiers (nearly 4 million in the city's metropolitan area).

Tunisia

Much of the western Mediterranean was once ruled from the great city-state of Carthage, whose ruins are located near Tunisia's capital, Tunis (population 700,000). Since gaining its independence from France in 1956, Tunisia has had only two leaders, both of whom have been moderate and basically pro-Western in foreign policy and repressive domestically. Islamist opposition parties are severely restricted, as are personal freedoms in general. On the other hand, the Tunisian government has been very progressive on women's issues and has done much to improve the social and legal standing of Tunisian women.

Tunisia has a relatively stable and diversified economy based on petroleum, mining (phosphates and iron ore), and agriculture. Its Mediterranean beaches and rich collection of ancient Carthaginian and Roman ruins and art are major tourist attractions. Of Tunisia's 9.8 million people, only 6 percent live below the poverty line, a very low figure for this region.

Libya

Libya became a monarchy after gaining its independence from Italy in 1951. The discovery of oil in 1959 turned the poverty-stricken county into an oil-rich nation, although much of its population has yet to benefit. In 1968, Colonel Muammar al-Qaddafi overthrew the king and established a radical, anti-Western, extremely anti-Israel, Islamic socialist regime. During the 1980s, Qaddafi diverted oil revenues to fund radical causes and terrorist attacks, most notoriously the downing of Pan Am flight 103 over Lockerbie, Scotland, in 1988. The United Nations imposed sanctions on Libya in 1992 that severely affected its economy. Those sanctions were not lifted until 2003, when Libya finally admitted its responsibility for the Lockerbie disaster and reached a financial settlement with the victim's families, part of a campaign on Qaddafi's part to improve Libya's international standing.

Libya's economy is all about oil, which accounts for almost all of its export income. The combination of oil revenues and a small population gives Libya one of the highest per capita gross domestic products in Africa. Its unemployment rate, however, is a whopping 30 percent. Libya has a population of 5.3 million people, 1.2 million of whom live in the capital, Tripoli.

Egypt and Its Lifeline, the Nile River

Egypt is arguably the leading nation in the Arab world. Among the many reasons are its long and glorious history, both ancient and Islamic; its rich culture; and its strategic location at the junction of Asia and Africa. Size counts, too: Egypt's population of nearly 71 million is the largest in the Arab world. Its capital, Cairo, with 8 million people (nearly double that in its metropolitan area), is the largest city not only in the Arab world, but in all of Africa.

The Land, the River, and the People

Egypt is bordered on the west by Libya, on the north by the Mediterranean, on the east by Israel and the Red Sea, and on the south by Sudan. Describing Egpyt's geography is simple: Think desert. Egypt is basically one big sandy plateau, dotted by a few oases in the west, and divided into two unequal parts by its famous lifeline, the Nile River and its delta. The Nile enters Egypt from Sudan in the south and flows north for about 960 miles to the Mediterranean Sea. At the southern border with Sudan the Nile feeds Lake Nasser, formed by the Aswan High Dam. One of the world's largest artificial lakes, Lake Nasser is 300 miles long and 10 miles wide; about two-thirds of it is in Egypt. Just north of Cairo, the Nile fans out to form its delta, which is about 155 miles wide when it reaches the Mediterranean Sea.

Terra-Trivia

Cairo is home to the Arab world's most important university, the 1,000-year-old Al-Azhar University. It is also a major center of Arab-language literature and media. The only Arab to win the Nobel Prize for literature so far is Egyptian novelist Naguib Mahfouz.

About 99 percent of Egypt's people live on 4 percent of the total land area—almost within sight of the Nile or its delta. The vast majority of the population is made up of ethnic Egyptians, Arabs, and Berbers, but in this highly cosmopolitan society there are also pockets of Greeks, Nubians, Armenians, and Europeans (mainly French and Italian). More than 90 percent of Egyptians are Sunni Muslims; the rest are mostly Coptic Christians. (The Coptic church is the ancient native Egyptian Christian church; it is independent of both Orthodoxy and Catholicism.) Arabic is the national and official language; most educated Egyptians also speak English or French.

Post-Colonial Egypt

At least on paper, Egypt became an independent monarchy in 1922, but British domination of its political life didn't fully end until 1952, when an army group led by General Gamal Abdel Nasser overthrew the king and established an Arab socialist republic. Nasser nationalized the Suez Canal in 1956, triggering an unsuccessful invasion by Britain, France, and Israel. His aggressively anti-Israel policies led to the disastrous 1967 Arab-Israeli War, in which Egypt lost the Sinai Peninsula and Gaza Strip (see Chapter 21) and much of its armed forces.

Nasser's successor, Anwar el-Sadat, launched the Yom Kippur War against Israel in 1973. Despite some early successes, the Arab side was again defeated by Israel. In 1978, U.S. President Jimmy Carter brokered the Camp David accords between Israel and Egypt, and the following year Sadat became the first Arab leader to sign a peace treaty with Israel. Although Sadat won praise and gratitude in the West (along with a Nobel Peace Prize), the reaction to the peace treaty at home and throughout the Arab world was far less positive. In October 1981, Sadat was assassinated by Islamic extremists opposed to the treaty. Sadat's successor, Hosni Mubarak, has since kept the peace with Israel, although it is an increasingly cold peace as the Israeli-Palestinian conflict has escalated in recent years. Although Murbarak has introduced a number of economic reforms, freedom of political expression is still restricted.

Egypt's main economic resource is the Nile and its fertile valley. It also has oil and natural gas, phosphates, iron ore, and hydroelectric power from the Aswan Dam. The threat of terrorist attacks at home and a post-September 11, 2001, decline in international travel and shipping have had a negative effect on two of Egypt's main income earners, tourism and Suez Canal revenues.

The Sahel, or Transition Zone, Countries

South of the Sahara is the Sahel, the semiarid transition zone where the alternately expanding and shrinking desert turns into the grassy plains known as *savannas*. This area is also a transition zone from Arab to black African and from Islam to Christianity and indigenous religions.

The Sahel comprises five countries located in a horizontal band just below the Maghreb and Egypt. All are primarily desert or flat plain, with mountains in north-western Chad and east-central Sudan. The other main physical features of the subregion are the Nile River and its two main tributaries, the Blue Nile and the White Nile, in Sudan; Lake Chad, shared by mainly by Chad and Cameroon; and in the west, the great Niger River, which flows through southeastern Niger and southern Mali. The only two Sahelian countries that aren't landlocked are Mauritania and Sudan.

Mauritania, Mali, Niger, and Chad all gained their independence from France in 1960. Sudan's independence from Britain and Egypt was proclaimed in 1956. In the years since independence, all five countries have known authoritarian rule, political and ethnic strife, and extreme poverty. With little arable land and, with some exceptions, few natural resources, these are among the poorest countries in the world. Soil erosion and desertification are major environmental issues throughout the region.

The Sahel.

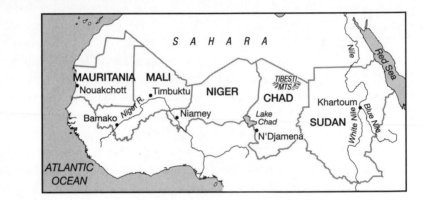

Mauritania

Although Mauritania has been holding relatively open, multiparty elections since 1992, its government is still largely controlled by one party. In 1976, Mauritania took over the southern third of Spanish Sahara, but withdrew in 1979 under mounting pressure from indigenous guerrillas. Ethnic tensions between the majority Maur (mixed Arab-Berber) and minority black populations are a persistent social and political problem. In terms of per capita gross domestic product, Mauritania is the most prosperous of the Sahelian countries. Subsistence farming, fishing, and iron ore mining are the bases of its economy. Offshore oil reserves have recently been discovered. Mauritania has a population of 2.8 million people, almost all of them Muslim. The capital, Nouakchott, has 661,000 people.

Mali

From the twelfth through the sixteenth centuries C.E., great trading empires flourished in what is now Mali. The fabled Malian cities of Timbuktu and Djenne grew vastly rich from the trans-Saharan caravan trade in gold, salt, and slaves and became major centers of Islamic culture. All that is long past. Despite recent political and economic reforms, Mali is desperately poor. About 80 percent of its workers live by farming and fishing in the only place where it's possible, in the southern region watered by the Niger and Senegal rivers. The vast majority of Mali's 11.3 million people are black Africans (mainly Mande); about 10 percent are Tuareg and Maur. The religious breakdown is Islam, 90 percent; indigenous beliefs, 10 percent. French is the official language; the majority African language is Bambara. The metropolitan area of Mali's capital, Bamako, is home to 1.3 million people.

Niger

The drop in world demand for uranium in the 1980s ended Niger's brief brush with prosperity. Since independence, Niger has experienced a succession of military governments, coups, assassinations, and ethnic unrest. In 1995, a longstanding conflict between nomadic Arab-Berber Tuaregs of the north and Niger's majority black population of the south ended in a cease-fire and later a peace accord. After two more coups in the late 1990s, democratic presidential elections were held in 1999. Niger is the poorest of the Sahelian countries, with, at last count, 63 percent of its population living below the poverty line. Substance farming, herding, and uranium mining are Niger's main economic activities. About 80 percent of Niger's 10.6 million people are Muslim; the rest practice indigenous religions or Christianity. French is the official language; Hausa and Djerma are the main African languages. Niger's capital is Niamey, with 748,000 people.

Terra-Trivia

Niger's uranium made headlines again in 2003, when an alleged attempt by Iraq to purchase uranium from Niger was used by the United States and British governments to bolster their case for invading Iraq in March of that year. The allegation later proved to be unfounded.

Chad

Chad's post-independence story has been particularly violent and chaotic, filled with coups, military regimes, ethnic civil wars, and foreign interventions. The primary fault line is between the Muslim north and east and the Christian or animist south, the traditional center of power. (The capital, N'Djamena, is in the south.) Libya's Muammar al-Qaddafi intervened throughout the 1970s and 1980s, either supporting Chadian factions or invading outright. Although Chad's current leader, in power since 1990, has attempted some political reforms, violence continues to flare in the troubled north.

Part of the problem in Chad is that its population of almost 9 million includes some 200 distinct ethnic groups. About half the people are Muslim, 35 percent are Christian, and the rest mostly *animist*. The economy is based on subsistence farming and herding, plus foreign aid. Oil, which is expected to start pumping in 2003, is a one bright spot in Chad's economic future.

Sudan

With an area of 968,000 square miles, Sudan is Africa's largest country; it also has a substantial population of 37 million. Geographically, Sudan's northern third is desert, its middle third is steppes (dry, level, treeless plains) and low mountains, and its southern third, called the Sudd, consists of vast swamplands. All the Nile tributaries flow through southern Sudan, with the two main ones, the Blue Nile and the White Nile, meeting near the capital, Khartoum (metropolitan area population: 5.7 million), to form the Nile.

In the nearly 50 years it has been independent, Sudan has been free of civil war for only 10, from 1972 to 1982. Its governments have alternated between shaky parliamentary coalitions and military dictatorships. In Sudan, the north/south divide pits the dominant Arab Muslim population of the north against the animist or Christian black Sudanese of the south. In 1983, the government, under the increasing influence of Islamic fundamentalist groups, incor-porated Shari'a (Islamic law) in the country's penal code. This action rekindled a civil war that over the last 20 years is thought to have cost more than 2 million people their lives. In addition, Sudan has come under international fire for harboring and supporting Islamic fundamentalist terrorist groups and for permitting the widespread enslavement of southern blacks.

GeoJargon

Animism is the belief that a universal force of life, or soul (*anima* is the Latin word for "soul"), is present in all things—human, animal, vegetable, or mineral—and endows them with power and energy. Animism is one of the indigenous religions of Africa, although it is also found in many cultures around the world.

Sudan's economy is still overwhelmingly agricultural and poor. But remarkably, since the mid-1990s, the government has implemented significant economic reforms that have helped fuel economic growth. In 1999, Sudan began exporting oil, which should help sustain growth over the coming years.

Eye on the Environment

Lake Chad, the Sahel's most significant body of water, is shrinking at an alarming rate. In the 1960s, with a surface area that fluctuated seasonally between 4,000 and 10,000 square miles, the lake was shared by Chad, Cameroon, Niger, and Nigeria. Today, at about 500 square miles, it is confined mostly to Chad and Cameroon. Although Lake Chad has been shrinking and expanding for millennia, weather changes and human activities seem to be accelerating the rate of shrinkage.

A series of severe droughts beginning in the late 1960s reduced the supply of water to the tributaries that feed the lake and increased the rate of desertification in the region. At the same time, expanding human and livestock populations had to rely more heavily on the water of the lake and its tributaries for consumption and irrigation. As the land grew drier, the need to irrigate increased as did pressure on the lake. The result is a lake that is now about one-twentieth the size it was 40 years ago and shows no signs of recovering.

The Horn of Africa

East of Sudan the Sahara Desert meets the high Ethiopian Plateau. This marks the beginning of the Horn of Africa, a huge eastern outcropping, shaped much like the horn of a rhinoceros, that juts up and out from the continent into the Indian Ocean.

The terrain of the Horn of Africa is dominated by Ethiopia's high central plateau, which is split diagonally by the Great Rift Valley, a fault system that starts in northern Syria, cuts through the Jordan Valley and the Red Sea, and in Africa, runs all the way down to Mozambique. In Ethiopia, the Rift Valley area is dotted by small volcanoes and hot springs in the north and by large lakes in the south. The central plateau is mountainous and has a generally temperate climate. It descends steeply to the hot, arid coastal plains of Somalia and the much narrower one of Eritrea. The Blue Nile, one of the Nile's main tributaries, rises in Lake Tana in northwestern Ethiopia.

Eritrea, Ethiopia, and the western leg of Somalia are all former Italian colonies. Djibouti was once a French colony; the north-western part of Somalia was formerly British. With the exception of Djibouti, the nations of the Horn of Africa have the unhappy distinction of being among the 10 poorest in the world in terms of per capita gross domestic product.

Terra-Trivia

The Great Rift Valley is the result of the pulling away of the Somali tectonic plate from the African plate. (For more on plate tectonics, see Chapter 2.) Eventually, as eastern Africa continues to split off from the rest of the continent a new sea or gulf will form in what is now the Great Rift Valley.

Eritrea

After the Italians were driven out of the Horn of Africa in 1941, Eritrea was administered by Britain until 1952, when it joined Ethiopia in a federation. In 1962, Ethiopia made Eritrea a province, triggering a disastrous 30-year civil war. Eritrea finally prevailed in 1991 and became independent two years later. In 1998, a vicious and costly border war broke out between the two neighbors. Although hostilities ended in late 2000 and United Nation peacekeeping forces are in place, tensions are still high between the two countries.

Eritrea's 4.5 million people comprise nine different ethnic groups; Tigrinya and Arabic are the most common languages, although English is widely spoken. The population is almost evenly divided between Christians and Muslims. Nearly 900,000 people live in the capital, Amsara, and its surrounding area. Eritrea's war-bankrupted economy is based on subsistence farming.

The Horn of Africa.

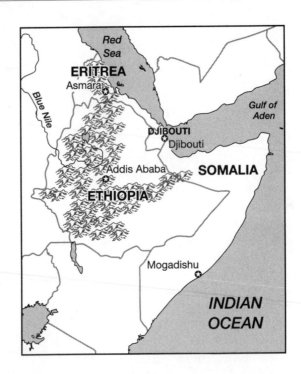

Ethiopia

Ethiopia is Africa's second oldest state after Egypt. According to legend, its first emperor was Menelik I, the son of Solomon and the Queen of Sheba. Except for the relatively brief period of Italian occupation from 1936 to 1941, Ethiopia has always been independent. In 1974, its last emperor Haile Selassie was deposed and later strangled in a military coup. The coup leaders set up a socialist military dictatorship, which became a full-fledged communist regime after Lt. Col. Mengistu Haile Mariam seized power in 1977. Supported by the Soviet Union, the Mariam government fought a civil war with Eritrean separatists and border wars with Somalia, while engaging in a reign of terror at home. After Mariam was forced into exile in 1991, Ethiopia's government eventually recognized Eritrea's independence and has made efforts to defuse ethnic tensions and establish a multi-party democracy in the long-troubled nation.

Ethiopia has more than 80 ethnic groups, although three of them—the Oromo, Amhara, and Tigrean—

Terra-Trivia

The coffee plant is thought to have originated in Ethiopia's central plateau, where it still grows wild. According to legend, the invigorating qualities of coffee were first discovered by an Ethiopian goatherd who noticed that his goats got particularly frisky after chewing on coffee beans.

account for about 75 percent are the country's 67.6 million people. About 2.7 million people live in the capital city of Addis Ababa. More than 70 languages are spoken in Ethiopia. Amharic is the official language; English is the most widely spoken foreign language. Nearly half the people are Muslim, more than a third are Ethiopian Orthodox Christians, and about 15 percent practice indigenous African faiths.

Coffee and foreign aid are the mainstays of Ethiopia's impoverished agricultural economy. The combination of war and recurring droughts over the last decades not only left Ethiopia's economy in shambles, but it also produced a series of horrifying famines.

Djibouti

Tiny Djibouti gained its independence from France in 1977. In 1991, mounting tensions between the dominant ethnic group, the Somali Issas, and the minority Afars led to the outbreak of a civil war that didn't fully end until 2001. The war, repeated droughts, and the influx of refugees from Ethiopia and Somalia have wrecked Djibouti's economy, although it remains the most prosperous in the Horn of Africa.

Djibouti has a population of 473,000 people, about two-thirds of whom live in the capital of the same name. The official languages are French and Arabic; the predominant religion is Islam. Djibouti's economy is all about its strategic location at the junction of the Red Sea and Gulf of Aden, not far from the Arabian oilfields. Its port and railroad line are the country's main source of income and employment.

Somalia

East of Djibouti and Ethiopia and bounded on the north by the Gulf of Aden and on the east and south by the Indian Ocean is Somalia, a country of some 7.7 million people, most of them Muslim Somalis. The former colonies of British and Italian Somaliland became the Republic of Somalia in 1960. The new country waged a costly territorial war with Ethiopia in the 1970s and faced mounting insurgencies in the 1980s. In 1990, Somalia began to disintegrate into hostile "ministates" controlled by rival warlords. War, famine, and mounting casualties led the United States to send in troops in 1992 to facilitate the delivery of food and assistance. U.S. forces withdrew in 1994 after the notorious "Blackhawk Down" incident in which the bodies of American soldiers were dragged through the streets of the capital, Mogadishu. As of mid-2003, Somalia has no national government and is a source of increasing concern as a likely base for al-Qaeda-linked terrorists.

In the News: Ethiopia—Birthplace of Modern Humans?

Ethiopia has long been fertile ground for scientists searching for the earliest human ancestors. One of the most famous and oldest of the potential human ancestors is Lucy, a 3.2-million-year-old female of the two-footed species *Australopithecus afarensis*, whose partial fossilized skeleton was discovered in 1974 at Hadar, Ethiopia. In June 2003, another landmark Ethiopian discovery made headlines around the world. Three fossilized *homo sapiens* (modern human) skulls discovered near the village of Herto in the Afar region of eastern Ethiopia were dated at 160,000 years old, making them the oldest anatomically human fossils ever found. What's remarkable about the Herto finds is that the three skulls (two adults and one child) date from the same time that biologists had predicted for the existence of a genetic "Eve," the common African ancestress of all modern humans.

The Least You Need to Know

- ◆ The Sahara Desert, the largest in the world, is the dominant physical feature of northern Africa.

- ◆ The Nile River, the longest in the world, is the source of life and civilization in an arid land.

- ◆ Ancient Egypt was one of the world's earliest civilizations.

- ◆ Islam is the predominant religion of Saharan Africa and acts as a unifying cultural link throughout the area.

- ◆ The nations of the Sahel and the Horn of Africa have had very troubled modern histories and are among the poorest countries in the world.

Chapter **20**

Sub-Saharan Africa: Cradle of Humankind

In This Chapter

- ◆ Overview of a troubled region
- ◆ The birthplace of humankind
- ◆ Ancient kingdoms and colonial masters
- ◆ Sub-Saharan Africa today
- ◆ Post-apartheid South Africa

In the previous chapter, you read about the Saharan countries that make up the northern part of the African continent. This chapter focuses on the vast and complex region of sub-Saharan Africa. About 70 percent of the more than 800 million Africans live in the 40 or so countries south of Saharan Africa.

Geographically, the sub-Saharan region is mainly one vast plateau covered with savannas (grasslands) in the north, tropical rain forests in the central equatorial belt, and more grasslands and deserts (the Kalahari and the Namib) in the south. Most of the region's lakes and mountains are located along the Great Rift Valley that cuts through eastern Africa from Ethiopia (see Chapter 19) to Mozambique. Sub-Saharan Africa's two great river basins are the

Niger Basin in the southern part of the continent's western bulge and the Congo Basin in equatorial west-central Africa.

Sub-Saharan Africa is a region of stunning contrasts. On the one hand, this is the Africa of the travel posters, with safari parks and nature preserves teeming with lions, elephants, giraffes, cheetahs, baboons, hyenas, rhinoceroses, and great herds of antelopes and wildebeests. The region is also vastly rich in natural resources: petroleum, natural gas, gold, silver, diamonds, copper, uranium, tin, zinc, coal, iron—if it's a metal or a mineral, sub-Saharan Africa probably has it. And yet, the reality of sub-Saharan Africa today is beyond grim. Along with the Sahel (see Chapter 19), sub-Saharan Africa is the least developed region on earth, with subsistence agriculture still the leading economic activity in most of the region. Of the 25 nations with the lowest per capita gross domestic product, 19 are in the Sahel and sub-Saharan Africa.

Ancient Roots

Humans have lived longer in Africa than in any other continent. It is here that fossil records indicate that the human species developed. The first members of our own genus, *Homo*, emerged in East Africa about 2 million years ago. *Homo habilis*, ("skillful human"), as these early tool-using proto-humans are called, developed into the larger-brained *Homo erectus* ("upright human") about 1.6 million years ago. Members of *H. erectus* eventually left Africa and populated Asia and Europe. Although the origins of modern humans, *Homo sapiens* ("wise human"), is still being debated, recent fossil finds (see Chapter 19) point to Ethiopia and East Africa as the likely birthplace of our species about 200,000 years ago.

As you read in the previous chapter, northern Africa gave rise to one of the first great human civilizations, ancient Egypt, and later to powerful empires and kingdoms in the northwest. Sub-Saharan Africa, however, remained isolated from developments in the Mediterranean world for thousands of years. By 1000 C.E., Muslim traders from the north were crossing the Sahara to trade salt for gold and slaves from West Africa and in the process brought Islam with them. The great Muslim trading empires of Ghana, Mali, and Songhai developed in the Niger River basin. Further east the Hausa city-states grew wealthy on the trans-Saharan trade.

Around 1100, Arab traders began establishing trading centers along the east coast of Africa from what is now Somalia down to Mozambique. By 1500, the states of Benin and Oyo had emerged west of the Niger delta, on what Europeans traders, now beginning to arrive in the region, would name the Slave Coast. South of the equator were the Bantu-speaking states, among them the mighty kingdom of Kongo, south of the Congo River delta, and the powerful trading empire centered on the city of Great Zimbabwe and its nearby gold mines.

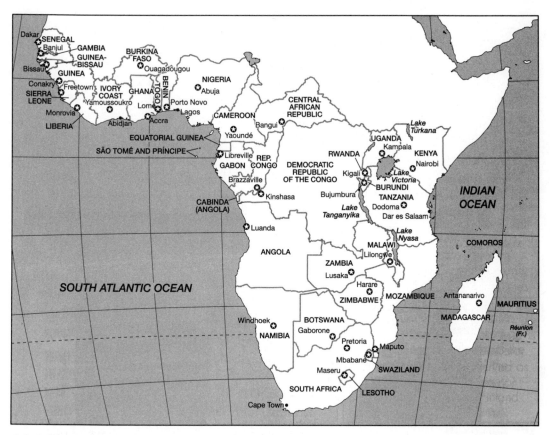

Sub-Saharan Africa.

Europe in Africa

Sub-Saharan Africa lacks good natural harbors, which explains in part why much of the region's early historical development was focused on overland trade routes in the interior and not on coastal settlements. The arrival of European explorers in the fifteenth century would force sub-Saharan Africa to turn its attention outward.

The Rise of the Slave Trade

Beginning in 1419, the Portuguese began sending expeditions down the west coast of Africa. The goal was to bypass Arab middlemen and find a safe water route around Africa to the riches of India and the Far East. In 1499, Vasco da Gama finally made it all the way around southern Africa to India. The Dutch, English, French and other European maritime powers soon joined the Portuguese in setting up coastal stations and ports along Africa's coasts. These eventually became centers of the slave trade.

Geographically Speaking

Although Africans and Arabs had been dealing in slaves long before the arrival of the Portuguese, the full-scale European slave trade is considered to have begun in 1444 when the first large group of slaves—about 235 kidnapped Mauritanians—were shipped to Portugal. But the slave trade really took off after plantation economies were established in the New World.

The first slaves in what would become the United States arrived in Virginia's Jamestown colony in 1619. The main New World destination for west African slaves was Brazil, which imported more than 4 million slaves between 1526 and 1870, when the slave trade finally came to an end. In one of the most shameful episodes in human history, Africa is estimated to have lost some 30 million of its people to bondage or to death during the horrifying trans-Atlantic journey known as the Middle Passage.

Slavery not only ravaged the interior of Africa, from which the slaves were "mined"—destroying families, entire villages, and even cultures—it also caused untold misery and suffering to the slaves themselves and their descendants for centuries to follow. Although European traders introduced New World foods to Africa, such as corn and cassava, that would become important crops on the continent, they also introduced guns, which helped to further destabilize African societies.

The European Takeover

For centuries, European interest in Africa remained focused on the coasts. It wasn't until the second half of the nineteenth century that European missionaries and explorers began to penetrate the interior. Their reports about the natural wealth of the "Dark Continent" soon caused the European powers to start establishing "spheres of influence" throughout Africa. In an attempt to impose order on what was becoming a frantic "Scramble for Africa," 13 European countries and the United States met in the Berlin Conference of 1884 essentially to plan the partition of Africa without the presence of a single African representative.

From the toeholds they already had on the coasts, the Europeans proceeded to carve up all of Africa, drawing and redrawing boundary lines through known and unknown expanses. The final lines took little account of African peoples, settled areas, hostile tribes, local cultures, languages, or in many cases even natural geographic barriers, such as rivers, mountains, deserts, and lakes. The arbitrary partition of Africa between 1880 and 1913 set the stage for many of the continent's current disasters.

Within 20 years of the Berlin conference, all of sub-Saharan Africa was under some sort of colonial rule, except for Liberia, whose ties with the United States kept it out of European hands. Britain carved out for itself a nearly uninterrupted 4,900-mile-long north-south corridor from Cairo to Cape Town. Only the presence of German East Africa (today's Rwanda, Burundi, and Tanzania) broke the continuity of British control (at least until after World War I, when Britain took over the former German colony as a protectorate). In West Africa, Britain also picked up resource-rich Nigeria, Ghana (then known as the Gold Coast), Sierra Leone, and Gambia.

Terra-Trivia

Liberia, Africa's first republic, was founded by the American Colonization Society in 1822 as a private colony (not part of the United States) for the settlement of freed slaves. Within 40 years, some 12,000 former slaves had settled what was then called Monrovia, after U.S. President James Monroe. In 1847, Monrovia became the Republic of Liberia, with the city of Monrovia as its capital.

The French claimed an enormous chunk of the continent—all of the western bulge of Africa except for Spanish Morocco, Spanish Sahara (now Western Sahara), the British west African colonies, independent Liberia, and the German colonies of Cameroon and part of Togoland, which France took over after Germany's defeat in World War I. France's African holdings also included what are now the central African nations of Gabon, Congo, and the Central African Republic, as well as the islands of Madagascar, Réunion, and the Comoros off Africa's east coast.

The huge and mineral-rich Belgian Congo (later known as Zaire and now as the Democratic Republic of the Congo) was ruled for a time as the personal property of the king of Belgium, who exploited it brutally. An international uproar led the Belgian government to take over the administration of the colony in 1908. Portugal clung to its ancient colonies of Angola, in the southwest, and Mozambique, in the southeast, and ruled both repressively. In addition to its previously mentioned colonies, Germany also controlled what is now Namibia in southwest Africa. After Germany's defeat in World War I, the British dominion of South Africa took over the former German colony.

Independent Africa: Instability and Turmoil

At the end of World War II, even the victorious European powers were too devastated to hold on to their empires. By the late 1960s, almost all the sub-Saharan nations had gained their independence, with some major exceptions. The Portuguese colonies of Angola, Mozambique, and tiny Guinea Bissau in west Africa fought long and bitter independence wars before Portugal finally withdrew from all its colonies in

1975. Britain didn't recognize a 1965 declaration of independence by Rhodesia's white supremacist regime until the country's black majority was given full voting rights and free elections were held in 1979. The following year Rhodesia became independent under the name of Zimbabwe. In 1990, Namibia finally gained its independence from South Africa after more than 20 years of armed conflict.

In the postcolonial era, sub-Saharan Africa has known much more than its share of violence, warfare, epidemic disease, environmental destruction, famine, and even genocide. From the 1960s through the 1980s, much of the region became a surrogate battleground for the cold-war struggles between the Soviet Union and the United States. Military coups, counter-coups, and brutal military regimes have been the norm across the region. Nationalist struggles and inter-tribal conflicts nearly tore apart Nigeria and the Republic of the Congo (now the Democratic Republic of the Congo) the 1960s.

Throughout the 1970s the east African nation of Uganda was in the grip of one of the continent's most bloodthirsty dictators, Idi Amin. In South Africa, decades of mounting violence, government repression, and international sanctions finally came to an end in 1992, with the dismantling of the apartheid system of white-minority rule. In 1994, long simmering ethnic hatreds between the Tutsi and Hutu peoples of Rwanda erupted in a genocidal massacre that left nearly a million Tutsi dead. In retaliation, Tutsi rebels overthrew the Hutu government, causing nearly two million Hutus to flee into neighboring Democratic Republic of the Congo, Burundi, Uganda, and Tanzania, further destabilizing an already shaky region. Most recently, in the summer of 2003, Liberia erupted in a frenzy of violence and chaos as rebel forces took over the capital of Monrovia and toppled the government of former rebel leader Paul Taylor.

An Overview of Sub-Saharan Africa

Sub-Saharan Africa has so many countries that a description of each one is beyond the scope of this book. This section divides the region into five subregions and describes the highlights of each.

Western Africa

This subregion, as shown on the following map, encompasses the southern part of the western African bulge, from Senegal to Nigeria. All the countries are coastal except for Burkina Faso. Because the area is tropical, it's hot year-round. Rainfall is heavy in the south and diminishes to the north. Tropical rain forests in the south give way to savanna in the northern half of the subregion.

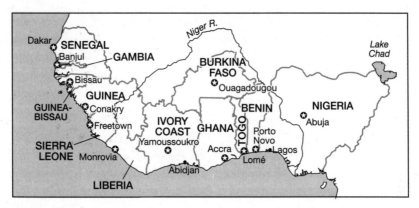

Western Africa.

The dominant physical feature of the area is the Niger River. The world's thirteenth longest river (2,600 miles long) rises in Guinea, flows northeast out of the subregion into the Sahelian countries of Mali and Niger (see Chapter 19) and then back down through Nigeria where it empties into the Gulf of Guinea. Lake Volta, one of the world's largest artificial lakes, is located in Ghana, and the northeastern corner of Nigeria borders on Lake Chad.

Western Africa has about 222 million people. Its largest country is Nigeria, with nearly 130 million. The cities along the southern coast of western Africa form one gigantic megalopolis, from Abidjan, the administrative capital of Côte d'Ivoire (Ivory Coast), with 4 million people, to Accra, capital of Ghana, with nearly 3 million people, and Nigeria's largest city, Lagos, with 9.5 million. To the north, Senegal's capital, Dakar, is also a major metropolis, with a population of nearly 2.5 million. Despite its number of large cities, western Africa is still heavily rural. Even in oil-rich Nigeria, nearly 70 percent of the workforce is engaged in agriculture.

The tribal, ethnic, and linguistic makeup of the subregion is extremely complex. Nigeria alone has more than 250 ethnic groups. Senegal and

Terra-Trivia

In the sixteenth century C.E. the great forest kingdoms of Benin and Oyo flourished west of the Niger delta in what is now southern Nigeria and Benin. It was here that some of sub-Saharan Africa's finest artwork was produced. Benin wooden masks and clay and bronze sculptures are famous around the world.

Terra-Trivia

Both Côte d'Ivoire and Nigeria have acquired new (and much smaller) capitals in their respective interiors: Yamoussoukro in Côte d'Ivoire and Abuja in Nigeria.

the Gambia are more than 90 percent Muslim; in the rest of the nations of the subregion, varying percentages of the population practice either Islam, Christianity, or indigenous faiths.

In size and population, Nigeria is the cornerstone of western Africa. It is also potentially one of Africa's wealthiest nations thanks to its oil reserves. An oil boom in the 1970s left Nigeria with the thirty-third highest per capita gross domestic product in the world. Today, thanks to corruption, mismanagement, and civil strife, Nigeria is one of the poorest countries in the world. Its stability is once again seriously threatened by mounting tensions between Christians and Muslims. The most prosperous countries in the subregion in terms of per capita gross domestic product are Ghana and Guinea; the poorest is Sierra Leone, which is tied with East Timor as the poorest country in the world.

Central Africa

Central Africa lies astride the equator in the west-central part of the continent. From north to south along the Atlantic coast are Cameroon, Equatorial Guinea, Gabon, Congo, a tiny separated bit of Angola (see the section on Southern Africa later in this chapter) called Cabinda, and the Democratic Republic of Congo's narrow outlet to the ocean. Inland, north of the Democratic Republic of Congo, is the Central African Republic. Off the western shore is the island nation of São Tomé and Príncipe.

Rainfall is ample (as you might expect) throughout this tropical subregion, although it's heavier along the coast and on the mountain slopes. West Cameroon's slopes get as much as 400 inches of annual rainfall! Two of the subregion's major rivers (it has hundreds) separate the Central African Republic from the Democratic Republic of Congo: the 2,720-mile-long Congo (the world's eighth longest) and its major tributary, the 660-mile-long Ubangi.

Geographically Speaking

Pygmies are short (average 5 feet tall) and dark-skinned hunter-gatherers who live in tropical rainforests in Africa, the Malay Peninsula, the Philippines, and New Guinea. Although questions about the origins of pygmies persist, modern blood typing and other studies seem to indicate that the various groups of pygmies are distinct from each other and had independent origins.

There are about 250,000 African pygmies, called Negrillos, living in the Congo River basin as well as in Rwanda and Burundi. Africa's pygmies are thought to predate their agricultural neighbors, with whom they trade and whose languages they've adopted.

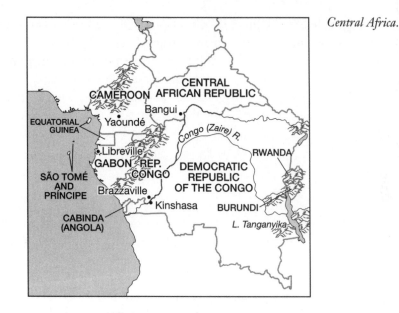

Central Africa.

Northern Cameroon and most of Central African Republic are covered by savanna and shrub growth. The rest of the subregion is mostly tropical rain forest. The heart of the subregion is the Congo Basin, shared by the two Congos. Most of the basin is surrounded by plateaus and mountains, the tallest being the Mitumba Range along the Democratic Republic of the Congo's eastern border.

Central Africa is relatively thinly peopled. Only 40 percent of the population is urban. The largest city is Kinshasa, the capital of the Democratic Republic of the Congo. This huge and troubled country in the heart of Africa is the giant of the sub-region, with more than half of the land and people.

Although Bantu-based languages prevail in most of the subregion, the legacy of France and (in the case of the Democratic Republic of the Congo) Belgium is evident in the fact that French is the official language of the two Congos, the Central African Republic, and Gabon. French and English are the official languages of Cameroon, while Portuguese is spoken in São Tomé and Príncipe. Equatorial Guinea is the only country in Africa whose official language is Spanish.

Most people in Central Africa farm, fish, herd, log, or mine for a living. The subregion includes the most prosperous sub-Saharan country outside of South Africa as well as one of the poorest. A combination of off-shore oil, a small population, a history of stable (if authoritarian) government, and more recently, a freer and more open political system explains the relative prosperity of Gabon today.

On the opposite side of the coin, a grim history of colonial exploitation, secessionist movements, the long and disastrous dictatorship of General Mobutu (1965 to 1997), civil wars, and ongoing ethnic conflicts both at home and with its neighbors has left the Democratic Republic of the Congo devastated and impoverished. Even though the country possesses vast amounts of natural and mineral resources—oil, gold, silver, diamonds, uranium, zinc, you name it—the DRC's per capita gross domestic product is just a notch above that of the poorest countries in the world: Somalia, East Timor, and Sierra Leone.

Eastern Africa

The subregion of Eastern Africa straddles the equator in the east-central part of the continent. Kenya and Tanzania are on the Indian Ocean coast; Uganda, Rwanda, and Burundi lie just to the west. The climate is tropical year-round but cooler in higher elevations. Although Uganda, Rwanda, and Burundi have ample rain, Tanzania is generally drier and Kenya considerably drier still. From west to east, this highland region ranges from light tropical forest vegetation to savanna lands.

The Nile rises in Burundi and flows north, but it's small and inconsequential in this area. A string of mostly elongated lakes marks the entire western edge of the subregion. These include Lake Tanganyika, the seventh largest lake in the world, and Lake Malawi, the tenth largest. These lakes and the surrounding mountains are all part of the Great Rift Valley system (see Chapter 19).

Eastern Africa is also home to two of the continent's great scenic wonders. Lake Victoria, the world's second-largest freshwater lake, lies on the equator where the boundaries of Uganda, Kenya, and Tanzania meet. Although eastern Africa is not as mountainous as Ethiopia to the north, northern Tanzania does contain Africa's highest mountain, the snow-capped Mount Kilimanjaro, at 19,340 feet.

The fertile Lake Victoria basin is one of Africa's major population centers. Almost half of eastern Africa's 106 million people live in the parts of Uganda, Kenya, and Tanzania surrounding the lake as well as in tiny Burundi and Rwanda. The population density here exceeds 500 people per square mile, compared to 65 for all of sub-Saharan Africa. With about 2.4 million people each, the largest cities in eastern Africa are Kenya's capital, Nairobi, and Tanzania's administrative capital, Dar es Salaam. (Tanzania has also moved its capital inland to the small city of Dodoma.) The subregion's next largest cities are Uganda's capital, Kampala, and Mombasa in Kenya.

Geographically Speaking

When you think of Africa, you probably think first of wildlife; its preservation has been a hot issue in Africa in recent years. As humans encroached on their territory, the big cats became endangered (one lion pride needs at least 15 square miles of living space). Elephants have long been targeted for their tusks, and rhinoceroses for their horns. Throughout the colonial era and beyond, all big animals were at risk from big-game hunters.

Drought, deforestation, and human poverty have also taken a tremendous toll on Africa's wildlife. But there is still a chance to preserve what's left. Western countries banned the import of ivory, and Kenya has outlawed hunting. Tanzania and Kenya have set aside huge tracts as wildlife preserves. Africa's magnificent national parks such as Serengeti, Ngorongoro, Mara, and Tsavo, are certainly part of the solution both for the wildlife and for struggling economies that depend on tourist income. But there's also a flip side. It's been estimated that 85 percent of Africans have never seen a wild animal. Relatively few can afford the parks' entrance fees, much less the car to get there.

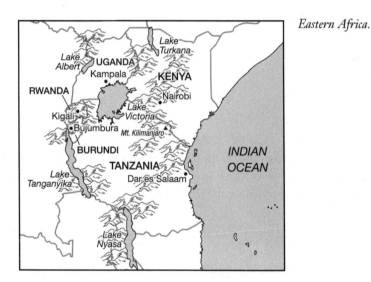

Eastern Africa.

Despite its large urban population, eastern Africa is still overwhelmingly rural. All of the countries in this subregion cluster at the bottom of the poverty chart. With a per capita gross domestic product of $1,200, Uganda is the best off of the group. It has achieved a measure of political stability recently and it has an abundance of fertile soil and regular rain to support its primarily agricultural economy (coffee is the main export). Uganda has also been one of the most successful countries in Africa in fighting AIDS, the lethal disease that is ravaging much of sub-Saharan Africa.

Kenya is the financial and business hub of eastern Africa, but its economy is plagued by deep-seated corruption and low prices for its main exports: coffee, tea, and fruits and vegetables. Tanzania had its first multiparty elections in 1995 and its heavily agricultural economy has been growing in recent years due to a pickup in the mining sector (diamonds and gold). Nonetheless, Tanzania remains one of the poorest countries in the world. In eastern Africa, only Burundi is poorer. The gruesome events in the recent histories of Burundi and Rwanda have been touched upon earlier in this chapter.

Southern Africa and Madagascar

Southern Africa includes everything south of the Democratic Republic of Congo and Tanzania. Heading south on the Atlantic coast on the west are Angola, Namibia, and South Africa (which surrounds Lesotho and Swaziland). Mozambique takes up the entire Indian Ocean coast on the east, north of South Africa. In the center are the landlocked countries of Malawi, Zambia, Zimbabwe, and Botswana. Madagascar, the world's fourth-largest island, lies off Mozambique's east coast, along with the island countries of Comoros and Mauritius and the tiny French island of Réunion. Because of its economic importance, South Africa is treated separately in the following section of this chapter.

Woodland forests dominate Angola, Zambia, Malawi, northern Mozambique, and eastern Madagascar. The Kalahari Desert covers western Botswana and eastern Namibia, and extends into South Africa. The Namib Desert runs all along the coast of Namibia into South Africa and north into Angola. Almost everywhere else, croplands and grazing lands prevail.

Although Northern Angola, Zambia, and Malawi are tropical, the rest of the region is temperate with mild winters and hot summers. The deserts are hotter and dry. The forest areas in the north get moderate amounts of rain; totals are lower elsewhere in the subregion.

Geographically Speaking

Most people in sub-Saharan Africa speak languages belonging to the Bantu subfamily of the Niger–Congo language family, with two interesting exceptions. On the island of Madagascar, people speak Malagasy, a Malayo-Polynesian language betraying the island's South Pacific roots. The San and Khoikhoi peoples of the Kalahari and Namib deserts speak a unique Khoisan "click" language that is spoken nowhere else on earth. (You might know the San and Khoikhoi better by the pejorative names the white settlers gave them: bushmen and Hottentots.)

The largest lake in the subregion is Lake Malawi, between Malawi and Mozambique. The region has several sizable rivers, including the Zambezi (Africa's fourth longest), which forms the border between Zambia and Zimbabwe; the Limpopo, which forms the eastern part of the border between Botswana and South Africa; and the Orange–Vaal, which flows through central South Africa before forming the border between South Africa and Namibia. The subregion is also home to another of Africa's great natural wonders: the Zambezi River's Victoria Falls on the eastern end of the border between Zambia and Zimbabwe.

In addition to the big cities of South Africa—Cape Town, Johannesburg, and Durban— southern Africa's other large cities include Angola's capital, Luanda, and Zimbabwe's capital, Harare, each with 2.3 million people, and Maputo, Mozambique's capital, and Lusaka, the Zambian capital, each with about 1.7 million people. Except for South Africa, the countries in this area are thinly populated: They have 25 percent of sub-Saharan Africa's land but only about 13 percent of its people.

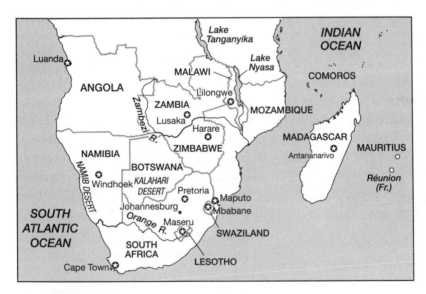

Southern Africa.

Excluding South Africa, Botswana is the wealthiest country in the subregion and one of Africa's rare economic and political success stories. Since gaining independence from Britain in 1966, Botswana has maintained a stable democracy (now Africa's oldest) and its diamond mining industry has fueled healthy economic growth and the second highest per capita gross domestic product in Africa. (Tourism and agriculture are its other main industries.) Unemployment and poverty rates remain high, however, and even more troubling, Botswana has one of the highest HIV/AIDS infection rates in the world.

In the rest of the subregion, the economic picture is grimmer. Violent independence struggles, communist dictatorships, and years of civil wars left both Angola and Mozambique in tatters. Although Mozambique has achieved a good measure of political stability and even economic growth since 1992, its already impoverished economy was dealt a heavy blow by severe flooding in 1999 and 2000. Thanks to its oil and diamonds, Angola is somewhat richer than Mozambique, but in many respects it is even worse off. From 1975 to 2002 the country was locked in a civil war that is estimated to have cost 1.5 million lives. Although the peace seems to be holding, Angola is struggling with food scarcities as thousands of people who had fled the country during the war return home. Life expectancy in both Angola and Mozambique is under 40 years.

Copper-rich Zambia has been hard hit by low copper prices, droughts, and the usual problems of political instability and corruption. Desperately poor Malawi has held multiparty elections in 1994 and 1999, but its economy is among the world's least developed. With agriculture accounting for more than 90 percent of the workforce, Malawi barely survives on tobacco exports and foreign aid. A growing HIV/AIDS epidemic is taking a heavy toll here as well.

Terra-Trivia

Madagascar is known as the "land of living fossils," as it is home to a huge number of wildlife species found nowhere else on earth. It is especially famous for its lemurs (primitive tree-dwelling primates) and chameleons. Deforestation is a serious and growing threat to the rainforests that support Madagascar's unique animals and plants.

The destructive economic policies of Zimbabwe's renegade president, Robert Mugabe, the only ruler Zimbabwe has known since independence, have caused its white farmers to flee and have wrecked the potentially healthy economy of this mineral-rich country. Inflation in early 2003 was 238 percent, economic growth declined more than 12 percent in 2002, and an estimated 70 percent of Zimbabweans live below the poverty line.

Madagascar held its first multiparty elections in 1992, but a contested election in 2001 nearly split the country apart. (A winner has since been declared and generally accepted.) Agriculture is the mainstay of this extremely poor country's economy.

South Africa: Challenges in the Post-Apartheid Era

Gold- and mineral-rich South Africa is the continent's most economically developed and wealthiest country. It is the world's biggest producer of gold, platinum, and chromium, and a leading producer of diamonds and a host of other metals and minerals. It has a diversified economy, with a well-developed infrastructure, a sound fiscal system, and a major stock exchange (one of the world's 10 largest). That's the good news. The bad news is that South Africa's economy continues to be weighed down by the high poverty rates and lack of equal economic opportunity for its nonwhite citizens—lingering legacies of South Africa's now dismantled system of racial separation known as apartheid.

South Africa's terrain is dominated by a high central plateau surrounded by mountains on the east and south; the highest elevations are in the Drakensberg Mountains in the east. Most of the plateau is rolling grassland, or *veld*, called highveld (over 5,000 feet).

To the west is the lower, semiarid middleveld, and between the mountains and the sea is the lowveld, bordered by a narrow coastal strip. The southern reaches of the Kalahari and Namib Deserts extend into South Africa from Botswana and Namibia. South Africa is the only truly temperate country in sub-Saharan Africa. Although temperatures vary by season and altitude, hot summers and mild winters typically prevail. Rainfall is moderate in the east but sparse in the west.

GeoJargon

Veld is a Dutch word meaning "field." In South Africa, it refers to the rolling grasslands found primarily on the central plateau. Elevation determines whether a veld is high, middle, or low.

South Africa has a population of 43.6 million. About 75 percent are black, 13.6 percent are white, 8.6 percent are Coloured, and 2.6 percent are of Indian descent. There are 11 official languages: Afrikaans (the first language of about 60 percent of the white population and most mixed-race people), English, and nine indigenous languages. About 68 percent of the population is Christian and more than 28 percent practice animism and other indigenous faiths. There are also small percentages of Hindus and Muslims. South Africa actually has three capitals: Pretoria (population: 1.5 million), Cape Town, the legislative capital (population: 3 million), and Blomfontein, the judicial capital (with 378,000 people). The population of Cape Town proper (as opposed to its metropolitan) is 2.7 million, making it South Africa's biggest city. Johannesburg, however, has the biggest metropolitan area population at 5 million.

Although the apartheid regime has been successfully dismantled and black-majority rule is firmly in place, South Africa still struggles to bridge huge economic and racial divides within its population. Although whites still dominate the economy, blacks, often poorly paid and ill-treated, make up more than three-quarters of the workforce. Despite the country's wealth, about 50 percent of South Africans live in poverty. South Africa also suffers from high crime rates and one of the world's worst HIV/AIDS crises. It is estimated that South Africa has more HIV-infected people (about 5 million) than any other country on Earth.

In the News: Africa's Health Crises

HIV/AIDS has been mentioned frequently in this chapter and for good reason. The disease today is the leading cause of death in sub-Saharan Africa. In 2001, 28.5 million people in the region were infected with HIV and 2.2 million died of AIDS. In a region where overpopulation was once the problem, population growth rates have either slowed dramatically or are declining in some places because of mounting infant and adult death rates due to AIDS. Life expectancy rates are dropping throughout the region to as low as 45 years in some countries and in others, like Botswana, to below 40. (In relatively wealthy Botswana, life expectancy would have been 72 years had it not been for AIDS.)

The consequences of AIDS in Africa go beyond individual suffering and loss of life. An entire generation is being targeted just at the age when it should be at or near peak productivity. The effects on the economies and social fabrics of many sub-Saharan nations threaten to be catastrophic. As of 2001, there were an estimated 11 million AIDS orphans in the region, and that number is expected to climb to 20 million by 2010. South Africa is projected to lose 11 percent of its workforce by 2005; Zimbabwe, about 20 percent. Compounding the crisis is the fact that the medications that have made AIDS in the West more of a chronic disease than an instant killer are either unavailable in Africa or out of most people's reach.

The Least You Need to Know

- In the Middle Ages, sub-Saharan Africa gave rise to wealthy and sophisticated trading kingdoms.

- The African slave trade inflicted unspeakable misery and suffering on the peoples of sub-Saharan Africa.

- In an unprecedented burst of imperialism, European countries carved up Africa into colonies in the late nineteenth and early twentieth centuries.

- Although all the sub-Saharan African countries are now independent, only a few have attained political stability and most are very poor.

- AIDS is taking a catastrophic toll throughout sub-Saharan Africa.

21

The Middle East and Central Asia: Crucible of Faiths

In This Chapter

- ◆ Untangling a tangled region
- ◆ Islam's heartland: The Arabian peninsula
- ◆ Regional tinderbox: The Middle East core
- ◆ Ancient Seats of Empire: Turkey and Iran
- ◆ East-West Crossroads: Central Asian 'stans

The region you're about to encounter has always been turbulent and conflict-prone, even in the best of times. But over the last several years it has turned into an even more violent and dangerous place. As of Fall 2003, the Middle East and Central Asia have become the main battlegrounds in the global war on terror. In the aftermath of the September 11, 2001, al-Qaeda-led terrorist attacks in New York City and Washington, D.C., the United States and its allies invaded Afghanistan in late 2001 and brought down the Taliban regime that had harbored al-Qaeda leader Osama bin Laden.

In March 2003, the United States targeted Iraq and toppled its murderous dictator, Saddam Hussein. The always-simmering conflict between Israelis and Palestinians over who will control what bits of the land that three religions call holy has erupted into a nightmare of hatred and violence. And in May 2003, a bombing in Riyadh, the capital of Saudi Arabia, killed 34 people and signaled that terrorism had arrived in the very heart of the Islamic world.

It all makes for pretty grim reading, but I hope you will stick with it because in so many ways what happens in the Middle East and Central Asia affects us all. That's why it's so important to know as much as possible about these places and what makes them and their people tick. But first we need sort out the individual parts of a very complicated regional puzzle.

Crossroads of Three Continents

In this chapter, we're actually dealing with two regions—the Middle East and Central Asia—that are linked by a common religion, Islam, and a lot of shared history. But even defining the term "Middle East" isn't that easy. Europeans came up with it as a name for the region they viewed as being midway between themselves and East Asia (the "Far East"). The term usually refers to the lands of southwestern Asia that lie west of Afghanistan and Pakistan—in other words, Iran and everything to the west (except for the Caucasus countries, which are not considered to be Middle Eastern). Egypt and even Libya are often included in the Middle East, but in this book we've covered them with the other Saharan countries of northern Africa in Chapter 19.

Depending on your definition, the Middle East either includes part of Africa (Egypt) or abuts it at the Israeli border. The region also includes a small part of Europe via Turkey, which straddles the Europe/Asia divide at the Bosporus and Dardenelles straits. This makes the Middle East the only region in the world that encompasses or at least abuts three continents.

Like Turkey, Iran is a Muslim country, but not an Arab one. So Iran's western border marks the end of the Arab Middle East, while its eastern border marks the beginning of Central Asia. (I never said this wouldn't be confusing!) The Central Asian region includes Afghanistan and all the former republics of the Soviet Union (see Chapter 15) whose names end in *stan* (meaning "place of").

Sorting Through the Subregions

To make this chapter a little more manageable, we've divided the dual Middle East and Central Asia region into five subregions:

◆ **The Arabian Peninsula:** Bahrain, Kuwait, Oman, Qatar, Saudi Arabia, United Arab Emirates, and Yemen

◆ **The Middle East core:** Israel, the West Bank and Gaza, Jordan, Lebanon, Syria, and Iraq

◆ **The Outer Middle East:** Turkey, Cyprus, and Iran

◆ **The Southern Caucasus:** Armenia, Azerbaijan, and Georgia

◆ **The Central Asian 'stans:** Afghanistan, Kazakhstan, Kyrgyzstan, Tajikistan, Turkmenistan, and Uzbekistan

The Middle East and Central Asia.

In a region with so many countries and so many crises, a detailed description of each nation is beyond the scope of this book. But a general overview of each subregion should give you a better understanding of the issues and forces at play throughout the Middle East and Central Asia.

The Middle East: Land of Faith, Oil, and Conflict

There are two key facts about the Middle East that are at the crux of its current turmoil. First, the region was the crucible for three of the world's great religions. Judaism, Christianity, and Islam all claim either spiritual or actual descent from Abraham and all hold the city of Jerusalem to be sacred. The other fact is oil. Petroleum was discovered in Persia (now Iran) in 1908 and in the Arabian peninsula in the 1930s. Today, about three-fifths of the world's known oil reserves are in the Middle East, mainly in Saudi Arabia, Iraq, Iran, Kuwait, and the United Arab Emirates. Control of the land that is now the state of Israel has been a source of conflict among Christians, Muslims, and more recently Jews for centuries. The struggle for control of Middle Eastern oil is not even a century old but is proving just as deadly.

Even before there was oil, the Middle East was a much sought-after prize. As mentioned earlier, this is the only region in the world that includes or borders on parts of three continents. Whoever controlled the Middle East controlled one of the world's great trade and conquest routes. Since the dawn of civilization the region has been conquered and lost countless times.

A Brief Look at a Long History

Settled agriculture in the Middle East dates back to at least 8000 B.C.E. Around 3500 B.C.E. one of the world's first great urban civilizations, that of Sumer, emerged in the fertile Mesopotamian plain between the Tigris and Euphrates Rivers in what is now Iraq. The Sumerians were followed by the Assyrians and Babylonians. The Hittites established a kingdom in Turkey in the 1600s B.C.E. Over the following centuries, Hebrews, Philistines, and Phoenicians settled in what would become Palestine and Lebanon. Among the waves of invaders who overran the region from 800 B.C.E. on were the Medes, the Persians, and in 331 B.C.E., the Greeks and Macedonians under Alexander the Great. The Romans were next, conquering much of the Middle East by around 30 B.C.E. About 30 years later, Jesus of Nazareth was born in Roman-ruled Palestine and the religion he founded spread throughout the Roman Empire. By 550 C.E., the western Middle East bordering the Mediterranean was part of the Christian Byzantine Empire.

The Prophet Mohammed, founder of Islam, was born in Mecca, in what is now Saudi Arabia, in about 570 C.E. Within 15 years of his death in 632, his followers had conquered a vast stretch of territory from Libya to Mesopotamia, spreading Islam as they went. By 711 the Muslim domains extended from Spain to what is now Pakistan. For the next several centuries, the Middle East would be ruled by Arab caliphates, the Umayyads, based in Damascus, Syria, followed by the Abbasids, based in Baghdad.

Geographically Speaking

Although the vast majority of the people of the Middle East are Muslim, there are many divisions within the Muslim world. The two main ones are *Sunni* Islam and *Shi'a* Islam. About 90 percent of Muslims are *Sunnis*. Lacking a centralized religious authority, Sunni Islam encompasses a wide array of beliefs, practices, and peoples. All Sunnis, however, recognize the legitimacy of the order of succession of the first four Caliphs (leaders or rulers of the entire Islamic community) who succeeded Mohammed after his death. Sunnis also believe in direct relations between human beings and God, with no intermediaries.

Shi'a Muslims, on the other hand, accept only the fourth Caliph, Ali, Mohammed's son-in-law, as the Prophet's legitimate successor. The majority of Shiites believe in a succession of twelve Imams (religious leaders and teachers), beginning with Ali and followed by his two sons and nine grandsons. The Twelfth Imam is believed to have been hidden by God and to be capable of sending messages to the faithful. Shi'a Islam places a great emphasis on martyrdom and suffering, and on visits to shrines of Imams.

There are Shiites throughout the Muslim world, but Iran is the only country that is almost entirely Shi'a (about 90 percent). Iraq's Muslim population is about two-thirds Shi'a, and some of Shi'a Islam's holiest shrines are located in Iraq.

Between 1096 and 1204, Christian Europe launched four Crusades against the Muslim Middle East in an attempt to regain the holy sites in Palestine. In the sixteenth century, two great empires arose in the Middle East, the Saffavids in Persia (and for a time, in Iraq) and the Ottoman Turks in the Mediterranean Middle East and North Africa. The Ottomans extended their power well into Europe, getting all the way to the gates of Vienna in 1529, but in the eighteenth and nineteenth centuries, the Ottoman Empire went into a long, slow decline. It sided with Germany in World War I and was finally dismantled at the end of the war.

The former Ottoman territories were divided into French and British spheres of influence, with Syria and Lebanon going to France, and much of the rest to Britain. The core of the old Ottoman Empire became the Republic of Turkey in 1923. The al-Saud family established an independent monarchy on the Arabian peninsula in 1926. The British mandate over Iraq ended in 1932. Syria and Lebanon gained their independence from France after World War II in 1946, and both the Kingdom of Jordan and the state of Israel were carved out of what had been the British mandates of Palestine and Transjordan. The proclamation of the state of Israel in 1948 immediately triggered the first Arab-Israeli War. Between 1961 and 1971, Britain withdrew from its protectorates on the Arabian peninsula: Kuwait, Qatar, United Arab Emirates, Oman, and South Yemen.

The Arabian Peninsula

The Arabian peninsula is bordered on the north by Jordan and Iraq, on the east by the Persian Gulf and the Gulf of Oman, on the southeast by the Arabian Sea, on the south by the Gulf of Aden, and on the west by the Red Sea and the Gulf of Aqaba. Seven countries share the peninsula. The subregion's giant and namesake is Saudi Arabia (capital: Riyadh), birthplace of Islam and the world's biggest oil producer. Going clockwise from the northeast, are the small Persian Gulf oil states: Kuwait (capital: Kuwait), whose invasion by Iraq in 1991 triggered the first Gulf War, tiny Bahrain (Al-Manamah), Qatar (Doha), and the United Arab Emirates (Abu Dhabi). On the peninsula's southern end are Oman (Muscat) and Yemen (San'a).

The countries of the Arabian peninsula are mostly desert, and it's the real-deal *Lawrence of Arabia*-type of desert, complete with rolling sand dunes. Although the Red Sea marks the end of Africa and the beginning of Asia, the Arabian desert is actually an extension of Africa's great Sahara desert. Along the coast of the Red Sea and the Gulf of Aden, mountain chains rim the peninsula, with some peaks rising higher than 10,000 feet; there are also mountains in northern Oman. As you might expect, most of the peninsula has a harsh, desert climate with huge temperature extremes. There are no rivers to speak of, but the mountainous southwestern part of the peninsula does get seasonal monsoon rains. Lack of arable land and a scarcity of fresh drinking water are major issues throughout the subregion.

The 50 million people of the peninsula are almost all ethnic Arabs, although there are also small South Asian and African minorities. Arabic is the main language of the peninsula; English is a fairly common second language. Some Indian languages are also spoken, mainly in areas with large numbers of workers hailing from South Asia.

In the homeland of the Prophet Mohammed, Islam is the nearly universal religion of the Arabian peninsula (the exceptions are essentially foreign workers). Mecca in Saudi Arabia is the birthplace of Mohammed and Islam's holiest site. One of the five basic obligations of every Muslim is to make a pilgrimage to Mecca, called the *Hajj*, at least once in a lifetime. Also in Saudi Arabia is the holy city of Al Madinah (Medina), where Mohammed died and was buried in 632 C.E.

With one exception, the countries of the Arabian peninsula are economically all about oil. The peninsula's Big Three oil producers (Saudi Arabia, Kuwait, and the United Arab Emirates) alone have more than 40 percent of the world's known oil reserves. The per capita gross domestic products of the oil-producing nations of the peninsula are all very high, with Qatar and the United Arab Emirates leading the pack. In contrast, Yemen, which has had a turbulent modern history and began producing oil only in the mid-1990s, is one of the poorest countries not only on the peninsula but in the world.

But even in oil-rich Saudi Arabia there are economic clouds overhead. A burgeoning population, high unemployment, and overdependence on world oil prices are increasingly serious problems for the Saudi economy. A long-term decline in oil revenues has caused the Saudi government to cut back on many of the perks its citizens once took for granted.

Although on the surface the Arabian peninsula seems more stable than the rest of the Middle East, the underlying reality is disturbing. In general, the governments of the region are autocratic monarchies or emirates with close political and economic ties to the West. Their populations, however, are increasingly anti-American and fundamentalist Islamic. The 9/11 mastermind, Osama bin-Laden, is the son of an extremely wealthy Saudi family, and 15 of the 9/11 suicide bombers were Saudi citizens. A major source of anger among conservative Saudis was the presence since the 1981 Gulf War of U.S. troops on Saudi Arabian soil. Although the United States withdrew its forces from the country in April 2003, the U.S. occupation of Iraq and the ongoing Palestinian-Israeli conflict will likely continue to fuel Islamic militancy on the Arabian peninsula.

The Middle East Core

One of the "cradles of civilization" and the wellspring of Judaism and Christianity, the Middle Eastern Core has been one of the world's most fought-over regions. Much of the contention surrounds the ancient city of Jerusalem, which is sacred to Jews, Christians, and Muslims alike. To Jews, it's the City of David, the ancient capital of Israel, and the site of Solomon's and later Herod's Temple. To Christians, it is the city where the crucifixion and resurrection of Jesus took place. Muslims regard Jerusalem as their third holiest city after Saudi Arabia's Mecca and Medina. The great Dome of the Rock, built on top of the remains of the Temple of Herod, enshrines the spot where Mohammed ascended to heaven. Today, both Israeli Jews and Palestinian Arabs claim Jerusalem as their capital.

The Middle East Core is bordered on the north by Turkey; on the east by Iran; on the south by a bit of the Persian Gulf, Kuwait, and Saudi Arabia; and on the east by Egypt and the Mediterranean. The subregion includes five countries and one entity whose status remains unresolved. The five nations are, clockwise from the north: Syria (capital: Damascus), Iraq (Baghdad), Jordan (Amman), Israel (Jerusalem), and Lebanon (Beirut). The status of the Palestinian territories of the West Bank (west of the Jordan River) and the Gaza Strip (the narrow coastal strip south and west of Israel) remains undecided and is currently (early 2004) being bitterly contested. Iraq's status is also uncertain. After the U.S. and British invasion in March 2003 overthrew Iraq's long-time dictator, Saddam Hussein, the country remains under mainly U.S. and British occupation and U.S. administration. A transfer of power to an Iraqi transitional administration is scheduled for July 2004.

The Middle East core.

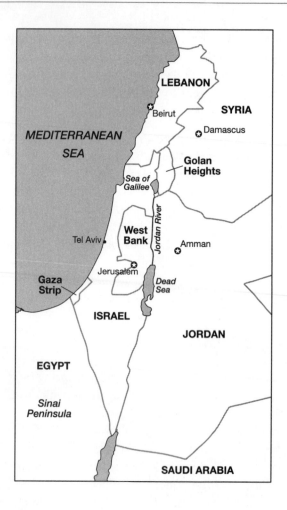

The Lay of the Holy Land

Syria, Lebanon, Israel, and the Gaza Strip all border the Mediterranean. Jordan and Iraq are almost landlocked except for Jordan's narrow outlet to the Gulf of Aqaba and Iraq's narrow but vital outlet to the Persian Gulf. Two deserts cover much of the subregion: the Negev in southern Israel and the much larger Syrian Desert, which encompasses southeastern Syria, western Iraq, and a good part of Jordan. Western and northern Syria and northeastern Iraq are somewhat mountainous. Inland from the Mediterranean coast, the Great Rift Valley (see Chapter 19) cuts through the subregion from Syria through the Jordan Valley before continuing down to Africa.

The Jordan River of biblical fame is the lifeline of Jordan, Israel, and the Palestinian territories. (Not surprisingly, control of scarce water supplies is a big bone of contention in this part of the world.) But far and away the region's most important river system is located in eastern Iraq. The Tigris and Euphrates Rivers both rise in Turkey and flow into Iraq (the Euphrates via Syria). The arc formed by Syria's Mediterranean coast, its part of the Euphrates valley, and the Tigris and Eurphrates river system in Iraq make up the famous Fertile Crescent, which nurtured great civilizations and mighty empires for millennia.

The climate in much of the subregion is hot and very dry. The coastal areas, however, have a typically Mediterranean climate, with long, hot, dry summers, and short, cool, rainy winters.

The overwhelming majority of the 57 million or so people of the Middle East Core are Arabs. The main exception is Israel, which is about 80 percent Jewish. Israel's non-Jewish population is predominantly Palestinian Arab. There are also Armenian minorities in Syria, Lebanon, Israel, and a tiny one in Jordan. Syria also has a small Kurdish minority, while Iraq has a sizeable one. Nearly 20 percent of Iraqis are Kurds, living in what currently amounts to an autonomous state of their own in northern Iraq.

Terra-Trivia

Despite their current conflicts, Jews and Arabs are both Semitic peoples. The term *Semitic* refers to a group of peoples who speak (or spoke) related languages all thought to stem from a common Semitic tongue. In addition to Jews and Arabs, the following groups, all familiar from the Bible, were also Semites: the Assyrians, Babylonians, Canaanites, Phoenicians, and Aramaeans (who included the Hebrews, or Jews).

Muslims live in every country of the Middle East Core and are the majority in most of them. Israel is the most notable exception. Only about 20 percent of its population is non-Jewish; about 15 percent are Muslim and 2 percent Christian. Although both Syria and Lebanon are more than 90 percent Arab, both have significant Christian minorities—10 percent in Syria and 30 percent in Lebanon. (Lesson: Not all Arabs are Muslim.) Iraq, on the other hand, is about 97 percent Muslim. The problem is that although Shi'a Muslims outnumber Sunni Muslims two to one in Iraq, the Sunni minority under Saddam Hussein ruled the country for more than 30 years and brutally oppressed the Shi'as. One of the unanswered questions at the moment in Iraq is whether or not Sunnis and Shi'as can put the past behind them and make a country together.

Israel, the Palestinian Territories, and Jordan

The escalating Palestinian-Israeli conflict, the global economic slowdown following the September 11, 2001, terrorist attacks, and the Iraq war and its aftermath have all badly hurt the economies of the subregion. Israel's economy, the most developed and technologically advanced in the Middle East Core, has been in a recession since the beginning of the second Palestinian *Intifada* ("uprising"). The reality or threat of suicide bombings in the buses and cafes of Israel's main cities, Tel Aviv and Jerusalem, have taken a heavy toll on the country, psychologically and economically.

On the other side of the Israeli-Palestinian fence, Israeli crackdowns and military attacks in response to suicide bombings have destroyed much of the infrastructure in the Palestinian territories and severely hampered freedom of movement and the ability to work. The result is that the Palestinian economy, fragile to start with, is in shambles. Foreign aid is all that is keeping it from collapsing.

Geographically Speaking

For a while in the early 1990s, it seemed possible that peace between Israelis and Palestinians might finally be at hand. The Oslo Accord signed by the Palestine Liberation Organization and the Israeli government laid out a five-year timetable for Palestinian self-government. It was not to be. In 1995, Israel's Labor party Prime Minister Yitzhak Rabin, who had negotiated the accords, was assassinated by a Jewish extremist. The hard-line Likud party government that succeeded him pulled back from the accords citing concerns about Israel's security.

Although some progress was made in talks between Israel and the Palestinian Authority in 1998 and again in 2000 at a meeting brokered by U.S. President Bill Clinton, both sides seemed intent on undermining the peace process: Israel, by its harsh treatment of Palestinians and its refusal to dismantle old settlements and stop building new ones on Palestinian territory; the Palestinians, by refusing to halt suicide bombings and put terrorist groups, like Hamas, out of business. Tackling the really contentious issues like final borders, water rights, and who gets what parts of Jerusalem seems pointless while the two sides are engaged in killing each other.

Jordan under King Hussein, and since 1999 under his son King Abdullah, has had to steer an often treacherous path between its Arab neighbors and its Western allies. Although King Hussein fiercely opposed the 1979 peace treaty between Egypt and Israel, he himself made peace with Israel in 1994. (Israel remains officially at war with Syria and Lebanon.) In part because of its dependence on Iraqi oil, Jordan sat out the first Gulf War in 1981, jeopardizing its relationship with the United States. The collapse of the Iraqi oil industry after the U.S–led invasion in 2003 has deprived Jordan of its main fuel source and put a heavy strain on its economy.

Lebanon and Syria

Since 1991, Lebanon has been putting itself together after a horrendous 16-year civil war, during which its capital, Beirut, once known as the Paris of the Middle East, was reduced more or less to rubble. Although Lebanon's many sects and factions have patched things up among themselves, repercussions of the civil war linger. Israel withdrew its troops from southern Lebanon in 2000, but Syria maintains a sizeable force in the country and the radical Shi'a group, Hezbollah, is firmly entrenched there. A highly entrepreneurial people, the Lebanese have successfully rebuilt much of their physical and financial infrastructure and brought inflation under control. Their economy, based primarily on banking and light manufacturing, has been growing slowly over the last several years.

A local branch of Saddam Hussein's Ba'ath Party still governs Syria. Since 1970 the country has been run dictatorially by Hafez al-Assad and after his death in 2000 by his son, Bashar. The long-standing conflict between Syria and Israel centers on Israel's capture of the Golan Heights from Syria in the 1967 Arab-Israeli War. Since the fall of Sadaam Hussein in 2003, Syria has been under increasing pressure from the United States for allegedly harboring members of Saddam Hussein's regime and developing chemical weapons. Syria's economy is still largely state-controlled and although the country is an oil exporter, its economy has been struggling with high unemployment rates, a fast-growing population, and a variety of internal and external crises.

Iraq

From 1958, when its king was assassinated in a military coup, until April 9, 2003, when the government of Saddam Hussein was overthrown by a U.S. and British invasion, Iraq had been a military dictatorship ruled by the Arab socialist Ba'ath party. From 1979 on, Iraq was under the thumb of the cruel and repressive regime of Sadaam Hussein. In 1980, Iraq launched an 8-year war against Iran that is estimated to have left some 1.5 million people dead. In 1990, Saddam Hussein invaded Kuwait and held it until January 1991, when his forces were routed by a U.S.–led United Nations coalition in under a week. Throughout the 1990s, U.N.–imposed sanctions curtailing Iraq's oil exports ruined the country's economy and caused great suffering to the Iraqi people. Yet Hussein would not cooperate with U.N. weapons inspectors seeking to verify the presence or absence of weapons of mass destruction in Iraq. Finally, in March 2003, U.S. President George W. Bush, alleging that Iraq did have weapons of mass destruction and was prepared to use them, launched an invasion of Iraq that succeeded in overthrowing the Hussein regime.

> **Terra-Trivia**
>
> With 6.7 million people in its metropolitan area, Iraq's capital, Baghdad, is the largest city in this subregion, It is also one of the most fabled in the world. It was founded in 762 as the capital of the Abassid caliphate and soon became one of the greatest cities in the Muslim world, a glittering center of scholarship, literature, and the arts.

Saddam Hussein's costly wars and the U.N. sanctions that followed the first Gulf War wrecked Iraq's oil-rich economy. (Iraq has the second largest oil reserves in the world.) Years of neglect under Saddam Hussein coupled with widespread post-war looting has seriously damaged much of Iraq's infrastructure. The U.S. occupation is proving considerably more difficult and costly than the Bush administration had estimated. Even after the capture of Saddam Hussein on December 13, 2003, insurgent attacks on U.S. and coalition forces have continued. And although no weapons of mass destruction have yet come to light, outside terrorist groups seem to be taking advantage of the instability in post-war Iraq. It remains to be seen whether Iraq will succeed in transforming itself into an open democratic society or become an even more dangerous threat to regional and global peace.

The Middle East—physical features.

Ottoman and Persian Domains

The subregion we're calling the "outer" Middle East lies north and east of the Middle East Core and consists of just three countries: Turkey (capital: Ankara), Iran (Teheran), and the part Turkish/part Greek island of Cyprus (Nicosia). The two big countries, Turkey and Iran, share a number of traits. Both are predominantly Muslim countries, but neither is an Arab country. (Cyprus is neither Arab nor predominantly Muslim.) Both touch other regions. Iran borders South Asia (Pakistan), Central Asia (Afghanistan and Turkmenistan), and the southern Caucasus countries. Turkey also borders the Causasus and it not only borders Europe, it is part of it. Most of Turkey consists of the great peninsula called Asia Minor or Anatolia, which is definitely in Asia. But a small, triangular part of Turkey lies in Europe, across the Bosporus Strait, the Sea of Marmara, and the Dardenelles Strait from Asian Turkey. European Turkey is bordered on the west by Greece and Bulgaria.

The nations of this subregion border many of the region's major bodies of water. Iran touches the Caspian Sea on the north, and the Persian Gulf, the Strait of Hormuz, and the Gulf of Aden, on the west and south. Turkey is washed on the north by the Black Sea, on the west by the Aegean, and on the south by the Mediterranean, which surrounds the island of Cyprus.

Both Turkey and Iran have long and glorious histories that have often intersected. Civilization spread very early on from Egypt and Mesopotamia (Iraq) into Asia Minor. The Turkish peninsula was part of the western frontier of the Persian empire in the sixth century B.C.E. and Alexander the Great added it to his empire in the fourth century B.C.E. The magnificent city of Byzantium, on the tip of what is now European Turkey, became the capital of the Roman Empire under the name Constantinople in 330 C.E. and remained the capital of the Byzantine Empire until 1453, when it was conquered by the Ottoman Turks and renamed Istanbul. After the fall of the Ottoman Empire in 1918 and the establishment of the Republic of Turkey in 1923, the capital was moved from European Istanbul to the city of Ankara near the center of Asia Minor.

Iran, known as Persia until 1935, was also the seat of ancient empires. Cyrus the Great founded the Persian Empire in 550 B.C.E. and made it the dominant power in the Middle East. The Sassanid Empire flourished from the early third century C.E. until it was brought down by Muslim Arabs in 636. Under the Safavids, who ruled from 1502 through 1736, Persian arts and culture reached one of many high points. The Iranian monarchy survived until 1979, when the repressive regime of the last shah (king), Reza Pahlevi, was overthrown by the Islamic Revolution of Ayatollah Khomeini.

The Rugged and Populous Subregion

All three countries in the subregion are mountainous. The western part of Asian Turkey consists mainly of the semiarid Anatolian Plateau, rimmed by mountains on the north and south. Eastern Turkey is the most mountainous part of the country. The country's tallest peak (17,000 feet) is Mount Ararat, famous as the place where Noah's Ark came to rest. It rises near the eastern border with Armenia in a part of Turkey that is historically Armenian. Turkey's lowland areas are mainly narrow strips along its coasts. The island of Cyprus has two east-west mountain chains with a wide plain in between. Iran also consists of a large plateau, in this case on the east spilling over into Afghanistan, with low mountains on the Iranian side of the border and tall, volcanic mountains in the north and west.

Most of Turkey is temperate, with hot, dry summers and cold, rainy winters. More extreme weather prevails in the interior. The northeastern and western parts of the country get the most rain. Cyprus enjoys a Mediterranean climate: hot, dry summers and cool winters. The climate in Iran's mountainous area is subtropical, with temperatures and rainfall depending on elevation. Elsewhere the climate is continental, with hot summers and cold rainy winters. The central plateau and southeastern coast plain are arid.

Turkey and Iran are the population giants of the Middle East and Central Asia region, with a combined population larger by far than any of the other subregions. There are 67 million Turks and 66 million Iranians (and only 767,000 Cypriotes). Turkey's capital, Ankara, has a population of 3.6 million, but its largest city, Istanbul, is really huge, with 10 million people in its metropolitan area. Iran's capital is even bigger—Teheran's 11 million inhabitants make it the largest city in the entire region.

Considering that Turkey was once the center of a vast, multicultural empire, it is a surprisingly homogeneous country today. Its population is 80 percent Turkish and 20 percent Kurdish. Turkish is the official language; Kurdish, Arabic, Armenian, and Greek are also spoken. The ethnic and linguistic breakdown in Cyprus mirrors the political division of the island. About 85 percent of Cypriots are Greeks, who live in the southern part of the island; about 12 percent are Turks, who live in the northern part.

Unlike Turkey, Iran is an ethnic and linguistic potpourri, reflecting its history as an international crossroads. Slightly more than half of Iranians are ethnic Persians; nearly a quarter are Azerbaijanis. Smaller minority groups include Kurds, Arabs, Lurs, Balochs, and Turkmen. About 58 percent of Iranians speak Farsi (Persian) or a Farsi dialect. Also spoken are Turkic dialects, Turkish, Kurdish, Armenian, Balochi, Arabic, and Luri.

Almost everyone in Turkey and Iran is Muslim, but as you've seen there are divisions within Islam and they are reflected here as well. Turkey is almost 100 percent Sunni Muslim, with tiny Christian and Jewish minorities. Cyprus, the exception in this sub-region, is 78 percent Greek Orthodox and 18 percent Muslim, with small Maronite and Armenian Christian minorities. Iran, the Shi'a heartland, is almost 90 percent Shi'a Muslim and only 10 percent Sunni. A tiny fraction of the population is Zoroastrian, Christian, Jewish, or Baha'i.

Turkey

Between 1923 and his death in 1938, Kemal Ataturk, the father of the Turkish republic, laid the foundations for a modern, secular Turkish state. Since 1938, Turkey has gradually developed into a parliamentary democracy despite periods of military rule and a lingering tendency on the part of the military to intervene in political affairs. Turkey has a troubled relationship with its own large and oppressed Kurdish minority. In 1984, a Kurdish separatist group launched a guerilla/terrorist campaign in southeast Turkey. Although a ceasefire was declared in 1999, skirmishes continue between government forces and Kurdish militants based in Iraq. The prospect of an independent Kurdistan in post-Saddam northern Iraq is a source of great concern to the Turkish government given the size and discontent of its own Kurdish minority.

Geographically Speaking

The Kurds are a nation without a country. A non-Arab, mainly Sunni Muslim people, the Kurds inhabit a region called Kurdistan that encompasses eastern Turkey, northeastern Iraq, northwestern Iran, and small parts of northeastern Syria and Armenia. Turkey has long oppressed its large Kurdish minority, although some of its harsher measures are being repealed as part of Turkey's bid to join the European Union. Saddam Hussein attacked Iraqi Kurdish villages with poison gas during the Iran–Iraq War in the 1980s and brutally put down a Kurdish uprising following the Persian Gulf War of 1991. A U.S.–enforced no-fly zone over northern Iraq enabled the Kurds to set up an autonomous region, but fighting soon broke between Kurdish factions over control of the area. A cease-fire was declared in 1999. Kurdish forces assisted U.S. and British forces in the seizure of Kurdish villages during the U.S.–led invasion of Iraq in 2003.

Turkey's economy is a mix of old and new. A modern, dynamic industrial and commercial sector coexists with a traditional agriculture sector, which accounts for nearly 40 percent of the Turkish workforce. Turkey has oil and gas reserves, but not on the level of Iran and the other Middle Eastern oil giants. Bicontinental Turkey continues to press its application to join the European Union.

Cyprus

Strategically located Cyprus has passed through many hands in its long history: Phoenicians, Assyrians, Egyptians, Persians, Greeks, Romans, Byzantines, the English (under Richard I in 1191), Venetians, and finally Ottoman Turks, who conquered it in 1571. At the start of World War I, Cyprus was annexed by Britain and gained its independence in 1960 after a long struggle. Tensions between Greek and Turkish Cypriotes continued to rise after independence, with the Greek side pushing for union with Greece and the Turkish side aiming for partition. With civil strife and violence mounting, Turkey invaded Cyprus in 1974 and took over more than 30 percent of the island, displacing some 200,000 Greeks. Cyprus remains a divided island with an internationally recognized government on the Greek side and a Turkish Republic of Northern Cyprus, recognized only by Turkey, on the northern end. Cyprus is scheduled to join the European Union in 2004, but after the collapse of United Nations–sponsored reunification talks in March 2003, it seems likely that it will join the E.U. as a divided country.

The Greek Cypriot side is far more prosperous than the Turkish side, which is hampered by its lack of international recognition. The Turkish side is heavily dependent on agriculture, whereas the Greek Cyprus has a more diversified economy oriented to industry and the service sector. Scarcity of drinking water is a problem on both sides.

Iran

After 25 years of restrictive religious rule in Iran, a younger generation is increasingly pressing for change. Nevertheless, attempts by Iran's relatively liberal current president, Mohammed Khatami, to curb the power of hard-line clerics have met with little success. And 25 years after the seizure of the U.S. embassy in Teheran and the taking of 52 American hostages, U.S.–Iranian relations remain troubled. In early 2003, President George W. Bush included Iran in his famous "axis of evil" along with Iraq and North Korea. After the Iraq invasion and the overthrow of Saddam Hussein, Iran is facing intense U.S. and international pressure to cooperate with U.N. weapons inspectors and shut down the nuclear weapons development program it is alleged to have.

The brutal Iran-Iraq War of the 1980s, in which both sides used chemical weapons, had a devastating effect on the economies and oil industries of both countries. Unlike Iraq, however, Iran has repaired much of the damage. Iran has abundant natural resources, but oil is king, accounting for 85 percent of the country's exports. Other major exports include petrochemicals, textiles, Iran's famous Persian carpets, fruits and nuts, and iron and steel. Inflation, unemployment, and poverty rates all remain high.

At the Edge of Asia: Transcaucasia

The Caucasus Mountains are an east–west chain that marks the dividing line between Asia and Europe. The greater Caucasus region straddles the mountains and is divided into two parts: North Caucasia, which remained part of Russia after the split up of the Soviet Union in 1991 (see Chapter 15) and Transcaucasia, the southern (actually southwestern) part of the region covered here. The three nations that make up this subregion—Armenia (its capital is Yerevan), Azerbaijan (Baku), and Georgia (Tbilisi)—are all former Soviet republics that spun off and formed independent nations in 1991.

Transcaucasia is dominated by the southern slopes of the Caucasus Mountains and the Azerbaijan Plateau in the east. The glacier-fed Kur River flows through the plateau into the Caspian Sea. Famous for its scenic beauty, the Caucasus region is prone to occasional earthquakes.

Georgia's climate along its Black Sea coast is subtropical with heavy rains. It's cooler and dryer in the eastern plains, with more extreme conditions in the eastern mountains. Bordering on the Caspian Sea, Azerbaijan has a dry subtropical climate with hot summers and mild winters. Totally landlocked Armenia has a highland continental climate with hot summers and cold winters.

Azerbaijan is the most populous of the three Transcaucasian countries with 7.8 million people. Georgia is second with 5 million, and Armenia third with 3.3 million. Azerbaijanis are about 90 percent ethnic Azeri, with small minorities of Dagestanis, Russians, and Armenians. Azerbaijani is the predominant language, followed in a small way by Russian and Armenian. About 94 percent of Azerbaijanis are nominally Muslim; small percentages are Russian or Armenian Orthodox. The Georgian population is about 70 percent ethnic Georgian; Armenians, Russians, Azeris, Ossetians, and Abkhaz are also present in smaller numbers. The predominant religion is Georgian Orthodox, with sizeable Muslim, Russian Orthodox, and Armenian Apostolic minorities. The people of Armenia are about 95 percent Armenian in ethnicity, language, and religion.

Like the other subregions covered in this chapter, the Caucasus (both south and north) have been invaded and settled by many different races and cultures over the centuries. Persians, Khazars, Arabs, Huns, Turko-Mongols,

Terra-Trivia

The Caucasus Mountains are no small barrier: Mount Kazbet in Georgia reaches a height of more than 16,500 feet. The Caucasus region is also known for the extraordinarily high percentage of its people who live past 100 years of age.

Ottomans, and Russians have all left their imprint on the Caucasus. Russia took over in the early nineteenth century, and while Christian Armenia and Georgia might have preferred Russian rule to domination by the Muslim Turks, Azerbaijan and the other Muslim states in North Caucasia fiercely resented and struggled against Russian authority.

The three Transcaucasian countries haven't had an easy time of it since independence. Azerbaijan has lots of oil and even greater untapped reserves, but is held back by Soviet-style inefficiency, corruption, and slow progress in opening up its economy. Another major problem is the ongoing struggle with Armenia over Nagorno-Karabakh, a small bit of Azerbaijani territory with a largely Armenian population, which Armenia seized in 1994, along with a chunk of Azerbaijan proper. Armenia's economy has also been badly affected by the conflict, especially as it has no oil and scant mineral deposits, and has to import food. If only by a hair, however, Armenia has the highest per capita gross domestic product in the subregion. Georgia comes in third, and it, too, is struggling to de-Sovietize and jumpstart its energy- and resource-poor economy. Georgia has also had to cope with ethnic and civil strife in two Muslim breakaway areas: South Ossetia and Abkhazia.

Terra-Trivia

Georgians are famous for their hospitality and for the artistic achievements in dance, theater, music, and design. The founder of the New York City Ballet, the great choreographer George Balanchine, was born in Georgia.

Central Asia: Afghanistan and the 'stans

The last subregion described in this chapter includes one remote country that has dominated headlines in recent years and five even more remote ones that are among the least known in the world. The famous country is, of course, Afghanistan, former Taliban stronghold and al-Qaeda terrorist training camp. The other five are the former Soviet republics of Central Asia: Kazakhstan (capital: Astana), Kyrgyzstan (Bishkek), Tajikistan (Dushanbe), Turkmenistan (Ashkhabad), and Uzbekistan (Tashkent).

One thing all these countries share is the syllable *stan* or *istan* at the end of their names. *Stan* is an Old Persian word for "place of," usually preceded by the name of a particular tribe (Kazakhstan, for example, is the place or homeland of the Kazakhs), which gives you an indication of the tribal nature of the societies in this area. "What about Pakistan?" you might ask. That's a "stan" country, too, but we've covered it in the South Asia chapter (see Chapter 22) because of its geographic, historic, and cultural ties to the Indian subcontinent.

The huge subregion of Central Asia abuts Russia to the north and China to the east. Pakistan and Iran are on the subregion's southern border; the Caspian Sea and Russia lie to the west. Although Turkmenistan and Kazakhstan border the Caspian Sea, all the Central Asian countries are landlocked.

Central Asia—physical features.

Mountains and Deserts

Beginning in the north, giant Kazakhstan (it's about four times the size of Texas) is mostly one vast, flat expanse of arid grasslands with high mountains rising in the southeast. The high, glacier-clad Tian Shan and Pamir Mountain ranges dominate Kyrgyzstan and Tajikistan. (At 24,590 feet, Pik Imeni Ismail Sameni in the Pamir range in Tajikistan is one of the world's highest mountains.) The Kyzyl Kum desert (*kum* is a Turkish word meaning "sand") covers most of central Uzbekistan; to the east and southeast are fertile valleys. Almost 90 percent of Turkmenistan is covered by the Kar Kum desert. From its center to its eastern border with Pakistan, Afghanistan is a very mountainous country. Its tallest peaks, some rising above 24,000 feet, rise up in the northeastern Hindu Kush range, home of the famed Khyber Pass (see Chapter 22). Southern and southwestern Afghanistan is mostly desert.

This immense area has a relatively small population. With almost 28 million people, Afghanistan is Central Asia's most populated country. Uzbekistan comes next with 25.5 million, followed by Kazakhstan with almost 17 million and Tajikistan with 6.7 million. Kyrgyzstan and Turkmenistan each have a bit under 5 million.

The Ethnic Web

The names of the countries in this subregion pretty much let you know who the predominant ethnic group is. In the five former Soviet republics, you're also likely to find a Russian minority of varying size. In addition, the population of each of these countries includes people from most of the other countries in the subregion as well as people from outside it. And finally, the majority Muslim populations of all of these countries are overwhelmingly Sunni. (Got all that?). Here are a few examples. The population of Kazakhstan is about 50 percent Sunni Muslim Kazakh and about 35 percent Russian Orthodox Russian, with smaller minorities of Ukrainians, Germans, Uzbeks, and Tatars. Uzbekistan is about 80 percent Sunni Muslim Uzbek and about 5 percent Russian Orthodox Russian, with tiny Tajik, Kazakh, and Tatar minorities.

Afghanistan deserves special mention because its ethnic profile is even more complicated. The largest group is the Pashtuns (ethnic Afghans), who live in the eastern and south central parts of the country. Baluchis inhabit the deep south. Tajiks live near the Tajikistan border, Uzbeks near the Uzbekistan border, and Turkmen near the Turkmenistan border. Hazaras, of Mongolian origin, live in the mountainous center. All these groups are mostly Sunni Muslim, except for the Hazaras, who are Shi'a.

Afghanistan

This is another country that has been crisscrossed, invaded, and migrated into for millennia. Persians, Greeks, Arabs, Mongols, Tatars, Russians, and the British all passed through, conquered, or tried to conquer Afghanistan. Russia and Britain fought three wars over Afghanistan in the nineteenth century, and although the British retained control of Afghanistan's foreign policy until 1919, they never fully colonized it.

Afghanistan was neutral in World Wars I and II and throughout the Cold War until the late 1970s. In 1979 the Soviet Union invaded Afghanistan to support the country's new and increasingly unpopular Marxist regime. The 10-year war that followed pitted Soviet forces and Soviet-backed Afghan government troops against United States-backed Afghan *mujahedeen* (Arabic for "holy warriors"). The war took a horrific toll on both sides, but the mujahedeen finally prevailed and the Soviets withdrew in 1989. The pro-Soviet government fell in 1992, and the chaos that followed gave an opening to a group of radical Islamic students called the Taliban. After taking control of Kabul in September 1996, the Taliban began imposing on Afghanistan a fundamentalist Islamic regime of unprecedented harshness. Under the Taliban, the country became a haven for the Islamist terrorist group al-Qaeda and its leader, Osama bin Laden.

After the September 11, 2001, al-Qaeda attacks on the World Trade Center in New York and the Pentagon in Washington, D.C., the Taliban refused to hand over bin Laden. On October 7, the United States and its allies launched an air war against the Taliban regime. Within five weeks, anti-Taliban Afghan forces on the ground had taken the key cities of Mazar-i-Sharif and the capital, Kabul. The Taliban government fell on December 7, but its leader, Mullah Omar and, more important, Osama bin Laden were not caught, and as of October 2003 remain at large, presumably in the lawless northeastern border zone between Afghanistan and Pakistan.

The Five 'stans

Like the former Soviet republics of eastern Europe, the Central Asian republics are struggling with varying degrees of success to reform and open their political systems and economies. Kazakhstan, the second largest of the former Soviet republics after Russia, is rich in fossil fuels and other natural resources and has a healthy agricultural sector and a growing industrial sector. Its per capita gross domestic product is the strongest in the region.

Turkmenistan also has major oil reserves and even more significant natural gas reserves (the fifth largest in the world). And even though it's a mostly desert country, Turkmenistan is the world's tenth largest cotton producer, thanks to intensive irrigation. Its economy, however, is seriously hampered by a repressive ex-communist government that doesn't tolerate opposition and has been reluctant to introduce much-needed market reforms.

Kyrgyzstan was the first of the Central Asian republics to adopt democratic institutions and the first of the ex-Soviet republics to join the World Trade Organization in 1998. It is a poor, predominantly agricultural country, with significant gold, mercury, uranium, and natural gas deposits. Election fraud, a high poverty rate, ethnic strife, and Islamic insurgencies are among the problems Kyrgyzstan is grappling with.

Uzbekistan is the world's second largest cotton producer, but that comes at a cost. Intensive irrigation and overuse of chemical fertilizers have poisoned an already arid land and dried up water supplies like the Aral Sea. Uzbekistan is also an exporter of gold and has other significant mineral and fossil fuel reserves, but its future is clouded by problems similar to those of Kyrgyzstan.

Despite healthy growth in recent years, Tajikistan is still the poorest of all the former Soviet republics. A civil war that lasted through most of the 1990s wreaked havoc with an already fragile, mainly agricultural economy. As elsewhere in the region, cotton is king in Tajikistan.

In the News: What's Next?

In a region that is so often in the news for all the wrong reasons, it's difficult to single out one particularly newsworthy issue. Let's just say that the world is watching intently to see what happens next in this volatile and dangerous land. And let's just hope that wise and courageous leaders will rise up in Israel, Palestine, Iraq, Iran, Afghanistan, and throughout the Middle East and Central Asia and do what it takes to defuse the hatreds in the region and spare future generations the suffering and violence of the present.

The Least You Need to Know

- The Middle East is at the crossroads of three continents: Europe, Asia, and Africa.
- The Middle East Core and the Arabian peninsula gave rise to three of the world's major religions: Judaism, Christianity, and Islam.
- Most of the world's oil comes from the Middle East Core and the Arabian peninsula.
- In Central Asia, the former Soviet republics are struggling to liberalize their economies and political systems.
- The entire Middle East and Central Asia region has become a battleground in the global war on terror.

22

South Asia: The Mighty Subcontinent

In This Chapter

- ◆ Looking back at South Asia's history

- ◆ Measuring the mountains and other natural features

- ◆ Discovering India

- ◆ The Perils of Pakistan

- ◆ Wading through Bangladesh

- ◆ Exploring Nepal and Bhutan

- ◆ Cruising through the island countries

The region of South Asia, as shown on the following map, includes the countries of Pakistan, India, Nepal, Bhutan, Bangladesh, Sri Lanka, and the Maldives, in other words, the Indian subcontinent and its immediate neighbors. The region is appropriately named, as it constitutes the south–central portion of the Asian continent.

South Asia is walled off on the north by the world's highest and mightiest mountain chain, the Himalayas. To the east is Myanmar (formerly Burma) and the Bay of Bengal; to the south, the Indian Ocean; and to the west, the Arabian Sea, Iran, and Afghanistan. The region resulted from the breakup of the ancient continent of Gondwanaland, which itself resulted from the breakup of the supercontinent of Pangaea (see Chapter 2). After the Indian plate broke off from eastern Africa and Antarctica, it began drifting northeast until it crashed into the Eurasian plate. This process caused the Himalayan upsurge and continues to make the Himalayas rise.

South Asia is small in area compared to other regions, but it more than makes up for this "shortcoming" in population. With more than 1.3 billion people, South Asia is running neck and neck with East Asia (China, Mongolia, the Koreas, and Taiwan) for the title of "most populous region in the world." And with its higher rates of population growth, South Asia is destined to win the race.

The Very Long Road to the Present

One of the world's four Bronze Age urban cultures (Egypt, Mesopotamia, and China are the other three), the Indus Valley civilization flourished in the fertile Indus River basin in the western part of South Asia (today's Pakistan and western India) from about 2600 B.C.E. to about 1600 B.C.E. Between 1600 and 1500 B.C.E. Aryan tribes from the northwest overcame what was left of the Indus civilization and settled in the Indus Valley region. The Aryans introduced the Sanskrit language and the Vedic religion, a precursor of Hinduism, and some time after 1000 B.C.E. the caste system was established.

Geographically Speaking

India's caste system is a hereditary social structure closely bound up with Hindu beliefs about personal conduct and reincarnation. The four main castes (each subdivided into many sub-castes) are, in descending order: Brahmins (priests and religious officials), Kshatriyas (warriors, rulers, and large landowners), Vaisyas (farmers, merchants, traders, and artisans), and Shudras (servants and peasants). At the bottom of the social order, beyond the caste system (or "outcastes," which is how the word originated) were the Untouchables, fit only for the most menial of tasks.

Under the caste system, people are born into the caste of their parents and are restricted to certain occupations. Marriage and intermingling across castes are forbidden. Over the centuries, the caste system has caused terrible injustice and human suffering in India. During the colonial period, the British tried hard to break down the caste system, as have post-independence Indian leaders, starting with Mahatma Gandhi, who referred to the Untouchables as *harijan*, or "children of God." Although India's constitution bans caste-based discrimination, the caste system, and its injustices, persist, especially in the countryside and rural villages that are the backbone of Indian life.

South Asia.

Buddhism was founded in India in the sixth century B.C.E. by Gautama Siddharta, who was born in what is now Nepal in 563 B.C.E. The new religion spread throughout northern India, especially during the reign of Ashoka, the great king of the Maurya dynasty, who unified most of India for the first time in the third century B.C.E.

The 33-mile-long Khyber Pass (elevation 3,500 feet) provides a passage through the rugged Hindu Kush Mountains that separate Afghanistan and Pakistan. Over the centuries, many conquerors and would-be conquerors have used the Khyber Pass to invade India: Cyrus the Great from Persia in 537 B.C.E., Alexander the Great from Macedonia in 327 B.C.E., the Huns in 430 C.E., various waves of Arab Muslim invaders beginning in the eighth century C.E., and the Mongols under Tamerlane in 1398. In 1526, another wave of Muslim invaders under Babur, King of Kabul (Afghanistan), established the great Mughal Empire. Although it gradually lost its power from the early 1700s on, the Mughal Empire would survive until 1858, when the British forced the last emperor off his throne.

In 1499, the Portuguese navigator Vasco da Gama became the first European to sail around Africa's Cape of Good Hope and reach India by sea. As a result of his remarkable achievement, Portugal enjoyed a near monopoly on trade with India for the next century. Founded in 1600 to conduct trade with the subcontinent, the British East India Company established its first trading post at Surat in 1615. From there, the Company's trading activities and eventually military and political influence began to spread throughout the Mughal Empire. By the end of the eighteenth century, the British had defeated their European rivals in India, most notably the French, and were in virtual control of large parts of the subcontinent. By the middle of the nineteenth century their domination would be complete. The greedy misrule of the East India Company, however, led to a violent uprising of Indian troops in 1857, known as the Sepoy Mutiny. After the revolt was suppressed and the last Mughal emperor deposed in 1858, the British Crown dissolved the East India Company and formally took over the administration of India.

Terra-Trivia

In 1633, the Mughal emperor Shah Jehan commissioned one of the world's most beautiful buildings and certainly the most famous structure in India today, the Taj Mahal, in Agra, near Delhi. A love letter written in gleaming white marble, the Taj Mahal was designed as a mausoleum for the emperor's beloved wife, Mumtaz Mahal, who had died giving birth to their fourteenth child.

Independence and Partition

Although the British could quell rebellions, they could not eliminate the nationalist feelings of the Indian people. Beginning in the 1920s, a Hindu lawyer, Mohandas K. Gandhi, known as the Mahatma, or "Great Soul," gave voice to India's aspirations and spearheaded an independence movement based on the principle of nonviolent civil disobedience and national self-reliance. Although British India was granted a measure of self-rule in 1935, it did not attain full independence until 1947.

Independence, however, was accompanied by a violent partition of British India into two nations: India, with a majority Hindu population, and the smaller Muslim state of

Pakistan, which consisted of two separate parts on either side of northern India. In one of the largest migrations in history, some 17 million people fled across the Indo-Pakistani borders in both directions to avoid sectarian rioting and bloodshed that was taking place on both sides. Millions died in the process.

A still unresolved conflict over the status of the northwestern Indian state of Jammu and Kashmir led to warfare between the two new countries in 1947 and again in 1965 and in 1999. In 1971, India supported an independence movement in East Pakistan, which, after yet another Indo-Pakistani war, became the independent nation of Bangladesh. Hostility between India and Pakistan and between Muslims and Hindus in both countries continue to make the subcontinent a dangerous tinderbox, especially since 1998, when first India and then Pakistan tested nuclear weapons.

Terra-Trivia

Mahatma Gandhi was one of the greatest visionaries and moral leaders of the twentieth century. He strongly opposed the partition of India, but once it became inevitable, he worked ceaselessly to promote peace and religious tolerance. In the end, Gandhi paid the ultimate price for his beliefs. On January 30, 1948, he was assassinated by a fellow Hindu and anti-Muslim fanatic.

From the Bay of Bengal to the Top of the World

South Asia lines up latitudinally with Saharan Africa and Middle America but has little in common with either. A region of great contrasts, South Asia has virtually every physical feature except major lakes. Below the Himalayas and their foothills, the Asian subcontinent encompasses the fertile Indus Valley and the Great Indian (or Thar) Desert in the west, the Ganges basin in the center, and the Brahmaputra River in the east.

The great Deccan Plateau dominates the center of the subcontinent peninsula. The Deccan is flanked by the Western and Eastern Ghats (steep slopes) down to the narrow coastal plains. Rugged and treacherous mountains separate the Indus from Afghanistan, and rain forest hills separate the Brahmaputra from Myanmar.

Although temperature varies with the seasons, it's colder in the mountains and plateaus.

Terra-Trivia

In the partition of British India, the northwestern province of Punjab was split between Pakistan and India. (The same thing happened to the eastern province of Bengal. East Bengal became East Pakistan and then Bangladesh in 1971.) The name *Punjab* means "land of five rivers," a reference to the five Indus tributaries—the Jhelum, Chenab, Ravi, Beas, and Sutlej rivers—that drain the region before joining the Indus.

The Himalaya Mountains average only 12 degrees F, whereas central India averages 79 degrees F. Rainfall also varies with topography. North of Bangladesh, Cherrapunji gets 450 inches per year (second only to Mount Waialeale in Kauai, Hawaii, with 460 inches). The Great Indian Desert receives only 4 inches; the rain forest in the east, 80 inches; the plain, 40 to 80 inches; and almost everywhere else, 20 to 40 inches.

The most famous of South Asia's many large rivers are the Indus (1,700 miles long), the main river of Pakistan; the Hindu holy river, the Ganges (almost 1,600 miles long), flowing from north central India into Bangladesh; and the Brahmaputra (also almost 1,700 miles long), in northeastern India and Bangladesh. All three rivers originate in the mountains of Tibet to the north and end in great deltas. The Indus River delta is on the Arabian Sea, and the Ganges and Brahmaputra join in Bangladesh to form a vast delta on the Bay of Bengal.

South Asia's Titan: India

India is a large country with a land area about one-third that of the United States. Its population of just over a billion, however, is almost four times the population of the United States. India today is home to about one-sixth of the world's population.

India has too many cities with more than a million inhabitants to list them all here. The two largest are the capital, Delhi, and the financial capital, *Mumbai*, whose metropolitan areas each number just over 17 million people. (Mumbai proper has 2 million more residents than Delhi proper.) *Kolkata's* metropolitan area has 14 million residents; *Chennai's*, 6.6 million.

GeoJargon

Mumbai? Kolkata? Chennai? Sound familiar? Probably not. In recent years, many Indian cities, including three of its megacities, officially replaced their colonial names with the local language version of their names. Bombay became **Mumbai** in 1995, Madras was changed to **Chennai** in 1996, and in 2000 Calcutta became **Kolkata**.

India's ethnic groups have so intermingled over the centuries that determining who's what is difficult. A widely accepted generalization is that about 72 per of the population is Indo-Aryan, 25 percent is Dravidian, and 3 percent is "other." Indo-Aryans are the result of the ancient intermingling between local populations and the Aryan tribes that migrated to India from the north around 1500 B.C.E. Indo-Aryans predominate in the north where Aryan influence was strongest. Darker skinned Dravidians, descendents of pre-Aryan Indian populations, predominate in the south.

If India's ethnic picture is complex, its language situation is even more so. Hindi is India's main official language, but it's the native tongue of only about a third of the population. In addition to Hindi and classical

Sanskrit, the Indian government recognizes 16 other regional languages as official. English enjoys "associate" status in the language hierarchy, but it is spoken by most educated Indians and is used for many national, political, and commercial communications. India's scores of languages and countless dialects fall into four main language families, with the majority of Indians speaking either an Indo-Aryan or a Dravidian tongue. (The linguistic north/south divide is much sharper in India than the racial divide.) The most widely spoken Indo-Aryan languages after Hindi are Bengali, Marathi, Urdu, Gujarati, Oriya, and Punjabi. Telugu, Tamil, Kannada, and Malayalam are the main Dravidian languages of the south. Many of India's languages are mutually unintelligible, which is why Hindi and English are so important.

India is the birthplace of two major religions, Hinduism and Buddhism, and is regarded by people around the world as a deeply spiritual place. Just over 80 percent of the Indian population is Hindu; about 12 percent is Muslim. Christians of all denominations account for 2.3 percent of the population. Less than 2 percent of Indians are Sikhs, practitioners of a monotheistic religion founded in Punjab in the fifteenth century C.E. Although both originated in India, Buddhism and Jainism (a reform movement in Hinduism that became a separate religion in the sixth century B.C.E.) today account for small fractions of the Indian population. Despite their small numbers, however, Jains play a disproportionately large role in the economic, political, and cultural life of India. Jain philosophy and ideals had a great influence on Mahatma Gandhi.

India After Independence

The miracle of modern India is that despite its many problems—massive overpopulation; widespread poverty; a hostile and nuclear-armed neighbor; and continuing ethnic, religious, and caste tensions within its borders—it has preserved the rule of law and its democratic institutions for more than a half a century. Except for a brief period in the mid-1970s, when prime minister Indira Gandhi (no relation to Mahatma Gandhi) declared a state of emergency and suspended civil liberties, India has been a functioning democracy, the largest in the world.

India Today

Behind the common perception of India as a place of grinding, if colorful, poverty lies a far more complex economic reality. By any definition, India is a very poor country. About a quarter of the population lives below the poverty line, and for India's poor, even finding adequate nutrition is often a daily struggle. Agriculture looms very large in India's economy, contributing about 25 percent of gross domestic product and employing a whopping 60 percent of its workers, most of them traditional village farmers. What is often overlooked, however, is that India also has the world's largest middle class, as many as 200 million people with good jobs and disposable income. One of India's most important resources is its large pool of well-educated, English-speaking workers.

India's main industries include textile manufacturing, chemicals, food processing, steel and heavy equipment manufacturing, coal and iron mining, and petroleum and natural gas production. In recent years, it has become a major exporter of software services and workers. The country has a lively press and publishing industry and the world's largest motion-picture industry (yes, bigger than Hollywood in terms of the number of films produced per year!).

Terra-Trivia

One of the great legacies of British rule in India is its remarkable railroad system. With nearly 39,000 miles of track, it is the largest railroad system in Asia, the second-largest state-owned system under a single management in the world, and the fourth most heavily used.

Despite impressive economic gains in recent years, however, India's economy is still weighed down by a poor infrastructure, inequitable income distribution, a cumbersome bureaucracy, and the enduring problems of overpopulation and Hindu-Muslim sectarian violence.

Pakistan: Born and Bred in Conflict

The easternmost of the "'stans" (see Chapter 21), Pakistan is bounded by India on the east, China and Afghanistan on the north, Afghanistan and Iran on the west, and the Arabian Sea on the south. Northern and western Pakistan are very mountainous; the Karakoram Range along the Chinese border contains some of the highest mountains on earth. In the southwest are the Baluchistan highlands, a forbidding landscape of deserts, rocky plains, and mountains. The eastern half of Pakistan is flat and dominated by the Indus River and its tributaries. This alluvial plain is shared by two provinces, the Sind in the south and the Punjab, the Pakistani heartland, in the north. East of the plain, straddling the Indo-Pakistani border is the Great Indian, or Thar, Desert. Southern Pakistan is very dry; the north, although far from lush, receives slightly more precipitation.

Pakistan's 147 million people make it the sixth most populous country in the world, and the second most populous Muslim country, after Indonesia. (India's Muslim minority is almost as large as Pakistan's entire population.) The vast majority of Pakistanis, about 77 percent, are Sunni Muslims; about 20 percent are Shi'a. (For more on Islam and its divisions, see Chapter 21.) Urdu is the official national language of Pakistan, although it is the native tongue of only about 8 percent of

Pakistanis. The language of Punjab province, Punjabi, is spoken by nearly half of the population. The other widely spoken languages are Sindhi, the language of Sind Province; Siraiki, a variant of Punjabi; Pashto, the language of the Northwest Frontier Province; and Baluchi, the language of Baluchistan. English is also an official language and the lingua franca of government ministries and the educated elite.

Pakistan's capital is Islamabad, a relatively small city (metropolitan area population of about 900,000) in northern Punjab. It was built in the 1960s to replace the old capital of Karachi located on the Arabian Sea near the Indus River delta. With a population 10.5 million, Karachi, the capital of Sind province, is Pakistan's largest city and main seaport and industrial center.

Post-Partition Pakistan

As mentioned earlier in this chapter, when British India became independent in 1947, it was divided into two states: predominantly Hindu India and Muslim Pakistan, the latter consisting of what are now the countries of Pakistan and Bangladesh. Pakistan was at first a dominion within the British Commonwealth, with the nationalist leader, Mohammed Ali Jinnah, as its first governor-general. A republic was declared in 1956, but military rule did not end until 1977, when Pakistan held its first elections under civilian rule. Pakistan's subsequent political history has been a complex and troubled one, with corruption-plagued civilian and military governments succeeding each other with alarming regularity.

The economic picture in Pakistan isn't much brighter. Despite the entrepreneurial skills of its people and its considerable natural resources—fertile land, natural gas, coal, some petroleum, and large hydroelectric potential—Pakistan remains poor and underdeveloped, its economy hampered by financial mismanagement, political instability, corruption, a low literacy rate, and the costly ongoing conflict with India. International sanctions imposed after Pakistan conducted nuclear bomb tests in 1998 were a further blow to its economy. Agriculture accounts for 24 percent of Pakistan's gross domestic product and employs about half its workforce. Textile and garment manufacturing is its major industry.

Terra-Trivia

The support of Pakistan's leader, General Pervez Musharraf, was crucial to the war on terrorism declared by U.S. President George W. Bush after the September 11, 2001, al-Qaeda attacks in the United States. To shore up that support, the nuclear test-related sanctions were lifted and Pakistan received a massive infusion of foreign aid.

The Kashmir Conflict

From the start it was clear that the partition of British India in 1947 was unfinished business. Tensions had existed all along between West and East Pakistan, with the East resenting the concentration of political and economic power in the West. In 1970, these tensions finally led to a civil war that resulted in the breakup of Pakistan and the creation of independent Bangladesh in the east.

Another major piece of still unresolved business is the contested northern Indian state of Jammu and Kashmir, renowned for its mountain beauty and cashmere wool. Before partition, Kashmir was one of India's so-called princely states, nominally ruled by a Maharajah. In this case, the Maharajah was Hindu, but most of the population was Muslim. When it came time to separate Pakistan from India, Pakistan invaded Kashmir and the Maharajah turned to India for help. Indian forces counterattacked and at the end of the war in 1949 Kashmir was divided between Pakistan and India, with India retaining the capital, Srinagar. In 1956, despite Pakistani protests, India-controlled Kashmir became the Indian state of Jammu and Kashmir, the country's only Muslim-majority state. The continuing dispute over Kashmir led India and Pakistan to war in 1965 and in 1999 and to the brink of war many more times, most recently in 2001, after a suicide bombing in the Indian Parliament killed 14 people, and in 2003, after 24 Hindus were killed in Indian-controlled Kashmir.

Disaster-Prone Bangladesh

When British India was partitioned in 1947, the historic region of Bengal in the northeast was divided into two parts: The Hindu west became the Indian state of West Bengal; East Bengal became the eastern part of the new nation of Pakistan. The cultural, economic, political, not to mention physical divide between East and West Pakistan eventually grew too great, and in 1971, East Pakistan declared its independence. After an extremely costly civil war (an estimated 1 million dead and 10 million refugees), East Pakistan prevailed and became the independent state of Bangladesh. Its first two presidents were assassinated, and the country was ruled by the military during much of the 1980s. In 2000, Sheikh Hasina Wazed, daughter of Bangladesh's first president, became the first prime minister to complete a full five-year term.

Bangladesh is bordered on the south by the Bay of Bengal and by India everywhere else, except for a small stretch in the southeast where it borders Myanmar (Burma). Aside from a few hills and forests along its perimeter, Bangladesh is a low-lying country, consisting mostly of the double delta of the Ganges and Brahmaputra rivers. Its climate is tropical monsoon, with a hot and rainy summer and a dry winter. Bangladesh is one of the wettest places in the world, with most of its rainfall occurring in the monsoon season from June through September. Since independence, Bangladesh has experienced catastrophic floods requiring emergency foreign assistance in 1974, 1984, 1987, 1988, and 1998. In the 1998 flood, nearly 70 percent of the country was under water for several months and some 1,200 people died.

Geographically Speaking

Like many Muslim countries, Bangladesh and Pakistan are highly patriarchal societies in which women's freedom of movement and expression is often severely restricted. But the problem of violence against women in South Asia is not limited to the Muslim countries. In India and Sri Lanka, as well as in Bangladesh and Pakistan, women are subjected to such atrocities as honor killings, dowry deaths (for insufficient dowries), disfiguring acid attacks, and female infanticide.

Ironically, all four countries have been pioneers in electing women to the highest offices in the land. Indira Gandhi was prime minister of India in the 1970s and '80s. Benazir Bhutto was Pakistan's prime minister in the 1980s and '90s. For most of the period from 1991 to the present (mid-2003), Bangladesh has been governed by two female prime ministers. In 1960, Sri Lanka became the first country in the world to have a female prime minister and since 1994 has had a female president.

Bangladesh is also one of the world's most cyclone-prone countries. Fierce tropical cyclones funnel up the Bay of Bengal to the coastal districts of Bangladesh at an average rate of 1.3 a year. In 1991, a cyclone killed almost 140,000 people and caused $2.7 billion in damages. Since then, however, Bangladesh has greatly improved its cyclone preparedness by building shelters and implementing evacuation plans.

Crammed into every inch of livable space are 133 million people, making Bangladesh the most crowded large country on Earth and the eighth most populous. Dhaka, the capital and largest city of Bangladesh, has more than 10 million people in its metropolitan area. The second largest city, Chittagong, has a population of about 2.6 million.

Bangladesh is an extremely poor and underdeveloped country. The statistics tell a grim tale. Bangladesh's per capita gross domestic product of $1,700 is in the sub-Saharan African range. Barely 45 percent of the total population is literate, about 40 percent is unemployed or underemployed, and an enormous 60 percent is engaged in agriculture. On the positive side, Bangladesh has plenty of fertile and well-watered (to say the least) land and is now self-sufficient in rice. Its other main crop and a major export

Eye on the Environment

As if Bangladesh did not already have enough water problems, naturally occurring arsenic is contaminating its groundwater supply, a primary source of drinking water. Inadequate well-water testing kits are compounding the looming public-health crisis.

product is jute, used to make burlap, twine, and rope. Bangladesh also has substantial natural gas reserves but has been slow to develop them. Despite determined domestic and international efforts to boost its prospects, the hurdles for Bangladesh remain high.

Mountain Monarchies and Island Nations

In addition to the three nations that came out of the partition of British India, South Asia includes two Himalayan kingdoms (it used to be three, until the tiny Kingdom of Sikkim became India's smallest state in 1975) and two tropical island nations. Following is a brief overview of the four states.

Nepal

This small landlocked country in the Himalayas is sandwiched between Indian and China's Tibetan Autonomous Region. Northern Nepal is dominated by the Himalayas: Eight of the ten highest mountain peaks in the world are in this part of the country, including the world's highest, Mount Everest (29,028 feet), which straddles the Nepal–China border. The center of the country is occupied by the Himalayan foothills; the southern Terai region is a flat and fertile plain. The mountains are snowy and cold year-round and the foothills are cool. The Terai is hot and humid in the summer and cool in the winter.

Nepal's 57,000-square-mile area supports a population of nearly 26 million. Its fabled capital of Kathmandu, located in a broad valley in the central hills, has a metropolitan area population of 1.2 million. Nepal is home to many different caste and ethnic groups (the two terms are often used interchangeably here). The Sherpas, a Tibeto-Nepalese people living in the sparsely populated mountain north, have gained worldwide fame as expert Himalayan expedition guides.

Terra-Trivia

Nepal's other famous ethnic group is the Gurkhas, a Tibeto-Mongolian people that conquered the small Nepalese state of Gurkha in the early 1500s and went on to unite the entire country (and to invade Tibet and China) in the 1700s. Gurkha soldiers are known around the world for their courage, loyalty, skill, and toughness.

Nepali is the official language of the country, spoken by about 90 percent of the population. About a dozen other languages and many more dialects are also spoken. Nepalese in business and government often speak English as well. Although Nepal was the birthplace of the Buddha (Siddhartha Gautama, believed to have been born in Kapilavastu around 538 B.C.E.), barely 8 percent of Nepal's people are Buddhist. The vast majority (more than 86 percent) is Hindu. There are also small Muslim and Christian minorities.

Nepal was originally populated by migrants from India, Tibet, and Central Asia. The modern kingdom dates back to the 1760s, when the Gurkha king Prithvi Narayan Shah united a number of independent hill principalities under his rule. By isolating itself from the outside world, Nepal successfully avoided being taken over by Great Britain during the colonial period, but at the cost of remaining economically backward. Nepal held its first democratic elections for a national assembly in 1959, but King Mahendra dismissed the government 18 months later and banned all political parties. In 1990, under pressure from a center-left pro-democracy movement, his successor, King Birendra, lifted the ban on political parties, and free and fair parliamentary elections were held the following year.

Since 1996, a Maoist guerrilla movement has been waging a growing and increasingly violent revolt against the Nepalese government. Peace talks in 2001 collapsed, but after another cease-fire in January 2003, negotiations resumed in April. In the meantime, Nepal was deeply shaken when on June 1, 2001, Crown Prince Dipendra shot and killed his father, King Birendra, his mother, and several other members of the royal family before turning the gun on himself. The slain king's younger brother, Prince Gyanendra, succeeded to the throne.

Nepal's per capita gross domestic product ($1,400) is even lower than that of Bangladesh. More than 80 percent of its people depend on agriculture for their livelihood, and about 42 percent live below the poverty line. In addition to farming, Nepal's main economic activities are tourism, textile manufacturing, and carpet production. The country's infant mortality rate is high and its life expectancy short (just 59 years).

CAUTION

Eye on the Environment

One of the most serious environmental problems facing Nepal, and much of South Asia, is deforestation. It has been estimated that between 1978 and 1994 Nepal lost nearly a quarter of its green forest. And if anything, Nepal's Maoist insurgency of recent years has accelerated deforestation by preventing forest department personnel from patrolling rebel-held forest hideouts. Lax security has given timber smugglers and rhino poachers a green light to step up their activities.

The consequences of unchecked deforestation are readily visible in Nepal and elsewhere in South Asia: severe soil erosion, water pollution, increased flooding and drought, and loss of wildlife. Nepal's forests shelter some of the world's rarest animals, including Bengal tigers, one-horned rhinos, and snow leopards, all of which are at risk.

Bhutan

East of Nepal, and separated from it by a small strip of India that was formerly the Kingdom of Sikkim, lies South Asia's other Himalayan kingdom, Bhutan. Like Nepal,

Bhutan is bordered on the north and east by China (Tibet) and by India everywhere else. It is mostly mountainous, with some fertile valleys and a small stretch of plain in the south called the Duars. The climate is tropical in the southern plain and gets progressively colder the higher up you go. Measuring only 18,000 square miles (less than a third the size of Nepal), Bhutan is home to some 2 million people. Its capital, Thimphu, has a population of 60,000.

Ethnically, the majority of Bhutanese are either Bhote (50 percent) or Nepalese (35 percent). The official language is Dzongkha, although a number of Tibetan and Nepalese dialects are also spoken. Three-quarters of the Bhutanese are Buddhist; the rest are Hindu.

Bhutan was largely isolated from the rest of the world until the 1960s, when it began opening up to the international community and launched a modernization campaign that included the abolition of slavery and the caste system. In 1998, Bhutan's king voluntarily transferred executive powers to an elected Council of Ministers. After nearly a century of absolute monarchy, Bhutan is slowly moving toward becoming a parliamentary democracy.

Agriculture and forestry are the mainstays of Bhutan's extremely small and under-developed economy. Most Bhutanese are subsistence farmers, and the country relies heavily on trade with and financial aid from India, which also handles its foreign affairs. Bhutan's resources include hydropower (it now exports electricity to India) and tourist appeal, which remains mostly unexploited.

Sri Lanka

Formerly known as Ceylon, Sri Lanka is the tear-shaped island off the southeast tip of India. The south-central interior of the island is mountainous; the rest is flat to rolling plains ending in beautiful palm-fringed beaches. Because Sri Lanka is only six degrees north of the equator, it has a tropical monsoon climate.

Sri Lanka is home to nearly 20 million people. About 2.5 million of them live in the metropolitan area of the capital city, Colombo. (Sri Lanka also has a legislative and judicial capital with a tongue twister of a name: Sri Jayewardenepura Kotte.) In recent years, Sri Lanka's economy has become much more focused on industry (textiles and garments) and services, and far less so on plantation crops (rubber and tea). Despite a disastrous ethnic civil war, Sri Lanka has experienced solid economic growth over the last decade and has South Asia's highest per capita gross domestic product.

The defining fact of Sri Lankan history has been the division and enmity between its two major ethnic groups. Nearly 75 percent of Sri Lankans are Sinhalese; one-fifth are Tamil. The Sinhalese are mainly Buddhist; the Tamils, mainly Hindu. Sinhala and Tamil are the national languages; Sinhala, however, is the official language. (English is often used for government business.)

The Sinhalese are thought to be descendents of Indo-Aryans who began arriving on the island, probably from northern India, in the late sixth century B.C.E., and then mixed with aboriginal peoples. The Tamils migrated to Sri Lanka from southern India beginning in the third century B.C.E. (It was at this time, too, that Buddhism was introduced in the island.) By the 1300s C.E., a Tamil kingdom had been established in the north, while Sinhalese power was concentrated in the southwest. The island was occupied by the Portuguese in the 1500s and by the Dutch in the 1600s. In 1796, the Dutch ceded Sri Lanka to Great Britain, who ruled it, as Ceylon, until 1948, when the island (renamed Sri Lanka in 1972) gained its independence.

In 1983, longstanding Tamil frustrations at the political and economic control of the majority Sinhalese reached a boiling point. A violent war broke out between government forces and Tamil rebel groups demanding a separate nation. The most notorious of the rebel groups is the Liberation Tigers of Tamil Eelam, or Tamil Tigers for short, pioneers in the use of suicide bombings. Finally, in December 2001, after tens of thousands of Sri Lankans had lost their lives (62,000 by some estimates), a ceasefire was declared. As of mid-June 2003, peace talks brokered by Norway seem to be making some progress.

The Maldives

The Maldives are an island republic in the north Indian Ocean, about 400 miles southwest of Sri Lanka. The country consists of about 1,190 coral islands on 26 atolls. (An atoll is a large ring of coral atop a defunct volcano surrounding a lagoon.) Although the total area of the islands is only about 115 square miles, they are spread out over 35,200 square miles. Only about 200 of the islands are inhabited by the Maldives' 320,000 people. (Another 87 "uninhabited" islands are set aside for tourist resorts.) Maldivians are of Sinhalese, Dravidian (South Indian), Arab, and African origin. All of them speak Divehi (a Sinhalese derivative), and all are Sunni Muslim. The capital, Male (that's its name, not its gender), has more than 80,000 residents.

The Maldives' first settlers are thought to have been Hindu and Buddhist seafarers who arrived from India and Sri Lanka around 500 B.C.E. According to tradition, Islam was introduced in the islands and a sultanate established in the twelfth century C.E. The Portuguese occupied the Maldives in the mid-1500s, the Dutch established a protectorate in the mid-1600s, and the British took over in 1796. The Maldives became independent in 1965 and replaced the Sultanate with a republic in 1968.

Given the Maldives' superb beaches, magnificent coral reefs, and tropical monsoon climate, it's not surprising that tourism is the country's main industry. With little arable land, the Maldives must import most of its food. In a country whose highest elevation is barely six feet, global warming and rising sea levels are a major concern. (For more on the possible effects of global warming on low-lying coral atolls, see Chapter 24.)

In the News: Pakistan's Northwest Frontier Province

Kashmir is not Pakistan's only international trouble spot. In many respects, its Northwest Frontier Province is an even more volatile and dangerous place. A wild and rugged area along the Afghan border, the Northwest Frontier Province is controlled in reality by local tribes and only nominally by the central government. As of mid-2003, it was widely believed that al-Qaeda leader Osama bin Laden was hiding in this deeply conservative Islamic region.

Before the terrorist attacks on New York City and Washington, D.C., on September 11, 2001, Pakistan had close ties with Afghanistan's radical Taliban government that had enabled bin Laden to carry out the attacks (see Chapter 21). Although Pakistan's leader, General Musharraf, ended up supporting the United States' 2001 invasion of Afghanistan and subsequent global war on terror (in return for billions in aid dollars), that support is weak to nonexistent among a large part of Pakistan's population, especially in the more radical west.

Western Pakistan's Islamic schools, or *madrassas*, have long been notorious for their anti-Western, pro-jihad (Islamic holy war) teachings, and have been fertile recruiting grounds for Islamic terrorist groups. In Fall 2002, fundamentalist Islamic parties swept elections in the Northwest Frontier Province, which voted the following June to adopt Islamic law, called *Shariah*. This makes it all the less likely that the central government will be able to follow through on its promises to curb and reform the madrassas.

The Least You Need to Know

- Rapid population growth means that South Asia will surpass East Asia as the world's most populous region.

- The world's largest democracy, India, has one-sixth of the world's people.

- Subject to devastating monsoon rains and cyclones, Bangladesh seems to be on nature's hit list.

- Nepal's 29,028-foot Mount Everest is the world's highest mountain.

23

Southeast Asia: Spice Islands and Golden Pagodas

In This Chapter

◆ Understanding the terminology: Indochina or Southeast Asia?

◆ Describing the countries: Peninsulas and archipelagoes

◆ Hacking through jungles and deltas

◆ Touring Southeast Asia's cities

◆ Getting to know the people

Southeast Asia, shown on the following map, has long been thought of as a remote and exotic region—a land of beautiful people, spectacular temples, spice islands, and tropical forests filled with rare wildlife. Behind all the magic, however, lies a tortured political history.

Much of that history has been determined by geography. Southeast Asia is squeezed between the two Asian giants, China and India, which have largely shaped the region's culture. Its natural resources and strategic location made Southeast Asia a magnet early on for European colonization. (It was to reach China and the "spice islands" of Southeast Asia that

Columbus sailed west from Spain and bumped into America.) In the decades after World War II, the region's struggle to put its colonial past behind it all too often resulted in war, revolution, more foreign intervention, genocide, repression, and lingering poverty. And yet, despite all the difficulties and traumas, the enchantment and resilience of Southeast Asia and its people are there for all to see.

What's in a (Place) Name?

The name "Southeast Asia" refers to the two peninsulas and two archipelagos that occupy the southeastern corner of the Asian continent—a region roughly bounded by China on the north; India, Bangladesh, and the Bay of Bengal on the west; the Indian Ocean on the south; and the Pacific Ocean on the east. What we now call Southeast Asia used to be referred to as the "East Indies" and sometimes "Indochina," although the latter term technically refers only to part of the region.

Southeast Asia.

Geographically Speaking

The term *Indochina* highlights the two major influences on Southeast Asia. The first part, *Indo*, refers, of course, to the region's western neighbor, India. From time immemorial, India has been one of the world's great cultural incubators. From it, Southeast Asia derived many elements of its culture, primarily its major religions. Buddhism and Hinduism both originated in India and have left a profound imprint on Southeast Asia. India's Muslim Mughal Empire served as a conduit for Islam.

The second half of *Indochina* is, of course, China, the world's most populous country. China has loomed large in the history, economy, and ethnic makeup of the region. For centuries, Chinese people have streamed south into Southeast Asia and today figure heavily in the region's population mix, especially in urban centers.

The Indochinese peninsula is the larger of the two peninsulas that make up the Southeast Asian mainland. It consists of the three countries that were once part of French Indochina—Vietnam, Laos, and Cambodia—as well as most of Thailand (the only nation in the region that was never a colony), and part of the former British colony of Burma, now known as Myanmar. The region's second peninsula is the Malay peninsula, which stretches south from the Indochinese peninsula much like the tail of a kite. It includes parts of Myanmar and Thailand, the peninsular part of Malaysia, and at its tip, the city-state of Singapore.

South and east of the peninsulas are two archipelagos—the Indonesian and Philippine. Indonesia occupies most of the Indonesian peninsula, except for the northwestern third of the island of Borneo, which is shared by Malaysia and the tiny state of Brunei, and the eastern half of Timor, which is now the new state of East Timor. The Philippine archipelago is occupied entirely by the Philippines.

Ancient Kingdoms

Southeast Asia has given rise to many great civilizations and mighty kingdoms. Among the most splendid was the powerful Khmer empire, centered in the city of Angkor in present-day Cambodia. Founded in the ninth century C.E., the Angkor dynasty controlled most of the Southeast Asian mainland from the tenth through the thirteenth century. Khmer rulers were passionate builders who left behind some of the region's most spectacular monuments. At the top of the list of Khmer architectural wonders is the temple of Angkor Wat in Cambodia. Built at the height of Khmer power in the midtwelfth century, Angkor Wat was dedicated to the Hindu god Vishnu and is generally considered to be one of the greatest religious structures in the world.

Southeast Asia's other great complex of ancient ruins is the temple-city of Bagan in modern day Myanmar (Burma). A major center of Theravada Buddhism (discussed later in this chapter), Bagan was the capital of the first Burmese Empire, founded in the mid-eleventh century C.E. Its kings lavished the city with so many temples and monuments that during its golden age, from the twelfth through the mid-thirteenth century, Bagan was known as the "city of four million *pagodas.*" In 1277, the Venetian merchant and explorer Marco Polo described Bagan's gold-sheathed towers as "one of the finest sights in the world."

A decade later, however, Kublai Khan's Mongol forces swept south and sacked Bagan, bringing the first Burmese Empire to an end. Today, only the remnants of Bagan's thousands of pagodas remain, lonely reminders of past greatness. Local villagers and farmers still worship at the many statues of the Buddha left in the temple niches.

GeoJargon

The term **pagoda** refers to a variety of tower-like religious structures that are usually part of Buddhist temple or monastery complexes. The characteristic *stupa* of Southeast Asia is a type of pagoda shaped like a Hershey's kiss and often sheathed in gold. Stupas usually house some revered relic, such as a hair from the Buddha.

Southeast Asia's colonial rulers.

European Colonization

The Mongols weren't the only foreign invaders to make their presence felt in Southeast Asia. Between the sixteenth and the nineteenth centuries, Europeans gained control over all of the region except for Thailand. First came the Portuguese and the Spanish, followed by the Dutch, the English, the French, and finally the Americans, who took over the Philippines from the Spanish in 1898, at the end of the Spanish American war. The preceding map shows who used to control what in Southeast Asia.

Within months of its attack on the U.S. naval base at Pearl Harbor, Hawaii, on December 7, 1941, Japan had overrun most of Southeast Asia. At the end of World War II, the displaced colonial powers returned to reclaim their former territories, but the days of foreign rule in Southeast Asia were numbered. Most of the region's countries gained their independence in the late 1940s and 1950s. Still, centuries of colonization left lasting imprints throughout the region. European administrative divisions, infrastructure, and even religions are evident in most Southeast Asian countries.

The Peninsular Countries

Peninsular Southeast Asia contains five countries and half of a sixth (Malaysia, which is split between the Malay peninsula and the island of Borneo). Following is a brief overview of each of the peninsular countries.

Myanmar

You get many strange looks when you mention Myanmar—even the geographically informed balk at this one. Not only is Myanmar off the beaten track, many people don't realize that the country once better known as Burma changed its name to Myanmar in 1989. Actually the name was imposed by the country's repressive military regime and isn't recognized by the United States State Department, which still uses the name Burma. When Burma became Myanmar, its capital and largest city Rangoon (population: 4.3 million) became Yangon (although that name isn't recognized by the United States either).

Diamond-shaped Myanmar occupies the northwestern portion of the Indochinese peninsula, with a narrow stretch of land jutting southward into the Malay peninsula. Bordered by Bangladesh, India, China, Laos, and Thailand, Myanmar is roughly the size of Texas.

Burma was colonized by the British in the nineteenth century and incorporated into British India. It became a separate, self-governing colony in 1937, only to be invaded by Japan in 1941. Burma was the scene of some of fiercest battles of World War II, many of them occurring along the famous Burma Road, the Allies' lifeline to China.

Terra-Trivia

Myanmar is the world's second largest producer of illegal opium, the source of heroin and other narcotics. The Myanmar, northern Thailand, and Laos border region is the notorious "golden triangle," an untamed area where drug lords and their personal armies control much of the countryside and where heroin and methamphetamines are the main exports.

Burma gained its independence from Britain in 1948, and has had a troubled existence ever since. A 1962 coup established a socialist military regime that continues to suppress dissent and restrict basic freedoms. In elections held in May 1990, the main opposition party won by a landslide, but the military government refused to recognize the results. The opposition leader, 1991 Nobel Peace Prize winner Aung San Suu Kyi, was held under house arrest from 1989 to 1995 and several times since, including in May 2003.

Myanmar was once Southeast Asia's major rice exporter. Although the country remains heavily agricultural, it now barely feeds itself. Despite Myanmar's agricultural and mineral potential, its 42 million people remain among the world's poorest.

Thailand

Thailand is located in the heart of mainland Southeast Asia, between Burma and the former French Indochina. Like Myanmar, Thailand has a thin, peninsular outcropping that reaches southward into the Malay Peninsula. Because of its unique configuration and position, Thailand has access to both the Adaman Sea and Indian Ocean on the west, and the Gulf of Thailand and South China Sea on the east. The country is about the size of France and has a population of some 62 million. The Thai capital, Bangkok, is known as the "Venice of the East" for its great waterway, the Chao Phraya River, and its many canals. This frenetic city is home to 6.6 million people (8.8 million in its metropolitan area).

In the mid-1400s, a unified Thai kingdom replaced Cambodia's Khmer empire as the dominant power in peninsular Southeast Asia. Siam, as the country was known until 1939, managed to make it through the nineteenth century without becoming a colony of either Britain or France. As a result of a bloodless coup in 1932, Thailand's absolute monarchy became a constitutional one. During World War II, Thailand was governed by a pro-Japanese puppet government and so was spared the full brutality of Japanese occupation. After the war, Thailand had its share of coups, attempted coups, and military governments. Since a violent pro-democracy uprising in 1992, Thailand has enjoyed more than a decade of free elections and no military interventions.

Thailand's diversified, free-market economy is one of the strongest in the region. Between 1985 and 1995, Thailand enjoyed one of the world's highest economic growth rates. And then disaster struck. The Thai financial crisis of 1997 caused its currency to plummet and sent shock waves through markets and economies around the world. Since then, the Thai economy has been making a comeback, thanks mostly to strong exports. Thailand's major industries include textiles, food processing, electrical appliances, computer hardware, and mining (Thailand is the world's leading producer of tungsten and third largest tin producer). Tourism is a major money-earner, and although agriculture accounts for about 11 percent of Thailand's gross domestic product, it employs more than half of its work force.

Geographically Speaking

The shape of a country is important because it can influence its politics, administration, and infrastructure. Politically speaking, the ideal nation shape is compact and circular with a central capital. Other shapes have inherent disadvantages and can lead to fracturing and the rise of autonomy movements.

Southeast Asia includes good examples of several nation shapes. *Compact* countries like Cambodia are roughly circular. Vietnam is an *elongated* country, long and thin. *Prorupt* countries are basically compact with a narrow elongation extending outward (Myanmar and Thailand, for example). *Fragmented* countries are part mainland and part island, like Malaysia, or are made up entirely of islands, like the Philippines and Indonesia.

Laos

Landlocked and mountainous, especially in the north, Laos borders on Myanmar, Thailand, Cambodia, and Vietnam, as well as on China in the north. This densely forested country about the size of Utah supports (barely) a population of about 5.7 million. Its capital, Ventiane, is home to 194,000 people.

A unified Laotian kingdom first emerged in the mid-1300s. By the late 1700s, most of what is now Laos was controlled by Siam (Thailand). Laos became part of French Indochina in 1893 and did not gain its full independence until France withdrew from Indochina in 1954 following its defeat by the Vietnamese at the famous battle of Dienbienphu. Laos was inevitably drawn into the U.S.–Vietnam War and sustained heavy American bombing during the late 1960s and early 1970s. Within months of the fall of South Vietnam and Cambodia in 1975, the communist Pathet Lao party took over the country, forced the king to abdicate, and established the Lao People's Democratic Republic.

Despite attempts since 1986 to decentralize and privatize the economy, Laos is one of the poorest countries in a poor region. Subsistence farming accounts for more than half of the country's gross domestic product and employs more than 80 percent of the work force. Electricity is scarce, roads are few, and railroads nonexistent.

Cambodia

Bordered by the Gulf of Thailand to the west, Thailand to the west and north, Laos to the north, and Vietnam to the east and south, Cambodia physically resembles a bowl, with a large alluvial plain in the center surrounded by mountains. Even though Cambodia is smaller than Laos, its population of nearly 13 million is more than double that of its neighbor to the north.

Once the heartland of the great Khmer empire, Cambodia became a French protectorate in 1863 and part of the union of French Indochina in 1893. Like Laos, Cambodia gained its full independence only after the end of the French–Indochina War in 1954. And also like Laos, but even more so, Cambodia was drawn into the U.S.–Vietnam War and became a major battlefield in it.

Although all three former French Indochinese colonies suffered terribly in the Vietnam War, Cambodia had the most horrific post-war experience. Communist Khmer Rouge forces under the monstrous Pol Pot took the Cambodian capital, Phnom Penh, in April 1975, and established the radical communist state of Democratic Kampuchea. During the Khmer Rouge reign of terror and the 1978 Vietnamese invasion that eventually ousted the regime, between 1.7 and 3 million people (out of a population of 7.3 million) were killed or died of starvation or overwork in the "killing fields" of Kampuchea.

Free elections in 1993 and the establishment of a constitutional monarchy under the country's longtime leader, Prince (now King) Sihanouk, seemed to signal an end to Cambodia's long nightmare. The years since then have been relatively stable for the country, but the threat of political violence is never too far away.

Cambodia's modern history has left it with a tattered economy. Although growth rates have been remarkably strong in the first years of this century, Cambodia is still plagued by extreme poverty, especially in the countryside, and by the lack of skilled workers and basic infrastructure. Except for tiny East Timor, Cambodia has the lowest per capita gross domestic product in the region. A strengthening tourism industry is a bright spot on the economic horizon.

Vietnam

Vietnam, the easternmost of the Southeast Asian peninsular countries, borders Laos and Cambodia to the west, China to the north, and the South China Sea to the east and south. This long, narrow country (31 miles wide at its narrowest) has a population of 81 million, making it the most populous country in peninsular Southeast Asia.

China ruled northern Vietnam for 1,000 years until the tenth century C.E., when a native Vietnamese dynasty took over and began expanding to the south. In the mid 1500s, Vietnam was essentially split between northern and southern dynasties, and by the time the French began taking over in the mid-nineteenth century, the country was divided into three parts. Southern Cochin China became a French colony in 1862; central Annam and northern Tonkin became French protectorates in 1884. In 1887, all three parts were united with Cambodia and later Laos in the single protectorate of French Indochina.

French rule in Vietnam lasted until 1954, when communist forces under Ho Chi Minh defeated the French. The country was divided into two parts: Communist North Vietnam, supported by China and the Soviet Union, and noncommunist South Vietnam, whose authoritarian regime was backed by the United States.

Full-scale war between north and south raged throughout the 1960s and into the 1970s. In an effort to shore up the South Vietnamese government, the United States became increasingly involved in the conflict. At the height of the war in the late 1960s, more than half a million U.S. troops were stationed in Vietnam. Facing mounting anti-war protests at home, U.S. President Richard M. Nixon agreed to a peace settlement in January 1973 and pulled out the last U.S. troops from Vietnam later that year. In April 1975, the South Vietnamese government collapsed and the two Vietnams were finally united under communist rule. Its involvement in the Vietnamese War cost the United States the lives of 58,000 soldiers and more than $140 billion. The Vietnamese, however, lost 1.3 million people in the struggle.

With the fall of South Vietnam, its capital, Saigon, ceased being a national capital and had its name changed to Ho Chi Minh City, in honor of the victorious North Vietnamese leader. But with a population of nearly 3.4 million (5.7 in its metropolitan area), the former Saigon remains Vietnam's largest and most prosperous city. Hanoi, the national capital (and former capital of North Vietnam and of French Indochina), is less than half the size of Ho Chi Minh City.

Like China, only much more slowly, Vietnam has been adopting a free-market economy while maintaining its communist political system. The results have been most impressive in the agricultural sector, with Vietnam now the world's second biggest rice exporter. The industrial sector, however, continues to stagnate. Vietnam is still a very poor country struggling to overcome the economic effects of years of war and destruction. Full diplomatic relations between Vietnam and the United States were restored in 1995, and in 2000 the two countries signed a bilateral trade agreement that is expected to bring significant benefits to Vietnam's economy.

Malaysia

Malaysia is both a peninsular and an insular nation. Most of the country's 22.6 million people live in West Malaysia on the southern end of the Malay Peninsula. Also located here is Malaysia's booming cosmopolitan capital, Kuala Lumpur, home to some 1.4 million people (3.7 million in its metropolitan area). Larger but much less populous East Malaysia occupies the northwestern third of the island of Borneo across the South China Sea. Both parts consist of coastal plains and interior mountains and are densely forested.

Much of Malaysia's history and economic development has been shaped by its location on the strategic Strait of Malacca, which very early on attracted the attention of European merchants and colonizers. First came the Portuguese in 1511, then the Dutch in 1641, and finally the British in the late 1700s. From their trading settlements on the Malay Peninsula—Malacca, Penang, and Singapore—the British went on to establish protectorates over the various peninsular Malay sultanates in the nineteenth and early twentieth centuries. After World War II, the British-held peninsular territories were granted semi-autonomy as the Federation of Malaya, and full independence in 1957. In 1963, the Federation of Malaysia was formed by the union of Malaya, Singapore, and the former British colonies on northern Borneo, Sabah and Sarawak (now the two states that make up East Malaysia). Two years later, Singapore pulled out of the federation and became an independent country (see Chapter 12).

Over the last 30 or so years, Malaysia has transformed itself from a producer of raw materials (rubber, tin, timber, petroleum, and natural gas) to a country with a diversified and prosperous economy, driven largely by exports of manufactured goods, particularly electronics. Aside from Singapore and Brunei, whose economies are in a different league from those of their neighbors, Malaysia has the highest per capita gross domestic product in Southeast Asia.

Island Countries

Southeast Asia has thousands of islands, including some of the world's largest and some of the most beautiful. The island nations are Indonesia, Singapore, Brunei, East Timor, the Philippines, and half of Malaysia. The highly developed country of

Singapore is covered with the other so-called Asian economic "Tigers" in Chapter 12. Because most of its people live on its peninsular half, Malaysia was covered with the other peninsular countries in the preceding section.

Indonesia

Indonesia's archipelago is the largest in the world, by number of islands, land area, and population. The country is made up of more than 17,000 islands straddling the equator; about 6,000 of them are inhabited. With a population of 231 million, Indonesia is the fourth most populous country in the world. Its capital, Jakarta, on the island of Java, is home to nearly 11 million people (more than 17 million in its metropolitan area), making it the largest city in Southeast Asia.

The fabled Spice Islands of Indonesia have a very long and rich history. Buddhist and Hindu empires and Muslim sultanates flourished here long before the arrival of the Portuguese in the early sixteenth century. The Dutch followed later in the century and gradually extended their control over all the islands except part of one: The east side of tiny Timor remained under Portuguese control until 1975.

The Dutch developed the Netherlands East Indies into one of world's most valuable colonial properties. During World War II, Indonesia endured a brutal Japanese occupation and afterward was briefly returned to Dutch control. After a bitter four-year war, Indonesia achieved its independence from the Netherlands in 1949. The Dutch colony in western New Guinea became part of Indonesia in 1963.

Much of Indonesia's post-independence history has been marked by separatist rebellions and authoritarian rule, first under founding president Sukarno and from 1967 through 1998, under his successor General Suharto. In 1965, a coup attempt against "president for life" Sukarno prompted a vicious anti-Communist purge in which an estimated 300,000 people were killed. Although free and fair national elections were held in Indonesia in 1999, the easing of authoritarian rule triggered violent separatist and sectarian uprisings in several parts of the country.

> ## GeoRecord
>
> Indonesia's thousands of islands include three of the world's largest. In this list of the largest islands in the world, square mileage is shown in parentheses and an asterisk (*) signifies an Indonesian island:
>
> 1. Greenland (840,004);
> 2. New Guinea* (309,000);
> 3. Borneo* (287,300);
> 4. Madagascar (226,658); 1
> 5. Baffin Island (195,928);
> 6. Sumatra* (182,860).
>
> *(New Guinea and Borneo are only partially occupied by Indonesia.)

GeoRecord

The world's largest Islamic country is not in the Middle East or in northern Africa. It's Indonesia, with more than 170 million Muslims.

Indonesia has significant oil and gas reserves, abundant minerals and timber, and a developing manufacturing base. Bali and many of its other islands are tourist paradises. Yet Indonesia's economy is bogged down by lingering political instability, corruption, ethnic and regional conflicts, and threats of terrorism.

Brunei

Brunei is a little different from any other country you've read about in this book. Located on the northwest coast of the island of Borneo, squeezed between the two East Malaysian states of Sabah and Sarawak, Brunei is a peanut of a country, measuring a little more than 2,200 square miles. The population is mite-size also (about 350,000 people). Those facts by themselves set this tiny country apart from the rest of the region, and the story gets stranger.

The Sultanate of Brunei had its first golden age about 500 years ago when it controlled all of Borneo and part of the Philippines. After a long period of decline beginning in the 1600s, Brunei became a British protectorate in 1888. Self-rule was restored in 1959, and the small country became fully independent in 1984.

Terra-Trivia

Scattered throughout the South China Sea are hundreds of tiny islands including two groups known as the Spratly Islands and the Parcel Islands. Although most of these islands are essentially uninhabited reef outcrops, China, Taiwan, Vietnam, Malaysia, the Philippines, and Brunei are all adamant about claiming them. Why? Whoever owns the islands of the South China Sea also owns the oil that lies beneath them.

Brunei might be very small, but it is also very rich, although less so than it was back in the 1970s and 1980s. Brunei's modern golden age has been all about oil and gas, and lots of it. Little Brunei is the third largest oil producer in Southeast Asia and the fourth largest producer of natural gas in the world. Although down considerably from its 1980 peak, Brunei's per capita gross domestic product ($18,600 in 2002) is second only to Singapore's in the region and far above that of most developing countries.

A hereditary constitutional sultanate, Brunei has been ruled by its current sultan, Haji Hassanal Bolkiah, since 1967. Oil revenues are used to subsidize education, health care, and even food and housing for the people of Brunei. The Brunei government also invests part of the country's oil revenues abroad as insurance against the day when the wells run dry.

East Timor

Southeast Asia's (and the world's) newest independent country is the former Portuguese colony of East Timor. This tiny (5,641 square-mile) nation of just under a million people is located on the island of Timor, one of the Lesser Sunda Islands at the eastern end of the Indonesian archipelago. In addition to the eastern half of the island, East Timor also occupies a small chunk of the northwestern coast and the two offshore islands of Pulau Atauro and Pulau Jaco. East Timor's capital, Dili, has a population of about 50,000.

The Portuguese first landed in Timor in 1521 and began colonizing the island in the mid-seventeenth century. (The western half was ceded to the Dutch in the next century.) After a revolution at home, Portugal withdrew from its remaining colonies, including East Timor, in 1975. The East Timorese declared their independence on November 28, 1975, but 10 days later Indonesia invaded the country and incorporated it as a province. Over the next 20 years, as many as 250,000 people are thought to have died in the violent "pacification" campaign that followed Indonesia's illegal takeover.

In August 1999, a referendum supervised by the United Nations was held in East Timor and the results were overwhelmingly in favor of independence. That triggered another wave of violence and killings on the part of anti-independence paramilitary forces backed by the Indonesian army. Some 200,000 East Timorese fled to the western side of the island as their country was reduced to ruins. Finally, under the auspices of a United Nations Transitional Administration, presidential elections were held in April 2002 and East Timor became fully independent the following month.

Much of East Timor and its economic infrastructure was destroyed in the violence following the 1999 independence vote. A major infusion of international assistance has helped repair some of the damage, and most of the people who fled the violence have returned. One bright spot for the future is East Timor's untapped offshore oil reserves. But for now, the country has the unhappy distinction of having the world's lowest per capita gross domestic product, a meager $500 in 2001.

The Philippines

Southeast Asia's other big archipelago is the Philippine Archipelago, a group of more than 7,000 volcanic islands off the east coast of Vietnam between the South China Sea and the Philippine Sea. The larger islands have mostly mountainous interiors surrounded by coastal lowlands. The entire group of islands is occupied by the nation of the Philippines.

Terra-Trivia

The Philippine capital of Manila is a sprawling, congested city of more than 10.2 million people (13.7 million in its metropolitan area). Its Intramuros section looks more like a Spanish colonial city in the Caribbean than anything else in Southeast Asia. Outside the colonial center, Manila is a dizzying mix of glittering office towers, huge port facilities, and urban blight.

Ferdinand Magellan claimed the islands for Spain in 1521 and for the next 377 years they remained Spanish territory. After its defeat in the Spanish American War in 1898, Spain ceded the Philippines to the United States. In 1935, the country became a self-governing commonwealth and was scheduled to become fully independent in 1946, but the outbreak of World War II threatened to derail the process. Japan invaded the Philippines in December 1941 and defeated the last American forces there in May 1942. Fulfilling his famous "I shall return" promise, General Douglas MacArthur landed in the Philippines on October 20, 1944, and after nearly a year of fierce and destructive fighting, it was Japan's turn to surrender. Despite the enormous devastation it had endured, the Philippines became independent on schedule in 1946.

After a series of orderly presidential successions during the post-war reconstruction period, Ferdinand Marcos came to power in 1965 and proceeded to rule despotically and corruptly for the next 21 years. The murder of opposition leader Benigno Aquino in 1983 sparked a "people power" uprising that eventually led to the ousting of Marcos in 1986 and the installation of Aquino's widow, Corazon, as the first post-Marcos president. Despite a number of attempted coups during Aquino's term, presidential successions have been more or less orderly in recent years. But as elsewhere in the region, corruption is a problem and the threat of instability persists. The country is also struggling to contain communist guerillas, Muslim separatist insurgencies in its southern islands, as well the growing threat of al-Qaeda-like terrorism.

Agriculture and light industries (mainly textiles, food processing, and electronics assembly) are the mainstays of the Philippines' economy. The country also has abundant mineral deposits, as well as significant natural gas, geothermal, hydropower, and coal reserves. Another asset is the high literacy rate (about 92 percent) of the Philippines' 84.5 million people. Among the problems holding the economy back are corruption, political instability, the threat of insurgencies and terrorism, inadequate infrastructure, high levels of government debt, and a weakening currency.

Physical Facts

The region of Southeast Asia straddles the equator and is located almost entirely within the tropics (only the northern portion of Myanmar peeks above the Tropic of Cancer). The typical Southeast Asian terrain consists of a mountainous interior with fertile coastal plains and river lowlands.

As you might expect, most of Southeast Asia is hot and wet. As is typical in tropical monsoonal climates, rainfall is heavy over the entire region between May and October, while northern areas receive considerably less rainfall between November and April. Tropical rain forests cover much of the region. Throughout Southeast Asia the main crop is rice; other important crops include rubber, tea, spices, and coconuts.

Indonesia and the Philippines are Pacific Ring of Fire countries (see Chapter 2), which means that they are subject to destructive earthquakes and volcanic eruptions. Despite the risks, people have been drawn to the these islands in great numbers over the centuries because of the fertility of their volcanic soils and their strategic trading location.

> **Terra-Trivia**
>
> In 1991, Mount Pinatubo violently erupted on the Philippine island of Luzon. The blast buried the United States' Clark Air Force Base and forced the evacuation of thousands of Filipinos. The eruption, the third largest of the twentieth century, affected weather around the world.

The Waters

Peninsular Southeast Asia is drained by several large and regionally important rivers. The Mekong River is Asia's sixth longest river and the thirteenth longest in the world. From its source in Tibet, the Mekong flows southward through China, along the Thai–Laotian border, through Cambodia, and on to its huge delta in southern Vietnam. Two of the region's most important rivers flow through Myanmar. Dominating the entire central portion of the country is the great Irrawaddy River, Myanmar's revered lifeline. Along its banks are the nation's two largest cities and biggest population concentrations. Originating deep in the heart of Chinese Tibet, the Salween River flows through eastern Myanmar, forming part of the Thai-Myanmar border. Thailand's Chao Phraya, famous for the golden pagodas that line its banks, flows down the western part of the country and through the capital city of Bangkok.

The sea is of vital importance to this region of peninsulas and archipelagoes. In the west, two arms of the Indian Ocean abut Southeast Asia—the Bay of Bengal and the Andaman Sea. The peninsular mainland is separated from the islands by the famous spice trade route, the Strait of Malacca. Amid the Indonesian islands are the Java Sea and the Banda Sea, while the Timor Sea and the Arafura Sea separate Indonesia from Australia. Between Indonesia and the Philippines is the Celebes Sea. East of Vietnam lies the South China Sea; east of the Philippines is the Philippine Sea and beyond that the vast Pacific.

Southeast Asia—physical features.

The Islands

As already noted, Southeast Asia includes literally thousands of islands—too many to mention here. The largest and most important are shown on the following map.

The region's largest island is New Guinea. Although Indonesia occupies half of it, geographically New Guinea is considered to be part of Melanesian group of islands in Oceania (see Chapter 24). The second largest island in Southeast Asia, and the largest in the Indonesian archipelago, is Borneo, which is shared by Indonesia, Malaysia, and Brunei. Borneo, Sumatra, Java, and Sulawesi (Celebes) make up the Greater Sunda Islands. One of the most densely populated places in the world, Java is Indonesia's core and the site of its capital, Jakarta.

East of Java are the Lesser Sunda Islands, extending from Bali to Timor. Between Sulawesi and New Guinea are the Moluccas, better known as the Spice Islands.

North of Indonesia are the thousands of islands that make up the Philippine archipelago. Although many of the islands are familiar to World War II buffs, we'll only mention the two largest here. Luzon in the north is the main population and industrial center and site of the Philippine capital, Manila. The Philippines' second largest island is Mindanao in the south, home to most of the country's Muslim population.

> **Terra-Trivia**
>
> The tiny islands of Komodo and Bali are part of the Lesser Sunda Islands. The feared Komodo dragon is a rare species of huge, meat-eating lizards found only on Komodo and a few neighboring islands. It is the world's largest living reptile. Bali, a Hindu island in predominantly Muslim Indonesia, is known famous for its spectacular scenery, beautiful people, and fascinating cultural traditions and architecture.

The Mix of Peoples

The mountainous nature of much of Southeast Asia has helped preserve the distinctiveness of many ethnic groups. On the mainland, the major groups roughly coincide with national borders: Burman peoples in Myanmar, Thai in Thailand, Lao in Laos, Khmer in Cambodia, Malay and Chinese in Malaysia, and Vietnamese in Vietnam. But the divisions aren't always this clear-cut, and grievances and aspirations of ethnic minorities continue to be a source of actual or potential conflict throughout the region.

The islands of Indonesia and the Philippines contain hundreds of smaller ethnic groups, but in general, the Indonesian mix consists of Javanese (45 percent) and smaller numbers of Sundanese, Madurese, coastal Malays, and in Indonesian New Guinea, Papuans. Filipinos are predominantly Malay.

The region's languages basically parallel its ethnic patterns. Some French lingers in the three former French Indochinese countries, although even in Vietnam English is becoming the more popular second language. English is spoken by business and government elites in many Southeast Asian nations and is an official language of the Philippines along with the national language, Pilipino. Although Spain bequeathed its Roman Catholic faith to the Philippines, the Spanish language has virtually disappeared from the islands.

Although Southeast Asia is also a land of many religions, its basic religious layout is not all that complex. The peninsular mainland is predominantly Buddhist; Indonesia is mainly Muslim, while Roman Catholicism prevails in the Philippines.

There are exceptions, however, to the basic rule. Although Buddhism is the predominant religion in Vietnam, the stronger influence there is Mahayana Buddhism. Indigenous religions, traditional Chinese religion (a combination of Confucianism, Taoism, and Mahayana Buddhism), and Roman Catholicism (a French colonial legacy) are also practiced in Vietnam. Malaysia, while partly a mainland country, is mostly Muslim. Indonesia is overwhelmingly Muslim, except for the island of Bali, which is Hindu. Thanks to their Iberian heritage, the Philippines and East Timor are the only two predominantly Christian (Roman Catholic) countries in Asia (outside of the former Soviet Union).

In the News: Indonesia—New Terrorist Haven?

The Indonesian island of Bali is one of the world's most idyllic spots, yet the bombing of a Bali nightclub in October 2002 served as a wake-up call to the world that even in this basically secular and tolerant country, Islamist terror groups were gaining ground. The Bali explosions killed more than 200 people, mostly tourists, making it the deadliest terrorist incident since the al-Qaeda attacks on New York and Washington, D.C. on September 11, 2001.

Although al-Qaeda has become a household word around the world, fewer, perhaps, are familiar with Jemaah Islamiyah ("Islamic Group"), an organization linked to al-Qaeda that seeks to incorporate Malaysia, Indonesia, and the heavily Muslim southern Philippines in an Islamist state. In August 2003, two members of the organization were sentenced to death for their role in the Bali bombing. That same month, in an incident also attributed to Jemaah Islamiyah, a car bomb partially destroyed Jakarta's Marriott Hotel, killing about 16 people.

What makes Indonesia potentially such fertile ground for terror groups are the lawlessness, sectarian violence, and separatist conflicts plaguing many parts of the country, especially outlying provinces like West Papua, on New Guinea, and the oil-rich "special region" of Aceh, at the northern end of Sumatra.

The Least You Need to Know

- ◆ Mainland Southeast Asia is a peninsular extension of Asia; the region also contains thousands of large and small islands.

- ◆ Southeast Asia is heavily influenced by its two large neighbors, India and China.

- ◆ Southeast Asia is hot and wet, with rain forests and mountainous interiors.

- ◆ Southeast Asia has many ethnic groups, numerous languages, and the world's largest concentration of Theravada Buddhists.

24

The Vast Pacific

In This Chapter

- ◆ Surveying the world's oceans
- ◆ Probing the Pacific
- ◆ Sorting through the 'nesias: Mela, Micro, and Poly
- ◆ Meeting the islands' peoples

The Pacific region consists mainly of ocean with a whole lot of islands. It has no continental landmass, and most of its islands are quite small. What dominates this region is not the sporadic scattering of land but rather the incredible expanse of water.

The planet Earth is often called the "blue planet"; 71 percent of its surface is water, mostly in the oceans. Early mariners used to refer to the world's oceans as the Seven Seas, by which they meant the North and South Atlantic, the North and South Pacific, and the Indian, Arctic, and Antarctic oceans.

Because the Pacific, Atlantic, and Indian Oceans all flow into each other around Antarctica, most geographers have abandoned the notion of a separate Antarctic Ocean. If you ignore the obscure north/south divisions of the Atlantic and Pacific, you're left with just four oceans. This chapter briefly examines the world's oceans and then looks more closely at the largest—the Pacific.

The World's Oceans

Although the earth's surface is mostly water, it is not evenly distributed. About 61 percent of the Northern Hemisphere is composed of water, while 81 percent of the Southern Hemisphere is water. Ocean water has an average depth of 12,450 feet, 4½ times the average height of dry land above sea level. The Pacific is the deepest ocean, with an average depth of 12,925 feet; the Indian and the Atlantic Oceans trail just behind it. The Arctic, however, is quite shallow—only about one-fourth as deep as the Pacific.

Geographically Speaking

Eighty-three percent of the ocean floor is between 10,000 feet and 20,000 feet deep. Only 1.3 percent is deeper, in areas called, naturally, *deeps*. Extremely deep deeps are called *trenches* (they're narrow), *troughs* (wide), or *basins* (broad). A *bank* is a shallow area that's safe for navigation. Both *shoals* and *reefs* are shallow areas that are unsafe for navigation; the difference between them is that reefs have rocks or coral, shoals don't.

A *ridge*, or *rise*, is an elongated underwater elevation or mountain chain; a *seamount* is an isolated elevation (a mountain whose summit is well below sea level), and a *plateau* is the upper part of an extensive flat-topped elevation.

Another indicator of ocean depth is the number of and depth of trenches. The Pacific leads again with sixteen trenches more than 20,000 feet deep; the Atlantic has five, the Indian four, and the Arctic none, although its Eurasian Basin is almost 18,000 feet deep. The Pacific's trenches are also much deeper than any others. The Pacific has three deeper than 30,000 feet: the Philippine Trench, just shy of 33,000 feet; the Tonga Trench, about 35,400 feet; and the deepest spot on earth, the Mariana Trench's Challenger Deep, at 35,840 feet. The Atlantic's deepest trench, the Puerto Rican Trench, measures about 28,200 feet.

The oceans also vary significantly in surface area. The Pacific Ocean, the largest, measures more than 64 million square miles and is almost as large as the other three oceans combined. The Atlantic measures 33 million squares miles; the Indian, 28 million; and the Arctic, only 5 million.

The Vast Pacific

The Pacific Ocean's maximum length is 9,600 miles from north to south and its maximum width is 11,000 miles from west to east at the equator. Another way to get a

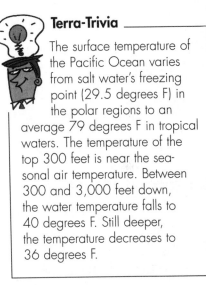

handle on the Pacific's size is to think of it in global terms. The Pacific covers one-third of the earth's surface and holds more than half of its free water.

Although ocean boundaries are often a matter of some debate, it's safe to say that the Pacific Ocean is bounded on the north by Russia and Alaska, on the east by the Americas, on the south by Antarctica, and on the west by Asia and Australia. The Pacific connects to the Arctic Ocean via the Bering Strait; to the Atlantic via the Drake Passage, the Straits of Magellan, and the Panama Canal; and to the Indian Ocean via various passages in the Indonesian Archipelago and between Australia and Antarctica.

The Pacific Ocean may be vast, but the Pacific region is very small in terms of overall landmass. The region consists of mostly tiny Pacific islands located south of the Tropic of Cancer and known collectively as Oceania. Although Oceania comprises as many as 30,000 islands, all of them, even including the regional giant, New Guinea, account for only about two-thirds of 1 percent of the Pacific realm. Other Pacific islands and island groups, such as Japan, Taiwan, the Philippines, and the Indonesian archipelago, are not usually considered part of Oceania and are covered elsewhere in this book (see index). New Zealand and even Australia are sometimes included in Oceania, but in this book they are treated separately in Chapter 13.

Terra-Trivia

The surface temperature of the Pacific Ocean varies from salt water's freezing point (29.5 degrees F) in the polar regions to an average 79 degrees F in tropical waters. The temperature of the top 300 feet is near the seasonal air temperature. Between 300 and 3,000 feet down, the water temperature falls to 40 degrees F. Still deeper, the temperature decreases to 36 degrees F.

The Not-So-Peaceful Giant

The name *Pacific* means "peaceful," but if you could take a look below the surface at the ocean floor, you'd see that the Pacific isn't really all that peaceful. From Antarctica, running north for about 6,000 miles, is the East Pacific Rise, which thrusts up about 7,000 feet. Along this rise, molten rock swells up from the earth's mantle and adds crust to the ocean plates on both sides. The plates are forced apart, causing them to collide with the continental plates all around them, as described in Chapter 2. The tremendous pressure generated by this process forces the continental plates to turn upward as mountains and the ocean plates to turn downward into deep trenches.

Island Origins

Tectonic pressures along the Pacific's perimeter have created the violent seismic and volcanic zone known as the Ring of Fire (see Chapter 2). Hot spots in the earth's mantle can cause molten rock to break through the surface. When this rock is beneath the ocean, a seamount (volcanic island) is born. Thousands of the "dots" that have appeared all over the Pacific, particularly in the southwest, were formed this way. Native inhabitants call these formations "high" islands.

Terra-Trivia

Coral reefs and islands form in shallow water or on top of nonactive volcanic seamounts. The typical Pacific coral island is an *atoll*, a large ring of coral above or below sea level, surrounding a lagoon. The coral above the surface supports tropical vegetation. From the air, an atoll looks like the rim and crater of a volcano, which in fact it grows on.

Thousands more of the Pacific dots are so-called "low," or coral, islands. Coral islands and reefs have been described as the earth's most complex ecosystems. They occur only in warm, sunlit water and result from thousands of live marine animals clustering in vast colonies. The skeletal deposits that remain after these creatures die eventually layer themselves and can reach a height of 300 or 400 feet.

Currents: Air and Sea

To best understand the Pacific Ocean's wind patterns, start at the equator and work north and south from there. Between 10 degrees north and 10 degrees south are the *doldrums*, a low-pressure area of light breezes. At 30 degrees north and 30 degrees south are the *horse latitudes*, belts of high pressure. The winds blowing from these high-pressure areas toward the low-pressure doldrums are deflected by the earth's rotation and become the tropical easterlies known as *trade winds*. Between 30 degrees north and 60 degrees north and between 30 degrees south and 60 degrees south, winds blow from the horse latitudes toward the poles and again are deflected by the earth's rotation to become the *westerlies*. The polar regions are high-pressure areas; the deflected winds blowing from them become the *polar easterlies*.

Although ocean currents in the Pacific are relatively complex, this chapter simplifies them with the usual broad brush. North of the equator, the ocean currents form a huge oval that moves clockwise. A similar oval current forms south of the equator, in which the current moves counterclockwise. Between the two ovals and running along the equator is a west-to-east movement of water, called the *equatorial counter current*.

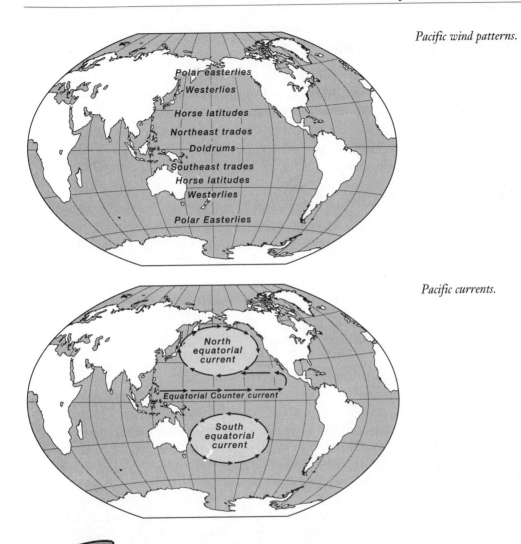

Pacific wind patterns.

Polar easterlies
Westerlies
Horse latitudes
Northeast trades
Doldrums
Southeast trades
Horse latitudes
Westerlies
Polar Easterlies

Pacific currents.

North equatorial current

Equatorial Counter current

South equatorial current

CAUTION

Eye on the Environment

Thanks to the sheer size of the Pacific Ocean, its air and water currents and surface temperatures play a major role in regulating global climates. This is never more evident than when normal Pacific air and water patterns are disrupted during the periodic phenomenon known as El Niño (Spanish for "the boy child," a reference to the Christmas season, when the disruption normally starts). During an El Niño period, which occurs at irregular intervals every two to seven years, the Pacific trade winds slacken or reverse direction, causing surface water temperatures to rise abnormally in the eastern equatorial Pacific off South America. All sorts of extreme weather phenomena are associated with El Niños, including flooding up and down the Pacific coast of the Americas; droughts in Australia, Indonesia, and Africa; and on the positive side, a milder Atlantic hurricane season.

Countries in an Endless Ocean

New Guinea and Oceania combined have almost 9.5 million people and encompass some 30,000 islands, which are divided into three major groupings—Melanesia, Micronesia, and Polynesia, along with a few scattered outliers. The groupings include 13 independent countries; one Indonesian province; one United States state; five other United States affiliates; one British, four French, one Chilean, and three New Zealand territories; and a part of Ecuador.

Islands of the Pacific.

Melanesia

The term *Melanesia* comes from two Greek words meaning "black islands"—a reference to the complexion of most Melanesians. Melanesia has nearly 80 percent of the region's population (about 7 million people) and 93 percent of its land area. The reason that Melanesia looms so large is that it includes the large and heavily populated nation of Papua New Guinea.

Papua New Guinea

The world's second-largest island, New Guinea, is divided almost evenl
Indonesia's West Papua (formerly Irian Jaya) province on the western ha
island and the Melanesian country of Papua New Guinea on the eastern
tion to its share of the big island, Papua New Guinea also encompasses s
smaller islands, including the Admiralty Islands, the Bismarck Archipelag
Bougainville. Papua New Guinea is home to nearly 5.24 million people. It
Port Moresby, has a population of almost 325,000, making it one of the lar
in the region.

Papua New Guinea has a hot and humid tropical climate and dense forests.
varies with topography, with some places receiving more than 230 inches per
The interior of the country is extremely rugged and the tribespeople who live
the Highlanders, were once among the most isolated people on Earth.

Before World War I, what is now Papua New Guinea was divided into a north
part controlled by Germany and a southern part controlled by the United King
and later Australia. Germany lost its New Guinea territory to Australia in Worl
I. Japan occupied part of the country during World War II. After the war, Austra
administered Papua New Guinea until 1975, when it gained full independence.
country's more than 700 Papuan and Melanesian tribes speak more than 700 languages. Papua New Guinea has abundant mineral and other natural resources, bu
has been slow to exploit them. Tribal people fish, hunt, and engage in subsistence
farming. In coastal areas, cocoa, coffee, and copra are grown for export.

The Solomon Islands

The Solomon Islands have seen their share of colonizing powers, including the
Spanish, Germans, British, Japanese, and Australians. The island group was also t
scene of some of World War II's most ferocious battles. A British protectorate si
the 1890s, the Solomon Islands became independent in 1978. The country is c
posed of 10 major volcanic, deeply forested islands (including Guadalcanal), 3
isles, and countless atolls. The approximately 495,000 Solomon Islanders sp
estimated 120 local languages. A Melanesian-English pidgin is used to brid
many language divides. The official language, English, is spoken by very
the presence of rich mineral resources, farming, fishing, and forestry ar
nomic mainstays of this chaotic country plagued by crime and intertri

Vanuatu

Formerly known as New Hebrides, Vanuatu was jointly administe
and French from 1906 until 1980, when the country became in

196,000 people inhabit Vanuatu's 70 mostly volcanic islands. The economy of the resource-poor nation is based on small-scale farming, fishing, offshore banking, and tourism.

New Caledonia

A French possession since 1853, New Caledonia became an overseas territory of France in 1956. Its population of nearly 211,000 is divided between Melanesians (about 43 percent) and French or other Europeans (37 percent). New Caledonians voted against independence in 1998. The country's economy depends on tourism and nickel mining. (New Caledonia has about 25 percent of the world's nickel supply.)

Fiji

After nearly a century of British colonial rule, Fiji gained its independence in 1970. The country consists of 300 islands (only 100 are inhabited) with a population of about 856,000. More than half the people are Fijian (mainly Melanesian); about 44 percent are Indian, descendents of contract laborers brought to work on Fiji's plantations in the late nineteenth century. Friction between the two groups has led to coups and political turmoil in recent years. Fiji's capital, Suva, has about 177,000 inhabitants. English is the official language, but almost everyone speaks Fijian; Hindustani is also widely spoken. Even though about 70 percent of its work force is engaged in agriculture, Fiji has one of the most developed economies in the region; tourism and sugar are the leading money earners.

Micronesia

The term *Micronesia* comes from Greek words meaning "small islands." The subregion's more than 2,000 islands cover only 1,000 square miles and support some 580,000 people. Although tiny coral atolls are the norm, the bulk of the population lives on the larger high islands.

Palau

The Palau islands, the westernmost archipelago of the Caroline chain, were controlled by Spain, Germany, Japan, and finally the United States until they became independent in 1994. They are still linked to the United States by a Compact of Free Association. Palau consists of more than 300 islands (half in Melanesia and half in Micronesia) divided into six groups. More than half of Palau's 19,000 people live in the capital, Koror. Tourism, farming, and fishing are the main ingredients of Palau's economy.

The Federated States of Micronesia

Formerly a United Nations Trust Territory administered by the United States, the Federated States of Micronesia became independent in 1986 but retained its ties to the United States under a Compact of Free Association. The country's four island groups contain more than 600 islands (many uninhabited), ranging from high volcanic islands to low coral atolls. Its population numbers 136,000 and its economy is heavily dependent on United States aid.

Nauru

With some 12,000 people packed on one 8-square-mile island, Nauru has a population density of about 1,500 people per square mile. Nauru has been ruled or administered by Britain, Germany, Australia, New Zealand, and Japan. After World War II, it became a United Nations trusteeship administered by Australia. Independence was achieved in 1968. Phosphate mining made Nauru relatively prosperous, but supplies are almost exhausted and the island has very few resources to compensate for the loss of the phosphate industry. Even fresh water has to be imported.

The Northern Mariana Islands

The Northern Mariana Islands are officially a "commonwealth in political union with the United States." The Marianas are a group of 14 islands, including Saipan, Tinian, and Rota, with a population of about 80,000. The chief industries are tourism, small-scale farming, and garment manufacturing. This last industry has come under severe criticism for using imported Chinese labor to produce "made in the U.S.A. garments" under deplorable sweatshop conditions.

> **Eye on the Environment**
>
> Phosphates have long been used around the world as fertilizers. Unfortunately, phosphate-enriched runoff causes algae to proliferate, choking lakes and drawing off oxygen that supports aquatic life. This process, called *eutrophication*, can result in fish kills and "dead" ponds and lakes. Ironically, the strip mining of phosphates has also wreaked havoc with Nauru's environment.

Guam

Large by Micronesian standards, the island of Guam measures 209 square miles and is home to nearly 164,000 people. Guam was ceded to the United States by Spain in 1898 and is today an unincorporated territory of the United States. About one-third of Guam's land is owned by the United States military, which maintains a highly strategic base on the island. Military expenditures and tourism are the mainstays of Guam's economy.

The Marshall Islands

Like Micronesia, the Marshall Islands became independent in 1986 under a Compact of Free Association with the United States. The country consists of two parallel chains of atolls and islands with about 74,000 inhabitants. The atolls include Kwajalein (the world's largest atoll, with 90 islets surrounding a 650-square-mile lagoon) and the former nuclear test atolls of Bikini and Enewetak. The capital, Majuro, is home to more than a third of the country's total population. The Marshall Islands are highly dependent on United States aid and continue to seek compensation for the damage done to Bikini, Enewetak, and other atolls by the U.S. nuclear testing program.

Kiribati

The Gilbert Islands, a group of 33 coral atolls straddling the equator and the international dateline, gained its independence from the United Kingdom in 1979 and changed its name to Kiribati. Its population is about 96,000. Its capital, Tarawa, is remembered as the site of a horrific World War II battle between the United States Marines and Japanese forces.

Terra-Trivia

The international dateline runs through the middle of the Pacific Ocean at 180 degrees longitude. Travelers are often perplexed by the dateline, so here's the short explanation: When you cross the line while traveling west, you add a day (6 P.M. Monday becomes 6 P.M. Tuesday). When you cross the line while traveling east, you subtract a day (6 P.M. Tuesday becomes 6 P.M. Monday).

Polynesia

The name *Polynesia* stems from a Greek word meaning "many islands." And indeed, Polynesia consists of many thousands of coral and volcanic islands in what is by far the most far-flung of the three major Pacific island groupings. Nearly two million people live in the various political configurations that make up Polynesia.

Tuvalu

Once part of the same British colony as the mostly Micronesian Gilbert Islands (now Kiribati), the mostly Polynesian Ellice Islands decided to become the separate colony of Tuvalu in 1974. Four years later, Tuvalu became independent. Tuvalu's 11,000 inhabitants are packed on 10 square miles of land over nine atolls in a 360-mile chain. "Global warming," which could cause sea levels to rise, is a nasty phrase in this tiny, remote, and very low-lying country. Tuvalu's highest elevation is only 15 feet.

Tonga

Tonga is made up of 170 coral and volcanic islands in a 289-mile-long north–south chain. Only 36 of the islands are inhabited by about 106,000 Tongans. Named the Friendly Islands by Captain Cook, Tonga was a British protectorate from 1900 to 1970, when it gained its independence as a hereditary constitutional monarchy (the only monarchy left in Oceania). The country's fairly stable economy depends on agriculture, tourism, and remittances from Tongans working abroad.

New Zealand Territories

Three Polynesian island groups have different levels of association with New Zealand. The Cook Islands (named for, you guessed it, Captain Cook) were a British protectorate from 1888 to 1900, when New Zealand took over. In 1965, the Cook Islands became a self-governing democracy in free association with New Zealand. Most of the 21,000 inhabitants live on the eight elevated volcanic islands of the southern Cook Islands; the northern atolls are sparsely populated. The remote island of Niue, one of the world's largest coral islands but with a population of only 2,100 and falling, is also a self-governing democracy in free association with New Zealand. The three coral isles that make up Tokelau cover four square miles and support 1,400 people. Tokelau is a self-administering territory of New Zealand, but is moving toward free association.

French Territories

French Polynesia consists of five archipelagos acquired by France in the nineteenth century: the Society, Tuamotu, Gambier, Austral, and Marquesas islands. The islands became an overseas territory of France in 1946, with Papeete, on the island of Tahiti, as the capital. This widely spaced collection of islands is home to more than 260,000 people and includes some of the most beautiful get-away spots on earth. The best-known islands are Tahiti (immortalized by the great French post-Impressionist painter, Paul Gauguin), Moorea, and Bora Bora. Tourism, not surprisingly, is the main industry, followed by pearl farming, agriculture, and deep-sea commercial fishing. The island groups of Wallis and Futuna became a French overseas territory in 1959. Most of the nearly 16,000 inhabitants earn their living by farming and fishing.

Samoa

Samoa (formerly Western Samoa) was German territory from 1899 to 1914, when it was taken over by New Zealand. In 1962, Western Samoa became the first Polynesian nation to regain its independence in the twentieth century. The "Western" part of its name was dropped in 1997. Samoa consists of two large volcanic islands, Savai'i and Upolu, and seven smaller islands that support a declining population of about 179,000.

Agriculture, development aid from New Zealand and Australian, and remittances from Samoans working abroad are the mainstays of the economy.

American Samoa

The seven eastern islands of the Samoan chain are an unincorporated territory of the United States. American Samoa's largest island is Tutuila, site of the capital, Pago Pago, one of the finest harbors in the South Pacific. The territory has a population of about 70,000. The economy of American Samoa depends heavily on tuna fishing and canning and on U.S. government subsidies.

Hawaii

Unlike the other United States possessions in the Pacific, Hawaii is a full-fledged state. The Hawaiian islands became a territory of the United States in 1900 and the fiftieth state of the Union in 1959. The 1,523-mile-long Hawaiian island chain consists of eight main volcanic islands and 124 islets and reefs. The largest island is Hawaii, the "Big Island." The state's capital and, with 370,000 residents, its largest city is Honolulu, located on the island of Oahu. The lushest island is Kauai, home of Mount Waialeale, the world's wettest place with an average annual rainfall of 486 inches. The island of Molokai is perhaps still best known as the site of Father Damien's famed leper colony. Hawaii's population of 1.2 million is very diverse: 42 percent Asian, 24 percent white, 9 percent native Hawaiian or Pacific islander, 7 percent Hispanic, and 2 percent black.

> **GeoRecord**
>
> The Big Island's Mauna Kea (13,796 feet) and Mauna Loa (13,678 feet) volcanoes are the state's two highest peaks. If these volcanoes were measured from their bases on the ocean floor, they would be 32,001 and 31,883 feet high, respectively. Both would be much higher than Mount Everest (29,028 feet) and therefore the two highest mountains in the world.

Hawaii's earliest settlers were Polynesians, probably from the Marquesas Islands, who arrived in outrigger canoes beginning in around 300 C.E. In the thirteenth and fourteenth centuries, new arrivals from Tahiti conquered and eventually absorbed the first Hawaiians. The first European known to have visited the islands was Captain Cook in 1778. He named them the Sandwich Islands, after the British Earl of Sandwich, and in 1779, was killed on the Big Island in a dispute with native Hawaiians.

Europeans began arriving in significant numbers in the 1790s, bringing with them diseases that decimated the native population. By 1810, King Kamehameha had unified the islands under his rule. Although his dynasty died out in 1872, the monarchy continued until 1893, when government officials, sugar planters, and American businessmen engineered its overthrow. A republic was declared the next year, and in 1898, the islands were annexed by the United States.

Sugar production was introduced in the islands in the 1830s and soon replaced whaling as the primary industry. In 1898, the U.S. naval base at Pearl Harbor was opened, adding a major element to Hawaii's economy—military expenditures. Today, tourism has eclipsed both agriculture and military spending as the base of Hawaii's economy. The post-9/11 decline in tourism and the global economic slowdown in the first years of this century have had a very negative impact on Hawaii's all-important tourism industry.

Miscellaneous Islands

A few outliers from the three main island groupings complete the Pacific survey. Wake Island (north of the Marshalls) and the Midway Islands (west of Hawaii) are unincorporated territories of the United States. Wake is run by the military; the Midway Islands are a wildlife refuge run by the U.S. Fish and Wildlife Service. Other outliers are France's uninhabited Clipperton Island (west of Costa Rica) and the Galápagos Islands (west of and part of Ecuador). Easter Island lies roughly halfway between Chile, its owner, and French Polynesia. The Galápagos are famous for their gigantic tortoises and iguanas (see also Chapter 18). Easter Island is renowned for its hundreds of mysterious huge-headed, long-eared, long-nosed stone statues.

The People of New Guinea and the Oceania Dots

Melanesians are believed to be the result of a mix between Papuan peoples and the black-skinned people of interior New Guinea, who then spread out to the islands between the equator and 20 degrees south latitude. More than 200 Malayo-Polynesian languages are now spoken in Melanesia, often several on the same island. Scattered throughout many of the southern Solomons and the northern Vanuatus are small-statured tribes who to this day practice polygamy and engaged in cannibalism in the not too distant past.

The Micronesians descend from a historic mixture of Asian, Indonesian, Filipino, and Papuan peoples. They are generally slender with straight hair and lighter skin than Melanesians. Micronesians were skilled seamen who navigated their craft between the islands to supplement their diets. The high islanders farmed, and the low islanders fished.

The Polynesians are thought to have come from the Malay Peninsula via Indonesia, probably in two major waves. The first wave settled in Samoa, Tonga, and Fiji; the second, a few hundred years later, migrated westward, northward, and southward (into New Zealand, whose Maori are thought to be of Polynesian descent). Polynesians have light brown skin and are known for their grace and beauty. The Polynesians were also renowned for their seamanship, both in small canoes for short interisland trips and in large craft that carried them over vast distances to discover, explore, and settle new islands thousands of miles away.

In the News: Rising Sea Levels

One of the predicted results of global warming is a rise in sea levels. Although this is of concern to all low-lying coastal areas, the Pacific island nations, with their large numbers of very low-lying atolls, are at particular risk. Abnormally high water temperatures also have a disastrous effect on coral reefs and their rich but fragile ecosystems.

Whether or not "greenhouse-gas" emissions are actually causing global warming is a matter of continuing debate (see Chapter 27), but there are troubling signs coming out of the Pacific region. High tides and unusually stormy weather have caused severe flooding in recent years in Kiribati, Tuvalu, Cook Island, the Marshall Islands, and the Federated States of Micronesia. Fiji, the Solomons, the Cook Islands, Tonga, and Palau have all reported coral bleaching, a discoloration of the coral reefs that are among the islands' main tourist attractions.

While the wealthy nations of the world debate energy policy and emissions controls, those most vulnerable to the possible consequences of doing nothing are the fragile economies of the developing world.

The Least You Need to Know

- The size of the Pacific Ocean almost equals the Atlantic, Indian, and Arctic oceans combined. It covers one-third of the entire earth's surface.

- The Pacific's entire rim, the Ring of Fire, has active volcanoes and is prone to earthquakes.

- The Pacific has almost 30,000 islands (mostly atolls) that are volcanic and coral.

- Oceania is divided into subregions called Melanesia, Micronesia, and Polynesia.

Chapter 25

The Frozen Poles

In This Chapter

- ♦ Identifying polar limits
- ♦ Divvying up ownership
- ♦ Highlighting polar similarities and differences
- ♦ Taking a closer look at the poles
- ♦ Keeping an eye on the ozone hole

The final regions in this part of the book are located at the far reaches of the earth, where ice and cold prevail and human beings are mere visitors. Many geographers don't consider the poles to be distinct regions because of their obvious dearth of human settlement. To help you in your quest for geographic knowledge, however, we've left no iceberg unturned and no region unexplored.

This book typically defines regions in terms of both physical and cultural characteristics. Because the poles don't fit this mold, however, this chapter will focus on physical features only.

Defining the Polar Regions

Determining the limits of the polar regions is not as clear-cut a proposition as you might expect. Some definitions are based on climatic conditions. For example, the Arctic and Antarctic regions can be defined as the parts of the globe in which the average temperature of the warmest month never exceeds 50 degrees F. Some geographers define the Arctic and Antarctic regions by vegetation: they are the parts of the globe above the latitude at which trees no longer grow (the tree line).

You might think that the logical boundary lines of the polar regions would be the Arctic and Antarctic circles, at about 66½ degrees north and south latitude, respectively. As tidy as those lines would be, they don't quite work.

Geographically Speaking

The earth's true, or geographic, poles are the ends of the axis around which the planet rotates. The magnetic poles are the ends of the axis of the Earth's magnetic field. The magnetic poles don't coincide with the earth's true poles, and they aren't stationary either. The north and south magnetic poles are actually average positions around which the poles wander on a daily basis. The average positions also shift over time. Currently, the north magnetic pole is drifting across the Canadian Arctic. The South Magnetic Pole is off the coast of Wilkes Land, Antarctica.

Although Antarctica is contained almost entirely within the Antarctic Circle, the continent peeks outside the circle in several locations. In the Eastern Hemisphere, parts of Wilkes Land and Enderby Land flirt with the Antarctic Circle, and bits of these areas poke out along the way. In the Western Hemisphere, Antarctica is contained entirely within the Antarctic Circle except for the northernmost portion of the Antarctic Peninsula.

The opposite situation exists in the Northern Hemisphere. The Arctic Circle encompasses parts of several other regions not considered part of the north polar region. A quick look at the map shows that Russia, Northern Europe, and North America all extend into the Arctic Circle. The north polar region includes everything north of these other landmasses, and therefore, contains no land itself, just water and ice.

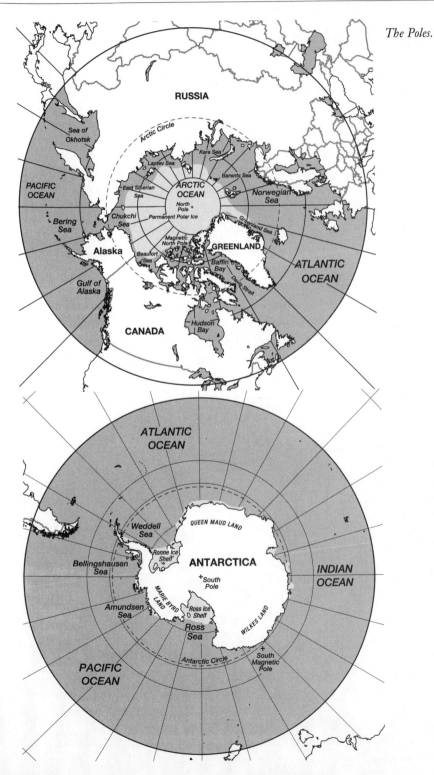

The Poles.

Making National Claims

Because the North Pole is located in the heart of the Arctic Ocean, there is no land to claim in the north polar region. Even with the 200-mile exclusive economic zone provided by the Law of the Sea Convention, no country's jurisdiction reaches the North Pole. At the South Pole, however, the situation is not so simple.

Antarctica is land, and although it has no permanent inhabitants and no local government, it has not escaped political squabbling. Seven different countries make claims to portions of this continent. No nation claims the part of Antarctica called Marie Byrd Land; this is the place to go if you're looking to claim a chunk of a continent.

The pie-shaped national claims all originate at the South Pole (as shown on the following map). Most of the claims have some logic behind them. Australia (which claims two slices of Antarctica), New Zealand, Argentina, and Chile—the countries closest to Antarctica—all make claims that are southern extensions of their national territories. Norway also makes a (more dubious) claim to a wedge of Antarctica, based on the fact that a Norwegian, Roald Amundsen, was the first person to make it to the South Pole. Because Robert Scott, from the United Kingdom, was a close second, the British claim also has some rationale. The other claimant is France, and if you don't get the logic on this one, join the club.

South Pole territorial claims.

Terra-Trivia

The northernmost point of land on Earth is on the tiny island of Oodaaq Qeqertaa. Located off the northern coast of Greenland, the island falls under Danish sovereignty. Because of the constant ice pack and the infrequent passing of visitors, the island was only recently identified as terra firma.

All these claims are something of a moot point anyway because of the Antarctic Treaty System, which came into effect in 1961. This treaty banned new territorial claims and prescribed peaceful, scientific use of the continent. A more recent agreement, signed in 1991, prohibits commercial mineral extraction on Antarctica. You don't have to look any further than the Antarctic Peninsula to see the potential for conflict: Three countries all claim the same area.

Looking at Polar Similarities and Differences

Even though the north and south polar regions are literally "poles apart," they do share certain similarities. Here's a list of the top nine similarities between the North and South poles:

- ◆ Both are located at 90 degrees latitude (ignore the fact that one's north latitude and one's south).

- ◆ During the winter, the sun never rises in either place.

- ◆ During the summer, the sun never sets in either place.

- ◆ Both have lots of ice and snow.

- ◆ If you stand on either pole, no matter which way you face, you can travel in only one direction (only south from the North Pole and only north from the South Pole).

- ◆ Seals live on both poles.

- ◆ Both are extremely cold.

- ◆ Neither pole has trees.

- ◆ You can forget about using your magnetic compass in either place.

But before you begin to think that the poles are almost identical, take a moment to peruse the top nine differences between them:

- One is north, and the other is south. The poles are 180 degrees apart, on opposite sides of the globe—as far apart as you can get on Earth.

- No polar bears live on Antarctica.

- The South Pole is on land; the North Pole is in the middle of an ocean.

- Although they're both cold, the South Pole is considerably colder than the North Pole.

- The South Pole is almost entirely covered by glacial ice; the North Pole has nary a glacier.

- The South Pole has penguins; the North Pole doesn't.

- Antarctica has a mountain more than 16,000 feet high and a depression more than 8,000 feet deep. The North Pole is pretty much at sea level.

- No national claims to the North Pole have been made, although many countries have claimed pieces of Antarctica.

- Several "permanent" scientific stations exist on Antarctica; only temporary floating stations exist at the North Pole.

Eye on the Environment

Another important similarity between the poles is that both are barometers of global warming. The discovery of open water at the North pole in 2000 raised worldwide alarms. Although the alarms might have been overplayed—arctic ice floats on top of the Arctic Ocean and is in constant motion, often exposing areas of open water, especially in the summer—the extent and thickness of Arctic ice are clearly declining. According to some projections, by the middle of this century, the Arctic Ocean might be ice-free during the summer.

The situation in Antarctica is no less disturbing. The process known as "calving," in which large chunks of the Antarctic ice shelves break off at the fringes and float off into the sea as icebergs, is a normal one. But the rate of disintegration and the size of the resulting icebergs have been increasing alarmingly in recent years. Although the cause of this phenomenon is still a matter of some debate, the evidence pointing toward global warming cannot be ignored.

Taking a Closer Look at the Poles

This section looks more closely at some of the similarities and differences between the North Pole and the South Pole. You might think that some of this information borders on the trivial, but the fact is that what happens to the polar regions has environmental repercussions for the entire planet.

Which One Is Land?

As mentioned previously, the North Pole is in an ocean, and the South Pole is in the middle of a continent. Of all the differences between the north and south, this one is the most important. This difference affects accessibility, geology, climate, and virtually everything else about the two locations.

The North Pole is in the Arctic Ocean. The smallest of the world's four oceans, it's also the shallowest and, obviously, the world's northernmost ocean. Most of the Arctic Ocean, including the North Pole, is always covered with ice. This ice is not a solid sheet; it's called *pack ice*—chunks of ice that slowly circulate about the pole.

The South Pole is on the continent of Antarctica. Even though almost its entire surface is covered with continental glacial ice, Antarctica is solid land. Antarctica is the fifth largest of the earth's seven continents. It's smaller than Asia, Africa, and North and South America, and larger than Europe and Australia.

Which One Is Colder?

No contest! The South Pole is the hands-down winner. Plateau Station on Antarctica is the coldest place on Earth, with an annual average temperature of –70 degrees F. The single coldest temperature ever recorded was also on Antarctica, at Vostok Station, where the thermometer dipped to –129 degrees F. These temperatures would be hideously cold even if the wind was calm on Antarctica, but it's not. Unfortunately for the Antarctic tourism bureau, Antarctica is also one of the world's windiest places. At Australia's Commonwealth Bay Station, cold air constantly rushes down from the ice cap, and wind speeds have been recorded at more than 180 miles per hour.

Gauging Arctic temperatures and winds is a little more difficult than in the Antarctic because no fixed, permanent stations exist. (In contrast, the United States Pole Station on Antarctica has been continuously occupied since 1956.) Although you would think that the coldest temperatures in the north would be at the North Pole, they're really farther south along the Arctic Circle. This phenomenon occurs because the North Pole is located over water, and the cold temperatures recorded along the Arctic Circle have been recorded over land.

Just outside the north polar region, at $76^{1}/_{2}$ degrees north latitude, is Thule Air Base in northern Greenland. Annual temperatures average about –11 degrees F—cold, but not approaching the –70 degrees F recorded at Antarctica. Verkhoyansk Station in northern Siberia has recorded January temperatures as low as –90 degrees F—again, it's cold, but not as bad as at Vostok Station in Antarctica.

There are several reasons for the temperature differences between the poles. At the height of the south polar winter, around July 1, the earth is at its *aphelion*, or farthest point in its orbit from the sun. Although that's not a huge factor in temperature, it does mean that the South Pole is about three million miles farther from the sun during its winter than the North Pole is during its winter.

> ### GeoRecord
>
> The highest point on Antarctica is the Vinson Massif, at 16,863 feet above sea level. The lowest surface point on earth is also in Antarctica, where a point on the continent drops to 8,327 feet below sea level. This low point on Antarctica's bedrock surface was caused by the weight of thousands of feet of ice pressing down on the continent.

The windier conditions at the South Pole make the windchill a factor to be reckoned with. To demonstrate the impact of the windchill effect: If the recorded temperature were –45 degrees F (a warm day at Plateau Station, Antarctica), winds of 45 m.p.h. would make the temperature feel like –125 degrees F. If you're considering a vacation in Antarctica, bring your long woollies.

Where Are the Polar Bears, Seals, and Penguins?

Aside from an occasional mite, there's no purely terrestrial life at the poles—all polar life is dependent on the seas. Under the ocean ice, algae and diatoms support small shrimplike creatures, called krill, that provide food for fish and larger aquatic life at the poles.

Penguins live only at the South Pole and are the only large animals that stick it out through the nasty Antarctic winter. Seals live on both poles. In the far north, polar bears as well as an occasional Inuit hunter have been known to venture out on the pack ice in search of seal meat. No such indigenous land predators exist on Antarctica to plague the penguins.

Where Are the Continental Glaciers?

When you think of the Ice Age, continental glaciers usually come to mind. The last ice age occurred during the Pleistocene era, when vast ice sheets spread to cover about one-fifth of the earth's continental surface. Only two remnants of the continental glaciers remain: One covers the island of Greenland, and the other covers Antarctica.

Continental glaciers form when climatic factors combine to allow a gradual buildup of snow and ice. When the accumulation of ice builds to a critical mass, the weight of the ice mass causes it to flow outward. As the glaciers advance, they grind and scrape the landscape with dramatic results.

Scientists don't agree on what climatic conditions must exist to cause glacial advance. Although you would suspect that colder temperatures would be a prerequisite, that's not necessarily the case. To build and flow, glaciers require snowfall in heavy amounts. Because cold air holds less moisture than warmer air, the coldest temperatures are often associated with lighter snowfall totals. Following this argument to its logical conclusion, warmer temperatures might be required to produce a new ice age.

Either way, glaciers have not historically formed over the North Pole. The continental glaciers of the Northern Hemisphere formed in subpolar areas and then spread northward toward the pole and southward toward more temperate climes. The ice cap on Antarctica is not advancing because of the scant amount of snowfall the continent receives. Nonetheless, if you want to trek across a continental glacier (even a semidormant one), head for the South Pole, not the North Pole.

Terra-Trivia

The Antarctic ice cap contains the world's largest supply of freshwater, approximately 70 percent of the total. More freshwater is contained in the continental glacier of Antarctica than the freshwater in all the world's rivers, lakes, atmosphere, groundwater, and other glaciers combined.

Night and Day

If you're standing on either the North Pole or the South Pole, you're in for a boring diurnal (daylight) pattern. When the sun rises at either pole, it stays daylight for six months. When the sun finally sets, nighttime settles in for six months (see "The Land of the North" section in Chapter 9).

The cause of this pattern is the $23^1/_2$-degree tilt of the earth on its axis. When it's summer in the Northern Hemisphere (March 21–September 23), the North Pole remains in the sunlight and the South Pole stays in the darkness. During the Southern Hemisphere's summer (September 23–March 21), the South Pole experiences constant daylight and the North Pole languishes beneath dark winter skies.

In the News: The Ozone Hole

The ozone layer in the upper atmosphere shields life on earth from the harmful effects of the sun's ultraviolet radiation. Normally, the ozone layer covers and protects the entire earth, but in 1985 scientists noticed a hole forming in the ozone layer directly over Antarctica. Since then, scientific observations have confirmed a 15 to 70 percent decline in ozone levels over the South Pole every spring.

As you'll see in greater detail in Chapter 27, the depletion of the ozone layer is an unfortunate result of human activities. The release into the atmosphere of certain chemicals, particularly a family of chemicals called chlorofluorocarbons, or CFCs, has broken down atmospheric ozone and will continue to do so for some time despite successful international efforts to ban CFC use and production. (See Chapter 27 for more on the international campaign to reduce CFC emissions.) Thanks to the circulation and concentration of cold air over Antarctica (a pattern called the *Antarctic vortex),* the problem is more severe over the South Pole. Although ozone depletion also occurs over the North Pole, the loss there has not been as great.

Because it's believed that excessive ultraviolet radiation can hinder plant development, Antarctica's ozone hole might be placing its basic food chain at risk. Leopard seals eat penguins and penguins eat krill, and krill eat algae. If algae are affected, so, too, are the higher vertebrates on the chain. Because algae also absorb carbon dioxide and produce oxygen, their loss might also have additional negative effects on the atmosphere. The Antarctic ozone hole reached a record size (about 11 million square miles) in September 2000, and although it shrank considerably in 2002, ozone concentrations in its interior continued to decline. As of September 2003, the ozone hole is nearing record size once again. The effects of global warming and atmospheric pollution on both poles will clearly be news stories well worth following over the coming years.

The Least You Need to Know

- The earth's true, or geographic, poles are the ends of the axis around which the planet rotates. They are located at 90 degrees north and 90 degrees south latitude. The geographic poles do not coincide with the magnetic poles.

- Although both poles are extremely cold, the South Pole is colder.

- The North Pole is over the Arctic Ocean, and the South Pole is over the continent of Antarctica.

- The poles have no permanent human inhabitants.

- The South Pole is covered with a continental glacier.

- A troubling ozone hole forms over the South Pole every spring.

Part 5

A Global Overview

Now that you have completed your tour of the world's developed and developing regions, it's time to stand back and look at the big picture. Although you can easily get caught up in the lines that define the boundaries of regions and countries, remember that if you take a stroll on the physical landscape, those lines don't really exist.

Part 5 of this book looks at issues that don't fit neatly into a discussion of a single country or even a single region. These global issues affect us all. More importantly, it will take international cooperation to solve some of the problems we have created for ourselves. We all have to be informed and take responsibility for our planet—the ultimate geographic task!

Chapter 26

Coping with Population

In This Chapter

- Understanding population rates
- Pondering the prospect of people, people everywhere
- Finding where the people are—and where they're not
- Coming to grips with the food dilemma
- Taking a step in the right direction: demographic transition

In your whirlwind tour through this book, you have visited every region on earth. Now that you know a little about the geography of each place, it's time to stand back and look at the entire planet. What problems are faced by every region on earth? What solutions can humankind turn to?

In many of the regions explored in this book, population pressures are straining the environment. Although some regions have relatively small populations, their land is typically unable to support large numbers of people. In some of the most fertile areas of the earth, millions of people pack into the usable land, creating massive clusters that dot the landscape.

At one time in human history, it was possible for overcrowded populations simply to move elsewhere. For centuries, crowded South Asian and East Asian populations emigrated to Southeast Asia. In the nineteenth and early twentieth centuries, crowded Europeans moved to distant lands in the New World, Africa, Australia, and New Zealand. Although emigration was once a solution to overpopulation, the earth is running out of places to put people.

As the human population continues to expand, every resource on earth is taxed. Many environmentalists consider the world's population problem to be the greatest environmental threat—because the sheer volume of the world's people lies behind virtually every other environmental problem humans face. Although the earth has shown remarkable resiliency, are humans stressing even the earth's ability to heal itself?

A Little About Demographics

Geographers look closely at *demographics*, the study of population, because it has so great an influence on the workings of the earth. They're particularly concerned with the distribution and movement of populations. One of the challenges of the twenty-first century will be to address the world's burgeoning population problem and come up with workable solutions.

GeoJargon

The study of population numbers, growth, distribution, and trends is called **demographics**. Demographers look at past and present population records to predict future population trends.

Birthrates and Death Rates

The field of demographics is based largely on rates: Birthrates and death rates are used to determine rates of population increase. The birthrate gauges the annual number of live births per thousand people; the death rates indicate the number of annual deaths per thousand people. To determine the natural growth rate, you simply subtract the death rate from the birthrate. The remainder is the increase (or, in rare instances, the decrease).

Terra-Trivia

In the year 1 C.E., about 300 million people were living on the earth. This figure did not double to 600 million until the year 1650, which means that it took 1,650 years for the world's population to double. The population took only 200 years to double again, and then again in under 100 years, and most recently in barely 45 years.

Population growth occurs when the birthrate exceeds the death rate. The greater the difference, the larger the growth. Population growth can occur because of an increase in the birthrate, a decrease in the death rate, or both. The only way to decrease population growth is to increase the death rate (not a people-friendly option) or decrease the birthrate.

Six Billion and Counting

Sometime in the fall of 1999, the world's population reached six billion, capping a century of unprecedented growth. You might say, "So six billion people live on the earth—what's the big deal? What are the consequences of having that many people, and should people be worried?" For comparison, a stack of $100 bills equaling $1 million would stand about $3\frac{1}{2}$ feet tall. A stack of $100 bills equaling $1 billion would measure about 3,583 feet tall—more than twice the height of the Sears Tower in Chicago.

Some people might say that although six billion people is a large number of people, the earth is a large place and can handle that number. Although the earth *is* large, according to the Population Education Training Project, the earth doesn't have that much space, after all. Here's the math:

♦ About three-fourths of the earth's surface is water; the remaining one-fourth represents the earth's land surface.

♦ About one-half of that land surface is uninhabitable, such as the poles, rocky and high mountains, swamps, and deserts. That puts the total at about one-eighth of the earth's total surface area.

♦ About three-fourths of the remaining one-eighth section can't be used to grow food because the land is too dry, too cold, too steep, too rocky, or too infertile, or it has been built over with urban areas, factories, parking lots, roads, houses, schools, and parks. All that remains to produce food is $\frac{1}{32}$ of the earth's surface.

♦ The top five feet of topsoil on this $\frac{1}{32}$ remnant of the earth's surface represents the entire amount of the earth's food-producing land for six billion people—and the topsoil is eroding and washing out to sea at an alarming rate.

Feeling cramped? You better get used to it. The world's population is expected to reach seven billion by 2013, eight billion by 2028, nine billion by 2054, and then perhaps stabilize at 10 billion by 2100. Will humans colonize another planet? Maybe. Will technology save the world? Hopefully. Without that technology in hand, however, humans have reason to be extremely concerned.

The Scattered Masses: Where the People Are

Geographers look at not only the numbers of people on earth but also their distribution. People are not distributed evenly across the face of the earth. The earth's largest population clusters are in South Asia and East Asia. The population of Asia represents more than half the world's total population. Although other, smaller clusters exist around the world, vast areas of the earth are essentially uninhabitable.

Just as population distribution varies dramatically around the world, so, too, does food production. The ability to grow food can depend heavily on the length of the growing season, rainfall, soil fertility, and available technology. Droughts, floods, disease, war, and pest infestation can decimate food production in an entire region. Although international aid can do much to help in times of food crises, it's often too little and too late.

In this century, horrible food shortages have caused starvation and suffering at previously unheard of levels. Each day, hundreds of millions of the earth's people (perhaps one-eighth of the world's population) go to bed hungry. The world's resources are being stretched, and desperate people are encroaching on marginal land. For example, steep, wooded slopes in Nepal are being terraced for marginal farmland, resulting in serious erosion and flooding. Despite the environmental consequences, rain forests are being cut as farmers attempt to scratch out a living in the nutrient-poor soil.

Although some countries, such as the United States and Canada, produce huge amounts of food for export, the food race also has its losers. Many countries in Africa, for example, already have more people than their land is able to feed. Even in a good year, these countries are dependent on imported food. Even more disturbing is that these same countries exhibit some of the world's highest growth rates.

Terra-Trivia

Industrialized countries shouldn't blame the world's woes on the expanding populations of developing countries. Not so long ago, industrialized countries were the global leaders in population growth. Today, these countries gobble up much more than their share of the world's resources. On average, one person in the United States uses the same amount of energy as 400 people in the African nation of Angola.

More People, Less Food

As you will see, changing the population-growth trends in a region takes years (except in places with a state-mandated control policy, such as China). Because of the time it takes to decrease growth rates, especially in developing countries, humans soon will face increasing food shortages unless they're able to produce more food on Earth. Is this possible?

As mentioned, the problem can't be solved by bringing new land under cultivation. Most of the world's arable land is already being used, and the remaining land offers little potential for any sizable agricultural yields. Although some people hope for the building of Martian colonies, bushel baskets of corn aren't likely to be shipping in from Mars for many more years.

With no more land to place under the plow, people must turn to methods of increasing production on lands already being used for farming, which is already being done in many areas. This section looks at some advantages and disadvantages of this effort.

Although this section focuses on agriculture in its discussion of food resources, keep in mind that another major source is available: fish. More than 90 percent of all fish consumed on earth come from the oceans and seas. Not so long ago, most people viewed the immense oceans as an unending resource—but not anymore. Industrial fishing techniques have taken a heavy toll on the world's supply of fish. It has been estimated that since the beginning of large-scale fishing in the 1950s the populations of such large ocean species as cod, halibut, tuna, and swordfish have been reduced by as much as 90 percent. Although the growth of aquaculture, or fish farming, can make up for some of the loss, the days when the oceans could be regarded as a limitless resource for the world's hungry may be history.

The Green Revolution

Beginning in the late 1960s, the world experienced a phenomenon known as the *green revolution*, a term that describes many scientific advances applied to agriculture to increase crop yields. Some technologies include bioengineered hybrid high-yield seeds, organic and chemical fertilizers, chemical pesticides, chemical herbicides, mechanization, high-tech irrigation systems, and multiple-cropping.

The application of green revolution techniques initially produced significant agricultural increases. Eventually, however, problems with the technology also became evident. One problem is expense: Hybrid seeds, irrigation systems, machinery, and fertilizers and other chemicals are all expensive—and energy intensive. Although the developing countries stand to gain the most from adopting green revolution techniques, those countries have the least amount of money to spend on expensive technologies.

Another problem is that large portions of the population in developing countries typically are involved in small-scale subsistence farming. Small fields can't handle the same kind of mechanization used on the great plains of North America. Green revolution techniques produce the highest yields when they're applied to large expanses of a single crop.

The introduction in the mid-1990s of genetically engineered (GE) crops opened a new chapter in the age-old struggle to increase agricultural yields. New genetic material can now be added to crop plants to increase their resistance to disease and pests, control weeds, and even extend shelf life. But this scientific breakthrough has been controversial, with many people questioning the safety of GE crops and their impact on the environment.

Disease, Rats, and Superbugs

One of the arguments in favor of genetic engineering is that it can increase a plant's disease- and pest-resistance and in some cases reduce the need for chemical pesticides. The opposite is true of hybrid seeds, which produce higher yields than indigenous crops, but are more vulnerable to local blights. Large areas of a single hybrid crop are much more susceptible to drought, pests, and disease than are small plantings containing multiple varieties of indigenous crops. To protect these high-yielding but vulnerable hybrids, farmers have turned to ever-increasing applications of chemical herbicides, pesticides, and fungicides.

In addition to the pollution and contamination these chemicals cause, they have only a short-term effect on the pest problem. The extent of the problem cannot be minimized: Each year, pests account for the destruction of an estimated half of the world's food production. To increase food production, the loss to insects, rodents, and fungi must be minimized—but are chemicals the answer?

Indiscriminate chemical spraying kills not only the targeted pest species but also the helpful natural enemies of that species. Although spraying typically produces a short-term benefit, the pests eventually return in increased numbers. After natural controls are destroyed, pests can decimate crops with abandon.

CAUTION

Eye on the Environment

One possible alternative to the widespread use of chemical spraying is *integrated pest management* (IPM). This system addresses the entire ecosystem and is sustainable over long periods while remaining environmentally friendly. In addition to its selective and limited use of chemicals, IPM relies on the use of natural pest predators, low-till agriculture (minimal soil disruption), and crop diversity to control pests.

A farmer using the IPM system accepts a small crop loss to pests in return for long-term sustainable yields. Many farmers would describe IPM as working with the ecosystem rather than against it. Although a farmer must be well educated about the subtleties of the local ecosystem, the lower costs of implementation are attractive to people in developing countries.

Although farmers can repeat the expensive spraying, of course, a higher dose is required the next time. Why? Because of the superbug phenomenon. Although most of the pests are killed during each spraying, some survive. The survivors have a natural immunity to the given level of toxin. These highly resistant survivors reproduce and breed a new generation of pests that are more resistant to spraying than the preceding generation. The only way to kill off these varmints is to hit them with yet a stronger dose of chemicals—and the cycle continues. The world eventually will become too toxic for human habitation—but will be just fine for the newest generation of superbugs.

A Glimmer of Hope: The Demographic Transition

This chapter has painted a rather bleak picture of a world straining under the weight of a population spiraling out of control and hungry people tapping marginal lands for minimal yields. You've read about food production that can't keep pace with an exploding populace and superbugs running rampant through the farmlands, laughing off all the poisons humankind can throw at them. Is there hope?

Yes, humans certainly have hope (I wouldn't leave you without at least a glimmer). Although population growth will continue to put pressure on the earth's resources in the short term, a phenomenon called the *demographic transition* could prove to be the answer for the long term, as shown in the following graph. To understand the demographic transition, you have to look at its three stages. The concept of demographic transition was developed around the population of industrialized Europe, which has passed through each stage.

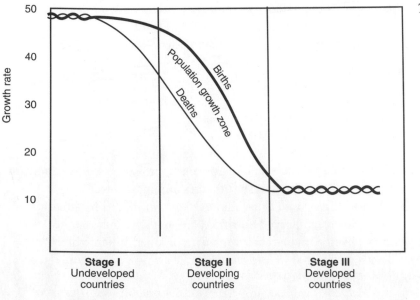

The demographic transition.

Stage I

Several hundred years ago, in preindustrial Europe, the population was relatively stable. The average woman in that society began having children early in life. If childbirth didn't kill her, she was likely to give birth to at least eight to ten children.

Why? One reason is that birth-control options were, at best, limited. Also, preindustrial Europe relied heavily on an agricultural economy, as do developing countries today. In this type of society, children are an economic asset: They don't eat much, and they can work in the fields and provide labor and income for their family. The more, the merrier.

The typical eighteenth-century peasant farmer didn't have a chunk of his paycheck deducted each week for Social Security. When he reached his golden years (his 30s), he depended on his family to sustain him if he could no longer work the fields. If he had no family, his prospects for a happy retirement were dim. The conclusion: The bigger the family, the better.

Terra-Trivia

In one of history's worst examples of a "destructive population check," Europe in the mid-fourteenth century was swept by the bubonic plague, otherwise known as the "black death." An estimated one-third to one-half of the continent's population was wiped out between 1347 and 1352. The death toll is believed to have exceeded 25 million people.

Although you might think that the high birthrates of preindustrial Europe would give rise to high population growth rates, it wasn't necessarily so. Don't forget the death rates. Infant mortality was also extremely high. Perhaps half the children who were born never reached adulthood. People also died young by today's standards. Anyone who made it to age 40 was doing quite well. Poor hygiene, disease, famine, and war all took their toll. Death rates were high. With both high death rates and high birthrates, population growth was slow and fluctuated with war or famine.

Stage II

As the industrial revolution created new jobs and opportunities in Europe, the region entered demographic Stage II. New crops (such as potatoes and maize) were imported from the New World and proved to be a boon to agriculture. The situation was looking up as medical advances, improved nutrition, and hygiene increasingly enabled people to live longer lives. Although the death rate plunged, the birthrate declined more slowly. (Death rates can decrease dramatically within just a couple generations, but birthrates take much longer.)

The Europeans still had no Social Security and no birth control. Children still helped on the farm and went to work in the new factories. People still wanted to have large families, and infant mortality had dropped. In a family with ten children, eight might have made it to adulthood. The result of high birthrates, lower infant mortality, and lower death rates were huge population growth rates. Europe's population exploded.

Europeans were lucky. Although the population explosion had made the region somewhat crowded, people could always take off for the New World, Africa, and Australia. Europe weathered its population explosion through emigration. Regions lagging 100 or 200 years behind Europe in the demographic cycle would not be as lucky.

Stage III

Europe is now a highly developed region. Few people farm the land, and most live in cities. Death rates are even lower now, with people living into their 70s. Social Security is available, and most people live in apartments, not on farms.

The net result: Raising children is expensive, and Europeans have stopped having large families. The economic tables have turned, and a large family is no longer an economic windfall; it's an economic liability, in fact. With effective birth-control methods readily available, the choice is sometimes between a new BMW or another baby.

In Europe, people are opting for the BMW. Families are small: One or two children is the norm. The result of low death rates and low birthrates is stable populations. Some European countries are even experiencing declining rates—which hasn't happened since the black death. The decline is steepest in Russia and parts of Eastern Europe. The disintegration of the Soviet Union in 1991 and the upheavals that followed led to a spike in death rates and a plunge in birthrates in many of the former Soviet republics and satellite nations. The resulting negative growth rate persists in the region despite increasing economic and political stability.

Slowing Growth

The force that drives the demographic transition is economics. The more economically advanced a society becomes, the lower its growth rate. Stable, and even declining growth rates, are possible under the proper economic conditions. Can this model be applied to today's developing countries? There are signs that this is already happening.

World population is expected to continue to skyrocket throughout this century—but at a slower rate than in the last century. It will take increasingly longer periods to add an additional billion people. After peaking in the late 1960s at 2.04 percent, the world's population growth rate fell to 1.2 percent by 2002.

Thanks to the increased availability and use of contraceptives, couples in developing countries are now having an average of three children, down from six in the 1970s. Even China, once the world's demographic time bomb, has brought its growth rate down to 0.87 percent, although the methods used to reach this level were often coercive and unacceptably harsh. An even grimmer factor in the slowing of world population growth is the rising death rate from HIV/AIDS, especially in sub-Saharan Africa where the AIDS crisis is most severe.

Slowing growth rates by no means signal the end of the population problem. The vast number of women of childbearing age in developing countries means that populations there will continue to rise, in many cases unsustainably. At the same time, the world's population is growing older, international migration from developing to developed nations continues to rise, and more and more people are concentrating in vast megalopolises. These demographic trends will have wide-ranging economic, social, and environmental consequences throughout this century.

Controlling the growth of the world's population is critical to ensuring the future security and well-being of humankind. However the population dilemma is resolved, it won't be by just a handful of countries, but by all the nations of the earth acting together.

The Least You Need to Know

- Demographics is the study of population numbers, growth, distribution, and trends.

- More than six billion people live on the earth; by the end of the twenty-first century that number will likely have climbed to 10 billion.

- Demographic transition describes the lower growth rates that come with improved economic conditions in an urbanized society.

- The world's population will continue to rise throughout the twenty-first century, but at slower rates.

Chapter 27

Healing the Atmosphere

In This Chapter

- ◆ Sorting through the layers of the atmosphere
- ◆ Keeping an eye on global warming
- ◆ Dealing with the sorry state of the upper atmosphere
- ◆ Realizing that the acid rain in Spain might very well stain
- ◆ Reeling from Chernobyl

Surrounding the earth is a thin layer of gases that's essential to all life. Humans unfortunately have not always treated that layer, the atmosphere, kindly. This chapter looks at some problems the atmosphere is experiencing and what we can do about them.

Each region on Earth has its own environmental problems, which have been described elsewhere in this book. When you're discussing the environment, it's easy to lapse into a lifeboat mentality (save yourself and disregard the plight of others). "It's a shame that the rhinoceros is becoming extinct in Africa—they really ought to do something." "You hate to see that deforestation in Haiti; I'm glad that we don't have the problem here."

And so it goes with many environmental issues in the news. What about the problems you can't ignore because they affect you directly?

The atmosphere is one of those universalizing agents. Many atmospheric problems that trouble Eastern Europe will also eventually trouble North America, for example. What's bad for South America might also prove to be bad for South Asia. The air on which all humans depend is constantly mixing as it swirls around the earth. In the process, human problems are also distributed.

For that reason, this chapter looks at the atmosphere outside the regional context. Some problems are certainly more acute in a particular region; all in all, however, humans are in this one together. Likewise, responses to atmospheric problems cannot be devised on a regional basis. If people hope to solve the array of atmospheric ills they face, they must learn to overcome their differences and work together.

Anatomy of the Atmosphere

The earth's atmosphere is made up mainly of nitrogen (78 percent) and oxygen (21 percent), as well as smaller amounts of other gases and water. It extends for nearly 350 miles from the surface of the earth and is divided into several distinct zones, or layers. The layer closest to the earth is the *troposphere*. "Tropos" is a Greek word meaning "turn," which is exactly what happens in this layer. The storms, temperature swings, winds, and other weather disturbances experienced on the earth's surface all take place in the troposphere. This zone encompasses the entire earth, although it's deeper at the equator and a little shallower at the poles. As you move higher in the troposphere, the temperature decreases.

About seven miles up, a thin transitional zone called the *tropopause* separates the troposphere from the next layer, the *stratosphere*, home of the famous ozone layer, which absorbs and deflects ultraviolet radiation from the sun (more about this later in this chapter). The stratosphere is drier and less dense than the troposphere and its temperature is fairly constant.

At about 30 miles up, the *stratopause* divides the stratosphere from the *mesosphere*. Within this layer, temperatures again decrease with height until, at 53 miles up, you reach another transitional zone, the *mesopause*, and beyond that, the final layer, called the *thermosphere*, or upper atmosphere. This is the deepest layer of the atmosphere, reaching some 372 miles above the earth's surface. Temperatures here remain constant for a while and then begin to increase with height due to the effects of solar energy.

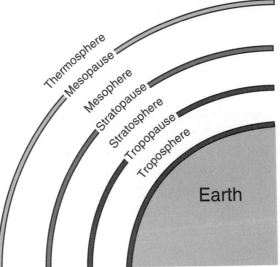

The earth's atmospheric zones.

Thermosphere
Mesopause
Mesophere
Stratopause
Stratosphere
Tropopause
Troposphere

Earth

Geographically Speaking

About 50 miles above the earth's surface begins a layer known as the *ionosphere*, composed of electrically charged particles, or ions. The ionosphere extends from the upper reaches of the mesosphere and all the way through the thermosphere.

The ionosphere is well known to ham radio operators as the layer that reflects radio waves and allows long-distance radio communication to take place. The ionosphere is also responsible for the famous aurora borealis, or northern lights, that grace the higher latitudes.

Turning Up the Heat: Global Warming

Global warming is one of today's most intensely debated environmental topics. Signs that the earth's climate is warming are there for all to see (and feel), and they are mounting. Global surface temperatures increased 1.1°F over the last century, and 0.4°F in just the last 25 years. Mountain glaciers are receding, polar icecaps are thinning, and sea levels rose at a faster rate during the twentieth century than in the preceding several *thousand* years. Pretty alarming stuff. The question is: What's causing it all? On one side of the debate are those who believe that this warm spell is just part of a natural and age-old cycle of global climate change. On the other side are those who point to human activity, specifically the burning of fossil fuels, as the primary culprit. If human activity is indeed to blame for global warming and nothing is done to reverse the trend, the results could be catastrophic to human life. This section looks at the probable causes and potential consequences of global warming, as well some possible solutions.

The Greenhouse Effect

Despite all its bad press, the process known as the *greenhouse effect* is a very natural and necessary phenomenon. To understand it, think about the way a greenhouse works, as shown in the following figure. Light from the sun passes through the glass of a greenhouse and is absorbed by the earth and plants within. The ground surface reradiates the energy as infrared radiation, or heat energy.

Although the short-waved sunlight initially passed through the greenhouse glass, the reradiated long-wave infrared heat energy is not able to pass back out efficiently. This heat energy becomes trapped and causes the greenhouse to warm. (The same process causes your car to get unbearably hot while sitting in the mall parking lot on a summer day.) The earth's atmosphere acts in much the same manner as a greenhouse.

The atmosphere contains gases that enable sunlight to pass through to the earth's surface but hinder the escape of the reradiated heat energy. Without this natural greenhouse effect, the earth would be much too cold to sustain life. The problem arises when too much of the heat-trapping gases are present in the atmosphere. Then temperatures begin to rise, and the troubles begin.

What Causes the Greenhouse Effect?

Naturally occurring gases in the atmosphere—carbon dioxide, methane, nitrous oxides, and water vapor—account for most of the greenhouse effect. The most abundant of these gases is water vapor, followed by carbon dioxide. Human activities such as agriculture, industry, mining, and especially the burning of fossil fuels, have increased the levels of some of these gases, and added others that are not naturally occurring, such as chlorofluorocarbons (CFCs). Although CFCs and methane are more efficient than carbon dioxide at trapping energy in the atmosphere, carbon dioxide is a greater cause for concern for two reasons: It is far more abundant in the atmosphere, and humans add so much more of it through the burning of fossil fuels. In fact, carbon dioxide is thought to be responsible for more than 50 percent of human-induced warming. This section will therefore focus on the carbon side of this very complex equation.

All life is based on carbon. (Remember the old sci-fi description of earth creatures as "carbon-based life-forms"?) The coal, oil, and natural gas that humans rely on so heavily to fuel their way of life were once the living carbon-based tissues of primeval plants and animals. Buried for eons under layers of sediment, these ancient organic remains were eventually transformed by heat and pressure into the fossil fuels people use today. When those fuels are burned, they release the carbon that had gone into their making back into the atmosphere.

CAUTION   Eye on the Environment

All sorts of human industries and activities result in the emission of greenhouse gases. Landfills, livestock, and rice paddies produce methane. A variety of industrial processes emit nitrous oxides. Chlorofluorocarbons (discussed in detail later in this chapter) are found in refrigerants, insulation, and other products. The burning of fossil fuels is the primary source of carbon dioxide emissions. Natural processes like evaporation and transpiration release water vapor, which is actually the most abundant of the greenhouse gases, accounting for as much as 85 percent of the total effect.

Carbon is also released when a forest catches fire, when you burn charcoal in a grill, or even when you take a breath. Humans survive by breathing in oxygen and releasing carbon dioxide. Plants do the exact opposite, taking in carbon dioxide and releasing oxygen through the process of photosynthesis. This mutually beneficial exchange between plants and humans has the added advantage of helping to control carbon dioxide buildup in the atmosphere. But plants can do only so much to absorb all the carbon dioxide generated by the burning fossil fuels.

Terra-Trivia

Atmospheric carbon dioxide levels are now more than 30 percent higher than they were in preindustrial times. By the end of the twenty-first century, the concentration of carbon dioxide in the atmosphere is expected to be anywhere between 75 and 350 (!) percent higher than it was in the 1850s.

Consequences of Global Warming

"So what if the temperature rises a few degrees; I hate shoveling snow anyway!" Why are people making a big deal about a slight rise in temperature? This question is a tricky one to answer because it's based to some degree on scientific conjecture. But projections based on computer climate modeling raise disturbing possibilities. Here are three of the most probable:

◆ **Rising sea levels:** In fact, this is already happening. A further increase in global temperature could cause the massive polar ice caps (especially in Antarctica) to melt. The huge influx of water from the ice caps would cause ocean levels to rise even faster than they are now. Ocean temperatures would increase, causing thermal expansion of the water molecules and an additional rise in sea level. Some of the world's largest population clusters exist along coastal areas, and rising sea levels would obliterate them. If you live in very low-lying countries like the Maldives Islands or the Netherlands, rising sea levels are especially bad news.

◆ **Changing climates:** Most people could survive if their hometown became a little hotter or drier. Unfortunately, agriculture around your town might be altered dramatically. Because temperature and precipitation would not be uniformly affected by global warming, the outcome of this scenario is difficult to predict (supercomputers are modeling at this moment). But if, say, the Great Plains of the United States became too hot and dry to grow grain, not only would residents be in trouble, the impact on the world's food supply would be severe. Although other areas farther north in Canada and Siberia might benefit from the climatic shift, their soils are not necessarily conducive to agriculture. Considering the earth's burgeoning population, climatic threats to the food supply are worth worrying about. And the problem isn't limited to farmlands. Some of the earth's most productive and fragile ecosystems—corals reefs, mangrove swamps, and tropical rainforests—are especially vulnerable to climate change.

◆ **Extreme weather:** Ten of the hottest years on record have occurred since 1987. The hottest year so far was 1998, which corresponded with the strongest El Niño on record and a host of weather related disasters: devastating floods, prolonged droughts, powerful hurricanes, and tropical forest fires. The El Niño phenomenon, the periodic warming of the ocean surface in the tropical Pacific that creates wide-ranging disruptions in normal weather patterns, is not caused by global warming, but has become more frequent and intense in recent decades. More persistent El Niños, more severe droughts and floods, and an increasing incidence of devastating weather events are all projected consequences of unchecked global warming.

Terra-Trivia

In another negative feedback scenario, warming atmospheric temperatures will cause ocean temperatures to rise, triggering an increased growth of algae. By removing carbon dioxide from the atmosphere, the algae initiate a global cooling. Will it work this way? Your guess is as good as any.

Is Global Warming Fact or Fiction?

Most scientists agree that the earth is experiencing a warming, but there is still no consensus on what's causing it or what to do about it. Some scientists assert that increases in greenhouse gases will eventually result not in a global temperature rise but rather in an *icebox effect*, in which global temperatures will drop.

According to this theory, as the earth warms, ocean water evaporates more quickly, resulting in an increase in cloud formation. A more extensive cloud cover will act to screen and reflect incoming sunlight, ultimately causing surface temperatures to drop. This is a classic example of negative feedback, in which a mechanism that alters the earth's climate equilibrium in one direction (e.g. increased warming) generates a response in the opposite direction.

Although scientists debate whether or not there is a global warming crisis, a consensus is building among the nations of the world that something needs to be done. The Kyoto Treaty, drawn up in Kyoto, Japan, in 1997, committed industrialized nations to reducing their overall emissions of greenhouse gases by an average of 5.2 percent below 1990 levels within 10 years—a far from adequate response according to some environmentalists, but a start nonetheless. Although as of May 2003, the treaty had been ratified by 109 nations, including the European Union, Canada, and Japan; the United States, the world's biggest producer of greenhouse gases, withdrew from the treaty in 2001, citing its negative economic impact.

So while we're still in global-warming limbo, is there anything we, as private citizens, should be doing? Absolutely. Whether the worst-case global warming scenarios occur or not, it is clearly in our interest to limit our consumption of fossil fuel, use clean-energy technology whenever possible, and encourage the conservation of our forests. These practices are sound, no matter what the outcome of supercomputer global-warming models might be. The conservation of fossil fuels will ensure an energy supply well into the future. Saving forests not only cleans the air but also helps to preserve biodiversity on the planet. Who knows? We might also just avoid the catastrophic consequences of global warming.

Ozone Depletion, or Pass the Sunblock!

Before looking closely at the atmospheric problem of ozone depletion, you should be able to distinguish between the ozone layer and ozone smog. In the lower layer of the earth's atmosphere (the troposphere), ozone is a pollutant. A by-product of auto exhaust and sunlight, ozone builds up to produce ozone smog that often menaces cities during the sunny and warm summer months. Although this form of ozone is a health threat, don't confuse it with the ozone in the upper atmosphere.

The ozone in the stratosphere is essential to life on this planet because it protects humans from the harmful effects of the sun's ultraviolet radiation. "Ozone depletion" refers to the degradation of this protective shield in the stratosphere. Unfortunately, humankind has devised several chemicals that, when released to the atmosphere, attack and degrade the protective ozone shield. The most notorious chemical family, responsible for the bulk of the world's ozone destruction, is the chlorofluorocarbons, or CFCs.

CFCs are found in foam packaging, refrigerator and air conditioner coolants, and insulation and are used as an industrial cleaner. (The use of CFCs as aerosol spray-can propellants was banned in the United States in 1978.) When CFCs are released, they break down the molecules of ozone in the stratosphere at an alarming rate. The reduction over the South Pole each spring is so severe, in fact, that it's called the "ozone hole" (see Chapter 25 for more details).

What are the consequences of ozone depletion? Without an ozone layer or with a reduced ozone layer, much more of the sun's ultraviolet radiation reaches the earth's surface. Ultraviolet radiation can be extremely harmful to life on earth in a number of ways. An excess of this radiation can damage plant life, which produces oxygen for humans. This radiation is also worrisome in its effect on the minute phytoplankton that abounds in the earth's oceans. Phytoplankton, which is at the base of the ocean's food chain, is a tremendous consumer of carbon dioxide and producer of oxygen. No one wants radiation messing with the world's life-giving phytoplankton.

The direct effect on humans has also become increasingly apparent in recent years. Increases in skin cancer (including the most severe, melanoma), eye cataracts, and damage to the human immune system are just a few of the consequences. Unlike the global warming issue, there is general scientific consensus on the link between CFCs and ozone depletion. The 1987 Montreal Protocol, which originally committed signatory nations to reducing their use of CFCs, was later amended to ban CFC production after 1995 in developed nations and by 2000 elsewhere. Unfortunately, these long-lived chemicals continue to deplete ozone long after their initial release to the atmosphere. Even with a total ban on CFC production, the ozone layer is not expected to fully recover until about 2050.

That Burning Sensation: Acid Rain

Often referred to as "death from the sky," acid rain is a broad term used to describe the various ways acid falls out of the atmosphere. In addition to acid rain, there's also acid snow and acid fog, and about half of the acidity in the atmosphere falls back to earth not as moisture but as particles or gases that are "dry deposited" on buildings, cars, and trees. In the past several decades, acid rain has become a persistent and global problem. Its primary causes are industrial sulfur dioxide emissions and nitrogen dioxide from automobile and truck exhausts. Placed into the atmosphere and mixed with water vapor, oxygen, and other chemicals, sulfur dioxide forms sulfuric acid, and nitrogen dioxide forms nitric acid.

The acidity that falls from the sky can result in a host of environmental problems. Ancient art and architecture exposed to the elements have suffered severely. The temples of the Athenian Acropolis and the statues of ancient Venice have been literally eaten away by the high concentrations of acid in the atmosphere.

When acidity levels rise in lakes and rivers, fish and aquatic life can no longer survive. The results are biologically dead lakes that are now prevalent in the northeastern United States and in eastern Canada. Acid rain, perhaps combined with other stresses such as insects, drought, disease, and other pollutants, is at least partially responsible for forest die-offs that are occurring worldwide. The eastern United States and Western and Northern Europe have been hit especially hard.

The solution, of course, is to reduce the emission of sulfur and nitrogen dioxides. This reduction is not easily accomplished, however. Because weather systems often move the problem far from its original source, identifying a specific polluter is difficult. It's also difficult to legislate mandates that would require one country to invest billions of dollars in a cleanup when the impact of its pollution occurs in its neighbor's backyard. In the meantime, Canadians are none too happy about their country's thousands of dead lakes that have resulted largely from pollution originating in the United States.

Terra-Trivia

Acidity is measured on a pH scale that ranges from 0 to 14. The 7 on the scale is considered neutral; lower values indicate increased acidity. Although rainwater is normally about pH 5.6 (slightly acidic), acid rain has been measured as low as pH 1.5, an acidity level that exceeds that of lemon juice, which is strong enough to etch a copper plate.

Nuclear Fallout

Nuclear energy is not generally regarded as an atmospheric problem. This section mentions it because of one specific event that made everyone aware of just how potentially dangerous the problem can be. That nuclear event now lives in infamy: Chernobyl.

On April 26, 1986, the number-four reactor at the Chernobyl nuclear power plant blew up. The roof of the containment building was blown off, and a deadly nuclear cloud rose into the atmosphere. The Soviets evacuated everyone within a 40-mile-diameter circle later known as the "zone of estrangement." The evacuation ultimately displaced about 116,000 residents of the area.

Although the accident occurred in what was then the Soviet republic of the Ukraine, the bulk of the radioactive fallout fell on nearby Belarus. As it circulated in the atmosphere, the radiation did not stop at national borders. Much of Western and Northern Europe received large doses of radiation from the accident. Even in the far Scandinavian north, the Lapp's reindeer milk was contaminated by radiation.

The accident is thought to have been directly responsible for as many as 5,000 deaths and 30,000 injuries and indirectly responsible for another 15,000 deaths. The number of people exposed to Chernobyl radiation in the Northern Hemisphere might have exceeded three billion. No longer can anyone think of nuclear accidents as a localized phenomenon. Everyone shares the same atmosphere, and you don't have to live next to ground zero to be affected.

By the way, you might not have heard the last of Chernobyl. Approximately 180 tons of uranium fuel and another 10 tons of radioactive dust remain sealed within a 24-story concrete-and-steel enclosure erected around the ruined reactor. The core of the reactor is still active, and by mid-2003 the enclosure had so deteriorated that it was in danger of collapsing. Equally disturbing is the fact that two of the four original Chernobyl reactors continued to operate until 2000 and that 13 other reactors of the same flawed design are still on-line in Russia and Lithuania.

The Least You Need to Know

- The health of the atmosphere affects not only individual nations or regions but the entire earth and all its inhabitants.

- The lowest layer of the atmosphere is called the troposphere. This is where humans live and weather happens.

- Global warming, resulting from the greenhouse effect, might already be causing global climatic change and rising sea levels.

- Ozone depletion is caused primarily by chemicals called CFCs. Loss of the ozone shield poses a health threat to all life.

- Acid rain is caused by industrial and vehicular emissions; dying forests and dead lakes are the result.

- The Chernobyl power plant disaster opened people's eyes to the potential global consequences of a nuclear accident.

Chapter 28

Energy for the Twenty-First Century

In This Chapter

- Coming to grips with the energy dilemma
- The true costs of fossil fuels
- Why renewable resources matter
- The promise of hydrogen

They say money makes the world go round, but what actually does the trick is energy. People need energy to run factories, drive cars, heat buildings, and operate just about every gizmo in the house. As society has become more industrialized, human energy needs have grown. The energy demands of the developed world far exceed those of developing countries, although the gap is narrowing. The United States is the most demanding country of all: With less than 5 percent of the world's total population, the United States consumes nearly 25 percent of the world's annual energy production.

World energy demand is expected to increase as much as 58 percent by 2025, with developing countries accounting for much of that increase.

How that demand is met and at what cost to world peace and the health of the environment are key questions confronting humankind in the twenty-first century.

Nonrenewable Resources: Here Today, Gone Tomorrow

Most of the world's energy needs come from *nonrenewable* sources—mainly fossil fuels—oil (petroleum), natural gas, and coal. Nuclear energy, derived from uranium, is not a fossil fuel, but it is nonrenewable. With all nonrenewable energy sources, the bottom line is this: Once you use it, it's gone for good. You can't replant it or regrow it, and supplies are not limitless.

The Old Fossils

The world today is largely powered by fossil fuels, which account for 85 percent of total energy consumption. The benefits of fossil fuels are easy to understand. They are still available in vast quantities and they are relatively cheap to extract from the earth. At one time, nations, industries, and to an extent, individual consumers made energy decisions on the basis of availability and cost. Today other factors have entered the equation and made it far more complicated. For one thing, supplies might be abundant today, but they won't always be. No one can say for sure when the world will literally run of gas, but that day will definitely come—and sooner rather than later if current consumption trends are not reversed. The energy shortages of the 1970s scared many people into conserving fuel, but the effects were short-lived, as oil prices dropped again in the mid-1980s and gas-guzzling cars came back into fashion.

GeoJargon

Nonrenewable resources, such as fossil fuels, metal ores, and uranium used for generating nuclear power, take so long to form that they're essentially irreplaceable. Although wood is generally considered a renewable resource, if a cleared forest is not replanted, its wood automatically becomes a nonrenewable resource.

Although extracting fossil fuels might be relatively cheap, there are many hidden and not-so-hidden costs involved in the process. Mining, drilling for, and transporting fossil fuels can all have devastating effects on the environment. Strip mining and oil spills are only two of the more visible examples. More than a decade after the Exxon Valdez oil spill in 1989, the beaches and wildlife of Alaska's Prince William Sound have yet to fully recover. And of course, once fossil fuels are out of the ground and in your gas tank or local power plant, they are adding greenhouse gases and other pollutants to the atmosphere, with all the unpleasant consequences you read about in Chapter 27.

```
CAUTION
```
Eye on the Environment

When it comes to energy matters, the world's big kahuna is without a doubt the United States. It is the largest producer, consumer, and net importer of energy, as well the biggest polluter and contributor of greenhouse gases (although China is quickly catching up). The United States ranks twelfth worldwide in oil reserves, sixth in natural gas, and first in coal. In 2002, the United States imported 58 percent of its total oil needs, of which 20 percent came from the Persian Gulf region. Canada, Saudi Arabia, Mexico, and Nigeria were the main U.S. oil suppliers. Under the administration of President George W. Bush, energy conservation and the development of renewable fuel sources have taken a back seat to securing overseas oil imports and finding new domestic sources of fossil fuels often in ecologically sensitive areas such as Alaska's Arctic National Wildlife Refuge.

Finally, there is the political side of the fossil fuel equation. The world is addicted to oil, and much of it is sitting under one of the world's most dangerous powder kegs. More than half of the world's estimated oil reserves are in the Middle East, mostly in the Arabian Peninsula, Iraq, and Iran. The Arab oil embargo of 1973, the Gulf Wars against Iraq of 1991 and 2003, and the escalating political instability and terrorist violence in the region are all clear indicators of the need to shake our global oil addiction and find alternative fuel solutions.

The Nuclear Card?

The newest of the nonrenewable energy sources, nuclear power plants, use the energy that is released by the splitting of uranium atoms to generate electricity. The first nuclear power plants went on line in the 1950s. Today, nuclear power accounts for about 16 percent of the world's electricity (20 percent in the United States). Once hailed as a safe, clean alternative to fossil fuels—the energy of the future—nuclear power is in decline. Few new plants are being built, and many are being phased out. The accidents at Three Mile Island in the United States and at Chernobyl in what was then the Soviet Union opened the eyes of many to the dangers of nuclear power plants. In addition to the threat of accidents, there's also the sticky problem of disposing of spent fuel and low-level radioactive waste. And in the aftermath of the September 11, 2001, terrorist attacks against the United States, nuclear power plants are being viewed with mounting concern as vulnerable terrorist targets.

Renewable Resources: Here Today, Here Again Tomorrow

Renewable resources continuously regenerate or can be recouped within a relatively short period of time. Tree stands that are harvested and replanted can regenerate within a lifetime. In addition to wood and other types of organic biomass, the main renewable energy sources are flowing water, wind, sunlight, and geothermal energy. This section looks at how renewable resources are being used to meet the world's energy needs.

Hydroelectric Power—Dam It!

Hydroelectric power, which uses the force of falling water to turn turbines and generate electricity, now accounts for about 20 percent of the world's electricity. Although hydropower would seem to be a safe, clean, and long-lasting solution to the world's energy problems, it presents its own problems.

First, potential sites for hydroelectric dams are limited and many have already been used. Second, huge reservoirs that build up behind hydroelectric dams can displace people and cause unforeseen ecological damage, as was the case in the Hydro-Quebec projects in northern Canada, which displaced the Cree people and contaminated one of their primary food sources with mercury poisoning. China's controversial Three Gorges Dam, the world's largest hydroelectric power project scheduled for completion in 2009, is expected to force between one and two million people out of their homes (see Chapter 16).

Rivers typically flood during their annual cycle, cleansing the land within their floodplain and revitalizing it with nutrient-rich silt. Damming a river prevents the annual flooding, with long-term consequences for floodplain fertility. Silt that no longer reaches farmlands downriver collects behind the dam, often clogging turbines and affecting future power generation. These problems have been evident on such large-scale projects as the Aswan High Dam on the Nile River in Egypt.

Dams also have a negative impact on wildlife habitats and migration patterns. In the western United States, for example, where most river systems have been extensively dammed, dams either have destroyed important salmon spawning habitats or keep the fish from reaching the spawning grounds that remain.

The Solar Solution

Every day the earth is bombarded with energy from the sun, so why not use it? It's clean, unlimited, and theoretically free. Well, at least the fuel is free. Converting it into usable forms of energy is still not cheap, although technological advances continue to bring costs down.

Energy from the sun is used in two principal ways: to generate electrical energy and to supply heat. Solar energy can be converted directly into electricity by means of *photovoltaic cells*. Even if you have not seen these cells at work on a rooftop or power station, you have undoubtedly seen them on your pocket calculator. Just give the cells a little light, and you're ready to start balancing your checkbook.

As new techniques for manufacturing these photovoltaic cells are developed, the cost might soon be competitive with traditional power sources. These systems are becoming sophisticated enough to work well with even ambient light and will therefore be adaptable to most places on earth (except maybe at the North and South poles).

Solar energy is also widely used for heating buildings. Many passive solar-heating techniques, such as orienting buildings to take maximum advantage of the sun's warming rays, are centuries old. Such traditional building principles have been adapted and combined with new techniques and materials to create energy-efficient designs at a minimal cost. In addition to passive solar techniques, *active solar systems* that collect and store solar power can also be used to heat buildings.

Terra-Trivia

Active solar systems became popular in the United States following the 1973–1974 Arab oil embargo. Spurred by tax credits offered by the United States government, the installation of active solar systems increased steadily until the mid 1980s, when oil prices fell and the government withdrew the tax credits. Without the tax incentive, the rate of installation markedly declined.

These active systems typically employ solar panels that absorb sunlight and heat fluids circulated through panels in pipes. The heated fluid is then used to warm living spaces or, more frequently, to heat water for the domestic hot-water supply. Solar panels work best, obviously, in areas that annually receive sufficient days of sunshine. Areas that are especially rainy or foggy, such as the Pacific Northwest in the United States or the British Isles, are worse candidates than drier areas.

Active solar systems also have industrial applications. Solar thermal concentrating systems work like super solar panels, using lenses and mirrors to concentrate solar radiation and produce extremely high temperatures that can be used to create steam and generate electricity.

Harnessing the Wind

People have used wind power for centuries to run mechanical equipment and pumps. The most notable examples are the beautiful old windmills of the Netherlands (Holland). The Dutch windmills were workhorses that constantly pumped water up and out of canals used to drain land reclaimed from the sea. Windmills (of higher-tech design than the old Dutch versions) are now used to turn generators for the production of electricity.

Some places are windier than others, of course. Windmills work well in the Netherlands because of the steady wind from the North Sea. Many places are simply too calm to make windmills a feasible option. They also tend to be noisy, and the whirling blades take their toll on the local bird population.

Although it still supplies only a small part of the world's electricity, wind power is the fastest growing source of electrical energy, and it's becoming increasingly cost competitive. The world's leading producer of wind-generated electricity is Germany, followed by Spain. Growth in the United States has slowed because of the uncertain status of Federal wind-power tax incentives.

Geothermal Energy and Biomass

Geothermal heat from deep within the earth can be tapped to create steam that turns turbines and generates electrical power. As you saw in Chapter 9, geothermal energy is an important source of power in Iceland and nearly every home there is heated with water from hot springs.

Unfortunately, geothermal energy is not easily accessible everywhere on earth. The largest geothermal facilities are located at geologic hot spots on the earth's surface, generally seismic areas, such as Iceland and along the Pacific Rim "Ring of Fire." The United States is the leading producer of electricity from geothermal reservoirs, with most of its plants concentrated in California and Nevada.

Another major renewable source of energy worldwide is *biomass*, which includes wood and other plant materials, organic garbage, and animal wastes. Biomass was the first source of energy used by humans and it remains a primary fuel in many developing countries. The most direct way to extract energy from biomass is to burn it for cooking, heating, or industrial purposes, but that releases carbon dioxide into the atmosphere, just like burning any other carbon-based fuel would, and it's inefficient. New technologies, however, are making it possible to use biomass in far more efficient and environmentally friendly ways. Biomass can be converted, for example, into cleaner burning combustible gases and liquid fuels, such as ethanol, which is being used to help reduce gasoline consumption. One promising scenario for the future is the large-scale farming of fast-growing grasses and trees as biomass "energy crops."

The Dawning of the Hydrogen Economy

It might still be only a dream, but a lot of scientists are dreaming it. What's the dream? A world fueled by a limitless energy source, powerful, efficient, and clean burning, whose most noxious emission is pure water. Sounds like a fantasy, right? Well it isn't that far-fetched. The magic fuel is hydrogen, the lightest, simplest, most abundant element in the universe. And right now, plying the streets of smog-filled cities around the world, are experimental fleets of buses running, pollution-free, on hydrogen fuel cells.

Hydrogen, Hydrogen Everywhere

Literally, hydrogen is everywhere. Although it doesn't exist naturally on earth as a gas, wherever there's water (H_2O), there's hydrogen (H_2). It is also a component of natural gas and can be extracted from biomass. So why isn't the world running on hydrogen? The main hitch, for now, is the cost of separating hydrogen from the compounds in which it is found. The technology for splitting ordinary water into hydrogen and oxygen is well-known, but it requires a lot of electricity and is costly.

> **CAUTION**
>
> ### Eye on the Environment
>
> Hydrogen power to the people! There is a definite populist appeal in the notion of a hydrogen-fueled future. According to Jeremy Rifkin, author of *The Hydrogen Economy* (J.P. Tarcher, 2002), hydrogen offers not only an alternative to fossil fuels but also a way to free power generation and delivery from the control of oil cartels and utilities. In the hydrogen economy, says Rifkin, energy users will be able to acquire and share energy by plugging in their fuel cells to local, regional, and national Hydrogen Energy Webs (HEWs), much as computer users today hook up to the Internet to exchange information.

Most of the free hydrogen used today is created by a process called "steam reforming," in which high-temperature steam is used to separate natural gas into hydrogen and carbon. But reforming is energy-intensive and reliant on fossil fuels. The challenge for all those scientists dreaming of a hydrogen future is to devise a cost-effective, environment-friendly way of liberating hydrogen.

Drinking from the Exhaust Pipe: Fuel Cells

Hydrogen can be used like gasoline to fuel a combustion engine, but it really comes into its own when it's used with a device called a *fuel cell*. Put very simply, a fuel cell is a sort of battery in which a catalyst causes hydrogen molecules to separate into protons and electrons. The electrons flow out as electric current, the protons combine with oxygen molecules from the air to form pure water, the only residue of the process. You might not actually want to drink from the tailpipe of a hydrogen-powered vehicle, but if you do, you'll live to tell the tale. In addition to powering vehicles, fuel cells can also be used to generate electricity and heat.

Although fuel cell technology is not new, the industry is still in its infancy and costs are high. But as the technology advances and hydrogen consciousness grows, costs are bound to come down to competitive levels. In a world faced with so many energy challenges, the potential of hydrogen is simply too great to ignore.

The Least You Need to Know

- ◆ Nonrenewable energy sources are finite in quantity, and their use presents pollution and disposal problems.

- ◆ Most of the world's oil reserves are in the volatile Middle East, where they are vulnerable to war, political upheaval, and terrorist threats.

- ◆ Renewable energy sources—water, sunlight, wind, geothermal heat, and biomass—offer a long-term alternative to fossil fuels.

- ◆ Hydrogen is an abundant, efficient, and clean-burning fuel with the potential to help the world finally break its fossil fuel addiction.

Chapter 29

Working to Save the Earth

In This Chapter

- Sailing an ocean of trouble
- The coming fresh water crisis
- What you can do to help heal the environment
- Looking back at your journey through this book

During the past several decades, people have become more concerned about the environment. They have generally become better educated about the problems they face and more active in their efforts to help save the earth. In the process, people have begun to grasp more fully the complexity and wonder of life on this planet.

The culmination of physical and human geographic studies is an understanding of how the two interrelate. How are human beings affected by the cycles of the physical world? How is the physical world, the environment, affected by human activities? The sad truth is that all too often human impact on the environment has had negative consequences for other species on this planet—as well as for our own.

This chapter looks at some of the problems humans face. Although Chapter 27 considers the atmosphere all humans share, this chapter focuses on the terrestrial realm. Despite the environmental problems that threaten the planet's future, hope exists. This chapter describes some of the ways that people, drawing on the remarkable human spirit, are solving problems and how their efforts are holding out the prospect of a brighter future.

Taxing the World's Oceans

In Chapter 26 you saw how overfishing has affected the world's oceans. Unfortunately, that is not all humans have done to the earth's most precious resource. All life on earth sprang from the oceans. Because oceans cover 70 percent of the entire Earth, it's difficult to imagine that something so huge could ever be threatened. But they are.

Terra-Trivia

You might suspect that the primary cause of oil pollution in the oceans is freighter accidents such as the *Exxon Valdez* disaster. However, most of the ocean oil pollution that now appears in even the most remote corners of the globe is caused not by accidents, but rather by conscious acts, mainly the routine flushing of oil tanker ballasts and tanks.

The world's creeks, streams, and rivers all comprise a huge highway to the sea. When you fertilize your lawn, for example, rain washes the chemicals into storm drains. From there, they flow into streams and then rivers, and eventually out to the sea. And so it is with sewage, industrial waste, and leaching landfills. Sometimes no one bothers with the indirect path: They just take it out to sea and dump it directly.

The old adage "The solution to pollution is dilution" summed up the conventional wisdom regarding the seas. People used to think that the oceans could never be mucked up because they were simply too big. That time has passed, however. The oceans are mucked up already and, unless humans act soon, will quickly get worse.

Many countries and municipalities simply cart their wastes out to sea and dump them. Although the type of waste varies, it might include treated or untreated sewage, industrial waste, garbage, construction materials, and even medical waste. Although most people want to forget about this refuse as it collects in the world's oceans, the stuff has been coming back to haunt us.

As bad as these wastes are, they can seem almost benign next to the really hideous materials humans spew into the oceans, such as toxic chemicals, radioactive waste, heavy metals, oil, and PCBs (polychlorinated biphenyls, which have been linked to birth defects, cancers, and immune system damage).

Although solutions are possible, they require a will and a wallet. Only 15 years ago, Boston Harbor was the cesspool of the East Coast. Sewage from Boston and its surrounding communities was dumped directly into the harbor with little or no treatment. Political clout was mustered, a budget was hammered out, and billions of billions of dollars and years of labor later, a huge new sewage-treatment facility stands on Deer Island at the mouth of the harbor. It's now possible again to find porpoises swimming in the clear blue waters of Boston Harbor.

Water, Water Everywhere and Not a Drop to Drink

Except for air, no other resource is as crucial to life on earth as fresh water. A reliable supply of clean fresh water is essential for drinking, cooking, and sanitation, but humans also rely on fresh water for transportation, agriculture, industry, and food (more than 40 percent of the earth's fish species live in fresh water). On a planet whose surface is mostly water, you might think that water scarcity would be the least of our problems. Not so. Only about 2.5 percent of the earth's water is fresh. More than two-thirds of that is frozen in glaciers and permanent snow cover, while the rest—the accessible water in lakes, rivers, and aquifers (ground water)—is very unevenly distributed. Some regions enjoy abundant supplies, others are water-stressed.

Dwindling Supply + Growing Demand = Crisis

Fresh water supplies are not only relatively restricted to start off with, they are becoming scarcer. Rising global temperatures, more frequent and intense droughts, and changing rainfall patterns are straining water resources in many parts of the world. An even bigger culprit is pollution. A toxic brew of industrial and agricultural wastes, chemical runoff, acid rain, and raw sewage takes large volumes of water out of the available supply every year, or even worse, causes sickness and death in those who have no choice but to drink tainted water. An estimated 1.1 billion people today drink water than is unsafe.

Geographically Speaking

The combination of water scarcity and burgeoning population can be politically explosive. Access to the waters of the Jordan River basin and its aquifers has long been part of the conflict between Israel and its Arab neighbors. Turkey, Syria, and Iraq are vying for control of the Tigris and Euphrates Rivers, which the three nations share. And far from the Middle East in North America, control of the waters of the Colorado River and the Rio Grande is creating tension not only between Mexico and the United States, but between neighboring states in the arid U.S. Southwest.

As water supplies decrease, population growth and industrial and agricultural expansion are rapidly increasing worldwide water demand. Agriculture is a particularly insatiable consumer of water, accounting for about two-thirds of the world's fresh water use. It is also highly inefficient. About 60 percent of the water used for irrigation is lost to evaporation before it ever gets into a plant. As with other resources, the United States leads the world in water use, to the tune of about 620 quarts per person per day. (The average African, by contrast, uses barely 50 quarts a day.)

Quenching the World's Thirst

Nearly half a billion people live in water-stressed areas today, and according to some projections, that number will climb to three billion by the year 2025. Although sea-water desalination (the process of removing the salt from salt water to make it potable) offers some hope for the long term, the worst-case scenarios could easily come true in the short term, unless the nations of the world get together to rethink water management policies, promote more efficient farm irrigation techniques, and curb the pollution that is poisoning the planet's lifeline.

Disappearing Forests: A Global Concern

Deforestation has received increasing press attention in the past several decades. Although much of the attention has focused on the world's tropical rain forests, temperate forests have also been under the ax. Although the global rate of deforestation slowed in the late 1990s, current estimates suggest that an area of forest the size of Maine is cut each year around the world.

The world's largest remaining concentrations of tropical rain forests exist primarily in three countries: Brazil, the Democratic Republic of Congo, and Indonesia. These forests are also being rapidly exploited. An area larger than France has already been deforested in the Brazilian Amazon. What are the causes of this rapid tree cutting, what are its consequences, and what can be done to save the vanishing forests?

> **CAUTION**
> ### Eye on the Environment
>
> In the Asian mountain country of Nepal, forests are being cut at an alarming rate. Extreme poverty has forced people to clear steep hillsides in search of marginal farmland and firewood. The result has been a devastating loss of topsoil. Without tree cover to control runoff, flash flooding has increased, and Bangladesh, downstream from Nepal's rivers, has been subjected to yet another source of flooding.

The Causes and Consequences of Deforestation

In different parts of the world, trees are cut for different reasons. As with most environmental ills, the reasons all stem from increasing pressures placed on the environment by the burgeoning human population. Those pressures include the need for more lumber and paper products, more farm and cattle ranching land, and, in the poorest countries, more firewood to heat homes and cook food.

Although other chapters have mentioned some of the consequences of deforestation, they don't even make a dent in the long list. The impact of deforestation is far-reaching:

♦ **Atmospheric carbon dioxide buildup:** Trees remove carbon dioxide from the atmosphere and add oxygen. When you remove trees, you also remove a major atmospheric cleanser. The burning of trees releases carbon dioxide into the atmosphere, adding to the greenhouse gases that are responsible for global warming.

♦ **Topsoil loss:** Trees hold precious topsoil in place. Without forests, the land is subject to erosion and soil loss. The quick release of runoff rainwater is responsible for flooding.

♦ **Climate change:** Forests play a vital role in the earth's hydrographic (water) cycle by taking in rainwater (absorption) and giving off water vapor (transpiration). The cutting of trees can result in altered patterns of precipitation.

♦ **Species decline:** Forests provide the habitat for an unrivaled diversity of plant and animal species. Cutting even small parcels of forest can result in species extinction. A loss of species results in diminished biological diversity on Earth and a loss of potential sources for future foods and medicines.

And finally, the aesthetic component of tree loss cannot be minimized. For most people, a world without forests would be a much poorer place.

What Can Be Done?

As with many global problems, it's easy to feel that, because you're just one person, you probably can't do much to help. This belief is far from the truth. The way you choose to live your life can not only affect the environment but also influence people around you. What can you do? You can support the many rain forest industries that help to make the forests economically viable without resorting to cutting. You can avoid using tropical hardwoods or woods (such as redwood) from old-growth forests. You can reuse or recycle wood-based products. You can plant a tree.

The Vegetarian Variation

In many parts of the world, especially Asia, diets are essentially vegetarian. The bulk of the diet consists of grains and vegetables with some fruit and little or no meat. Despite the damper this diet puts on backyard barbecues, it's part of an environmentally friendly lifestyle.

A person's diet can affect the environment in many ways including changing how land is used. For instance, a fertile plot of land can support 20 times as many people on a vegetarian diet as on a typical American diet. Similarly, fewer tropical forests in Central America would be cut down to support cattle grazing if people reduced the amount of beef they eat. And did you know that more than half of all water consumed in the United States is used in the livestock industry?

Putting Your Money Where Your Mouth Is

Money not only talks, it influences behavior. One way to curb environmental ills is to make them costly for the perpetrators through the use of the tax code. In a movement that is gaining ground in Europe and elsewhere in the world, governments are experimenting with "environmental tax shifting," which penalizes pollution and resource waste while reducing the tax burden on workers and investors. The result can be a win-win for society: economic stimulus and a cleaner environment. Even in the United States, where tax policy under the administration of President George W. Bush is trending in a completely opposite direction, "green" taxes have been used successfully to phase out ozone-destroying CFCs.

A Final Word

As you've toured the entire globe in this book, you have learned a thing or two about geography. You've explored the physical world and the human world and looked at how the two interact and are dependent on one another. The best way to understand geography is to live it. Perhaps this guide has whetted your appetite and provided a spark. The entire world awaits you, filled with wonders to see, people to meet, and adventures to experience. Pack your bag, set your course, and, with a map in hand—discover your world!

The Least You Need to Know

◆ The world's oceans are being contaminated at an unprecedented rate, but with the will and the funding to do it, the damage can be reversed.

◆ The increasing scarcity of fresh water might be one of the biggest environmental challenges of this century.

◆ Forest cutting has global consequences that will take a universal effort to abate.

◆ "Green" taxes help clean up the environment by increasing the cost of pollution and resource waste.

Appendix A

Country Statistics

The following population data is for each country's territory as of 2003. Population estimates were derived from information available as of early 2003. A minus sign (–) indicates a decrease.

Country and Capital	2003 Population (1,000)	Annual Growth (%)	People per Square Mile
World	6,233,822	1.2	117
Afghanistan, Kabul	27,755	3.4	111
Albania, Tiranë	3,545	1.1	334
Algeria, Algiers	32,278	1.7	35
Andorra, Andora la Vella	68	1.1	393
Angola, Luanda	10,593	2.2	22
Antigua and Barbuda, St. John's	67	0.7	397
Argentina, Buenos Aires	37,813	1.1	36
Armenia, Yerevan	3,330	–0.15	290
Australia, Canberra	19,547	1	7
Austria, Vienna	8,170	0.2	256
Azerbaijan, Baku	7,798	0.4	233
Bahamas, Nassau	300	0.9	77
Bahrain, Manama	667	1.6	2,596
Bangladesh, Dhaka	133,377	1.6	2,580

continues

continued

Country and Capital	2003 Population (1,000)	Annual Growth (%)	People per Square Mile
Barbados, Bridgetown	277	0.5	1,666
Belarus, Mensk	10,335	–0.1	129
Belgium, Brussels	10,275	0.15	878
Belize, Belmopan	263	2.65	30
Benin, Porto–Novo	6,788	3.0	159
Bhutan, Thimphu	2,094	2.15	116
Bolivia, La Paz	8,445	1.7	20
Bosnia and Herzegovina, Sarajevo	3,964	0.8	201
Botswana, Gaborone	1,591	0.2	7
Brazil, Brasília	176,029	0.9	54
Brunei, Bandar Seri Begawan	351	2.1	175
Bulgaria, Sofia	7,621	–1.1	178
Burkina Faso, Ouagadougou	12,603	2.6	119
Burundi, Bujumbura	6,373	2.4	644
Cambodia, Phnom Penh	12,775	2.2	187
Cameroon, Yaounde	16,185	2.4	89
Canada, Ottawa	31,902	1.0	9
Cape Verde, Praia	409	0.85	255
Central African Republic, Bangui	3,643	1.8	15
Chad, N'Djamena	8,997	3.3	19
Chile, Santiago	15,499	1.1	54
China, Beijing	1,284,304	0.9	357
Colombia, Santa Fe de Bogotá	41,008	1.6	102
Comoros, Moroni	614	3.	838
Congo, Brazzaville	2,958	2.2	22
Costa Rica, San José	3,835	1.6	196
Côte d'Ivoire, Yamoussoukro	16,805	2.45	137
Croatia, Zagreb	4,391	1.1	201
Cuba, Havana	11,224	0.35	262
Cyprus, Nicosia	767	0.6	213
Czech Republic, Prague	10,257	–0.1	337

Country and Capital	2003 Population (1,000)	Annual Growth (%)	People per Square Mile
Democratic Republic of Congo, Kinshasa	55,225	2.8	63
Denmark, Copenhagen	5,369	0.3	327
Djibouti, Djibouti	473	2.6	56
Dominica, Roseau	70	–0.8	242
Dominican Republic, Santo Domingo	8,722	1.6	466
East Timor	998	2.13	172
Ecuador, Quito	13,447	2.0	126
Egypt, Cairo	70,712	1.7	184
El Salvador, San Salvador	6,354	1.8	794
Equatorial Guinea, Malbo	498	2.45	46
Eritrea, Asmara	4,466	3.8	95
Estonia, Tallinn	1,416	–0.5	81
Ethiopia, Addis Ababa	67,673	2.6	157
Fiji, Suva	856	1.4	121
Finland, Helsinki	5,183	0.1	44
France, Paris	59,766	0.35	284
Gabon, Libreville	1,233	1.0	12
Gambia, Banjul	1,456	3.1	373
Georgia, Tbilisi	4,961	–0.55	184
Germany, Berlin	83,252	0.3	616
Ghana, Accra	20,244	1.7	228
Greece, Athens	10,645	0.2	211
Grenada, St. George's	89	0.02	682
Guatemala, Guatemala City	13,314	2.6	318
Guinea, Conakry	7,775	2.2	82
Guinea–Bissau, Bissau	1,345	2.2	125
Guyana, Georgetown	698	0.2	9
Haiti, Port–au–Prince	7,064	1.4	666
Honduras, Tegucigalpa	6,561	2.3	152
Hungary, Budapest	10,075	–0.3	282

continues

continued

Country and Capital	2003 Population (1,000)	Annual Growth (%)	People per Square Mile
Iceland, Reykjavik	279	0.5	7
India, New Delhi	1,045,845	1.5	911
Indonesia, Jakarta	231,328	1.5	329
Iran, Tehran	66,623	0.8	105
Iraq, Baghdad	24,683	2.8	143
Ireland, Dublin	3,883	1.1	146
Israel, Jerusalem	6,029	1.5	773
Italy, Rome	57,716	0.1	509
Jamaica, Kingston	2,680	0.6	638
Japan, Tokyo	126,975	0.2	833
Jordan, Amman	5,307	2.9	150
Kazakhstan, Astana	16,741	0.1	16
Kenya, Nairobi	31,139	1.15	142
Kiribati, Tarawa	96	2.3	348
Korea, North, Pyongyang	22,244	1.1	478
Korea, South, Seoul	48,324	0.85	1,275
Kuwait, Kuwait	2,111	3.3	306
Kyrgyzstan, Bishkek	4,822	1.45	63
Laos, Vientiane	5,777	2.5	65
Latvia, Riga	2,366	−0.8	95
Lebanon, Beirut	3,678	1.4	943
Lesotho, Maseru	2,208	1.3	189
Liberia, Monrovia	3,288	1.9	88
Libya, Tripoli	5,368	2.4	8
Liechtenstein, Vaduz	33	0.9	528
Lithuania, Vilnius	3,601	−0.25	143
Luxembourg, Luxembourg	448	1.25	449
Macedonia, Skopje	2,055	0.4	208
Madagascar, Antananarivo	16,473	3.0	73
Malawi, Lilongwe	10,702	1.4	295
Malaysia, Kuala Lumpur	22,662	2.0	179

Country and Capital	2003 Population (1,000)	Annual Growth (%)	People per Square Mile
Maldives, Male	320	2.95	2,764
Mali, Bamako	11,340	3.0	24
Malta, Valletta	397	0.7	3,207
Marshall Islands, Majuro	74	3.9	1.054
Mauritania, Nouakchott	2,829	2.9	7
Mauritius, Port Louis	1,200	0.9	1,681
Mexico, Mexico City	103,400	1.5	139
Micronesia, Palikir	136	1.1	501
Moldova, Chisinau	4,434	0.1	341
Monaco, Monaco	32	0.45	41,423
Mongolia, Ulan Bator	2,694	1.5	4
Morocco, Rabat	31,168	1.7	181
Mozambique, Maputo	19,607	1.1	65
Myanmar, Yangon	42,238	0.6	166
Namibia, Windhoek	1,821	1.2	6
Nauru, Yaren	12	2.0	1,521
Nepal, Kathmandu	25,874	2.3	490
Netherlands, Amsterdam	16,068	0.5	1,227
New Zealand, Wellington	3,908	1.1	38
Nicaragua, Managua	5,024	2.1	108
Niger, Niamey	10,640	2.7	22
Nigeria, Abuja	129,935	2.5	369
Norway, Oslo	4,525	0.5	38
Oman, Muscat	2,713	3.4	33
Pakistan, Islamabad	147,663	2.1	491
Palau, Koror and Melekeok	19	1.6	110
Panama, Panama City	2,882	1.3	98
Papua New Guinea, Port Moresby	5,172	2.4	30
Paraguay, Asuncion	5,884	2.6	38
Peru, Lima	27,950	1.7	57
Philippines, Manila	84,526	2.0	734

continues

continued

Country and Capital	2003 Population (1,000)	Annual Growth (%)	People per Square Mile
Poland, Warsaw	38,625	−0.02	328
Portugal, Lisbon	10,084	0.2	285
Qatar, Doha	793	3.0	189
Romania, Bucharest	22,318	−0.2	251
Russia, Moscow	144,979	−0.3	22
Rwanda, Kigali	7,398	1.2	771
St. Kitts and Nevis, Basseterre	39	0.01	279
St. Lucia, Castries	160	1.2	679
St. Vincent and the Grenadines, Kingstown	116	0.4	889
Samoa, Apia	179	−0.25	162
San Marino, San Marino	28	1.4	1,197
São Tomé and Principe, São Tomé	170	3.2	460
Saudi Arabia, Riyadh	23,513	3.3	28
Senegal, Dakar	10,590	2.9	143
Serbia and Montenegro, Belgrade	10,657	−0.1	293
Seychelles, Victoria	80	0.5	455
Sierra Leone, Freetown	5,615	3.2	203
Singapore, Singapore	4,453	3.5	18,482
Slovakia, Bratislava	5,422	0.1	288
Slovenia, Ljubljana	1,933	0.1	248
Solomon Islands, Honiara	495	3.0	47
Somalia, Mogadishu	7,753	3.5	32
South Africa, Pretoria	43,648	0.02	93
Spain, Madrid	40,077	0.1	208
Sri Lanka, Colombo	19,577	0.85	783
Sudan, Khartoum	37,090	2.7	40
Suriname, Paramaribo	436	0.55	7
Swaziland, Mbabane	1,124	1.6	170
Sweden, Stockholm	8,877	0.02	56
Switzerland, Bern	7,302	0.2	474

Country and Capital	2003 Population (1,000)	Annual Growth (%)	People per Square Mile
Syria, Damascus	17,156	2.5	241
Taiwan, Taipei	22,548	0.8	1,804
Tajikistan, Dushanbe	6,718	2.1	122
Tanzania, Dodoma and Dar es Salaam	37,188	2.6	109
Thailand, Bangkok	62,354	0.9	316
Togo, Lomé	5,285	2.5	252
Tonga, Nukualofa	106	1.85	383
Trinidad and Tobago, Port–of–Spain	1,164	–0.5	582
Tunisia, Tunis	9,816	1.1	164
Turkey, Ankara	67,309	1.2	226
Turkmenistan, Ashkhabad	4,689	1.8	25
Tuvalu, Funafuti	11	1.4	1,110
Uganda, Kampala	24,699	2.9	320
Ukraine, Kiev	48,396	–0.7	208
United Arab Emirates, Abu Dhabi	2,446	1.6	76
United Kingdom, London	59,778	0.2	641
United States, Washington, D.C.	280,562	0.9	79
Uruguay, Montevideo	3,387	0.8	51
Uzbekistan, Tashkent	25,563	1.6	148
Vanuatu, Port Vila	196	1.7	34
Venezuela, Caracas	24,288	1.5	71
Vietnam, Hanoi	81,098	1.4	646
Yemen, San'a	18,701	3.4	92
Zambia, Lusaka	9,959	2.0	35
Zimbabwe, Harare	11,377	0.05	76

Source: CIA, The World Factbook 2002; World Almanac 2003

Index